Basic Health Science Chemistry

A Review and Workbook

Joan Creager
Marymount University

Basic Health Science Chemistry

A Review and Workbook

WCB **Wm. C. Brown Publishers**

Dubuque, Iowa • Melbourne, Australia • Oxford, England

Book Team

Editor *Colin H. Wheatley*
Developmental Editor *Elizabeth M. Sievers*
Designer *Mark Elliot Christianson*
Permissions Editor *Vicki Krug*
Art Editor *Rachel Imsland*

Wm. C. Brown Publishers
A Division of Wm. C. Brown Communications, Inc.

Vice President and General Manager *Beverly Kolz*
Director of Sales and Marketing *John W. Calhoun*
Marketing Manager *Christopher T. Johnson*
Advertising Manager *Amy Schmitz*
Director of Production *Colleen A. Yonda*
Manager of Visuals and Design *Faye M. Schilling*
Design Manager *Jac Tilton*
Art Manager *Janice Roerig*
Publishing Services Manager *Karen J. Slaght*
Permissions/Records Manager *Connie Allendorf*

Wm. C. Brown Communications, Inc.

President and Chief Executive Officer *G. Franklin Lewis*
Corporate Vice President, President of WCB Manufacturing *Roger Meyer*
Vice President and Chief Financial Officer *Robert Chesterman*

Cover photo © Human Touch Jeon/The Image Bank

The credits section for this book begins on page 246 and is considered an extension of the copyright page.

A Times Mirror Company

ISBN 0–697–15276–6

Printed in the United States of America by Wm. C. Brown Communications, Inc., 2460 Kerper Boulevard, Dubuque, IA 52001

10 9 8 7 6 5 4 3 2 1

Contents

7

Digestion, Absorption, and Regulation of Metabolism *166*

8

Nutrition *197*

9

Excretion; Fluid, Electrolyte, and Acid-Base Balance *218*

This book provides a thorough review of basic chemistry relevant to the life sciences, with concepts presented in the context of human physiology. It will serve as a study aid for students in undergraduate curricula in the health sciences and as a review tool for those entering graduate or professional schools.

Each chapter begins with a correlated layout of outline and objectives that give a quick overview of the chapter and allow the reader maximum flexibility in selecting appropriate sections for study or review. Questions in the chapter text ask for information provided in the preceding major section of the chapter. In the Study and Review section, "Did you get the essentials?" requires that boldface terms, which appear throughout each chapter, be used to fill in the blanks. The combined glossary-index gives a definition and page reference for every boldfaced term in the text.

Acknowledgments

In a way, all the people who contributed to the second edition of *Human Anatomy and Physiology* also contributed to this book and I thank them again for their contributions. I am especially indebted to reviewers who considered this manuscript for its usefulness as an aid to mastering basic concepts of chemistry. Those reviewers follow:

Anne E. Lesak
  Moraine Valley Community College
Virginia Rivers
  Truckee Meadows Community College

Working with the WCB staff continues to be a rewarding experience. Elizabeth Sievers and Donna Slade have been especially helpful in getting this book into production. No matter how many good helpers I have had, errors may have slipped by and for those I accept responsibility and extend my apologies for any inconvenience to users.

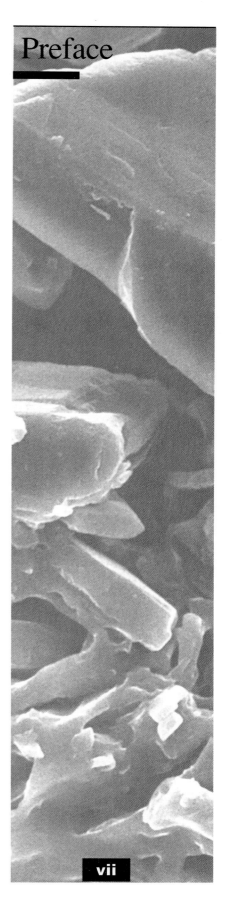

Preface

vii

Fundamentals of Chemistry

OBJECTIVES

Chemical Building Blocks and Chemical Bonds

1. Describe how chemical building blocks and chemical bonds contribute to the structure and function of biological molecules.

Mixtures, Solutions, and Colloidal Dispersions

2. Define the terms solvent, solute, and solution.
3. List the properties of water and explain how water contributes to body functions and homeostasis.
4. Distinguish between a solution and a colloidal suspension.

Acids, Bases, and pH

5. Define acid, base, and pH and relate the characteristics of acids and bases to their physiologic effects.

Complex Molecules

6. Briefly describe the properties of organic molecules and name four kinds of complex biological molecules.

Carbohydrates

7. Describe the properties, categories, and physiologic importance of carbohydrates.

Lipids

8. Describe the properties, categories, and physiologic importance of lipids.

Proteins

9. Describe the properties, categories, and physiologic importance of proteins.

Nucleotides and Nucleic Acids

10. Describe the properties, categories, and physiologic importance of nucleotides and nucleic acids.

Bioenergetics

11. Briefly explain the laws governing chemical changes.
12. Describe and give examples of coupled reactions.

Enzymes

13. Relate the properties of enzymes to the regulation of chemical reactions.
14. Explain how temperature, pH, and the concentrations of substrate and enzyme affect the rate of enzyme reactions.

Radioactivity

15. Define radioactivity and describe its properties.

Study and Review

1

Chemical Building Blocks and Chemical Bonds

Chemistry is concerned with matter—the properties and interactions of matter. **Matter** is anything that occupies space and has mass (substance), including air, water, rocks, and living things. Many physical and chemical properties of living organisms can be described in terms of physical and chemical properties of matter. Matter is composed of basic chemical building blocks. Just as letters of the alphabet can be combined in different ways to make thousands of words, chemical building blocks can be combined to make thousands of different substances, many of them more complex than words. Although few English words contain more than 20 letters, complex chemical substances can contain more than 20,000 building blocks.

In the body, various chemical substances undergo changes in chemical reactions. The sum total of all the body's chemical reactions is called **metabolism** (met-ab'o-lizm). Metabolism includes the breakdown of nutrients for energy and the making of body substance. Other chemical changes occur as muscles contract and nerves and hormones send signals.

Homeostasis is the maintenance of internal conditions within a narrow, normal range. It is possible only when each of a large variety of chemical processes is occurring at the right time, in the right place, and at the proper rate. Understanding the fundamentals of chemistry is important, not only because of these chemical processes, but because maintaining homeostasis often depends on events at the chemical level.

Particles of matter are too small to be seen even with the strongest microscopes. By using various experiments chemists have deduced certain characteristics of the particles. They have identified the **atom** as smallest chemical unit of matter. Matter consisting of one kind of atom is called an **element.** Each element has specific properties, for example, the graphite in a pencil consists of a vast number of carbon atoms. The earth's atmosphere consists of gaseous elements such as oxygen and nitrogen. Sodium is a soft, metallic element.

Atoms of an element combine with other atoms of the same or different elements. The ability of carbon atoms to form long chains is important in the structure of living things. Chemists use letters to designate elements— C for carbon, N for nitrogen, Na for sodium (*Natrium* is the Latin word for sodium), and subscripts to indicate how many atoms of the element are present. Though oxygen and nitrogen can occur as paired atoms, O_2 or N_2, most atoms combine with atoms of other elements. One atom of carbon combines with two atoms of oxygen to form carbon dioxide (CO_2) and two atoms of hydrogen combine with one atom of oxygen to form water (H_2O). Two or more atoms combined chemically form a **molecule.** A few

Table 1.1
Properties of Atomic Particles

Particle	Relative Mass	Charge	Location
Proton	1	+	Nucleus
Electron	1/1836	−	Orbiting the nucleus
Neutron	1	None	Nucleus

molecules consist of atoms of the same element, N_2 for example, but most consist of atoms of different elements like CO_2. Molecules that contain atoms of two or more elements are called **compounds.** CO_2 and N_2 are molecules; CO_2 also is a compound.

More than 85 percent of the human body weight consists of four elements—carbon, hydrogen, oxygen, and nitrogen. A molecule of the sugar glucose contains 24 atoms, $C_6H_{12}O_6$. Carbohydrates, proteins, fats, and nucleic acids can contain thousands of atoms.

Atoms are the smallest particles that retain the properties of an element, but their subatomic components contribute to those properties. Among the many subatomic particles now known, we are concerned with only **protons, neutrons,** and **electrons** (Table 1.1). The mass of a proton and a neutron is arbitrarily designated as equal to one mass unit, and by comparison, electrons have a much smaller mass. Charges are also arbitrarily designated as negative for electrons and positive for protons. Neutrons are neutral—they have no net charge. Protons and neutrons occupy the nucleus of the atom, whereas the electrons move in orbits around the nucleus. All atoms have an equal number of protons and electrons and are electrically neutral; the atoms of a particular element have a specific number of protons that determines the element's **atomic number.** Atomic numbers range from 1 to over 100.

Electrons, being in constant motion, form an **electron cloud** around the nucleus. Some electrons display more energy than others, and their motion can be represented by concentric circles to suggest different energy levels within the cloud (Figure 1.1). Electrons with the least energy stay in orbits near the nucleus and those with more energy move to orbits farther from the nucleus. The negative charge of electrons holds them close to positively charged protons in the nucleus, and orbital motion moves them away from the nucleus. Orbital motion is analogous to the circular path made by swinging an object tied to a string.

Depending on the size of an atom, its electrons occupy one or more concentric circles or shells. An atom of hydrogen has 1 electron located in the innermost shell. An atom of helium has 2 electrons, the maximum number in the innermost shell. Atoms with more than 2 electrons

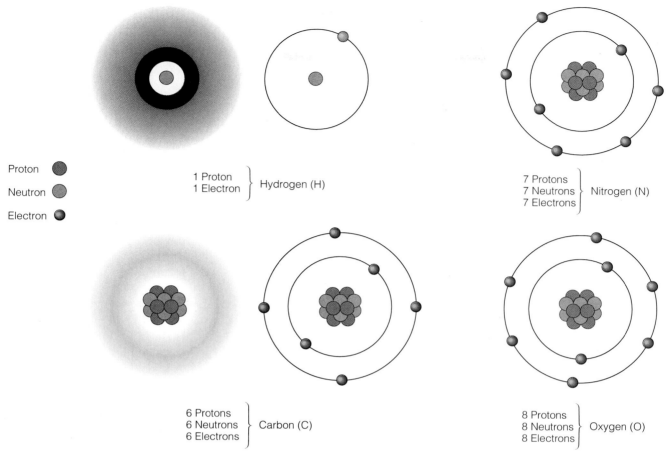

Proton ●
Neutron ●
Electron ●

1 Proton
1 Electron } Hydrogen (H)

7 Protons
7 Neutrons } Nitrogen (N)
7 Electrons

6 Protons
6 Neutrons } Carbon (C)
6 Electrons

8 Protons
8 Neutrons } Oxygen (O)
8 Electrons

Figure 1.1

The structures of some atoms commonly found in the human body.

always have 2 electrons in the inner shell and up to 8 additional electrons in the second shell. The inner shell is filled before electrons occupy the second shell, the second shell is filled before electrons occupy the third shell, and so on. Very large atoms have several more electron shells, and some shells can contain more than 8 electrons. However, for the elements of physiologic importance, the outer shell is considered filled if it contains 8 electrons.

Atoms with outer electron shells that are nearly full (have 6 or 7 electrons) or nearly empty (have 1 or 2 electrons) have a tendency to form ions. An **ion** is a charged particle produced when an atom gains or loses electrons (Figure 1.2). When a sodium atom with 1 electron in its outer shell loses this electron it has 1 more proton than electrons and becomes a positively charged ion called a **cation** (kat'i-on). (An easy way to remember that cations are positively charged is to notice that the top of the "t" in cation makes a "plus sign.") When a chlorine atom with 7 electrons in its outer shell gains an electron it becomes a negatively charged chloride ion called an

Table 1.2 Ions Commonly Found in the Body	
Cations	**Anions**
Sodium (Na^+)	Chloride (Cl^-)
Potassium (K^+)	Hydroxyl (OH^-)
Calcium (Ca^{2+})	Bicarbonate (HCO_3^-)
Hydrogen (H^+)	Phosphate (PO_4^{3-})
Magnesium (Mg^{2+})	

anion (an'i-on). Sodium or chloride ions are chemically more stable than corresponding atoms because their outer electron shells are full.

Many elements occur as ions in the body (Table 1.2). Those with 1 or 2 electrons in their outer shell tend to lose electrons and form ions with charges of +1 or +2,

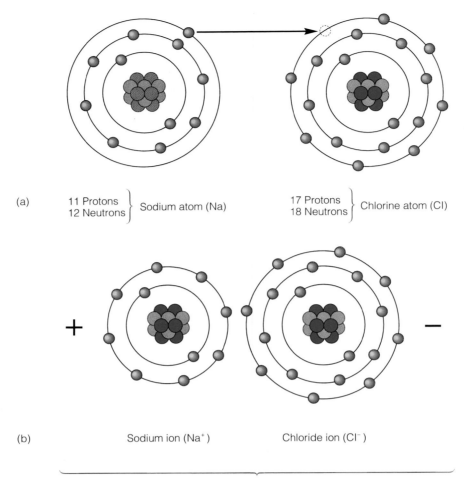

(a)   11 Protons
       12 Neutrons } Sodium atom (Na)

       17 Protons
       18 Neutrons } Chlorine atom (Cl)

$+$ (b)   Sodium ion (Na$^+$)   Chloride ion (Cl$^-$) $-$

Sodium chloride

Figure 1.2

The formation of sodium and chloride ions: (*a*) When sodium loses the single electron in its outer shell, it becomes a positively charged sodium ion. When chlorine, which has seven electrons in its outer shell, gains an electron, it becomes a negatively charged chloride ion. (*b*) The oppositely charged sodium and chloride ions attract each other electrically and form a crystalline molecule of sodium chloride.

respectively; those with 6 or 7 electrons in their outer shell tend to gain electrons and form ions with charges of -1 or -2. Some ions, such as the hydroxyl ion (OH$^-$), contain more than one element.

All atoms of the same element have the same atomic number, but they can have different atomic weights. **Atomic weight** is the total number of protons and neutrons in an atom. For example, carbon atoms with 6 protons and 6 neutrons have an atomic weight of 12. Some naturally occurring carbon atoms have 7 or 8 neutrons and atomic weights of 13 or 14. Such atoms can also be created in the laboratory. Atoms of a particular element that contain different numbers of neutrons are called **isotopes.** For an element that has naturally occurring isotopes, the atomic

weight is the average atomic weight of the mixture of isotopes. Atomic weights can be decimal numbers, but a particular atom has a specific number of whole neutrons. Some isotopes are stable, while others are not. Unstable isotopes emit radiation from their nuclei. Such emissions can be used to follow chemical processes in living things, but they also can harm them. Properties of elements found in living things are summarized in Table 1.3.

Chemical bonds are forces between the outer shell electrons that hold the atoms of a molecule together. Three kinds of bonds commonly found in living organisms—ionic, covalent, and hydrogen bonds—vary in strength and help to determine how molecules behave in living organisms.

Table 1.3
Some Properties of Elements Found in Living Organisms

Element	Symbol	Atomic Number	Atomic Weight	Electrons in Outer Orbit	% of Body Weight	Biological Occurrence
Carbon	C	6	12.0	4	18	Forms "backbone" of all organic compounds
Hydrogen	H	1	1.0	1	10	Found in most biological molecules; H^+ important component of solutions
Oxygen	O	8	16.0	6	65	Found in most biological molecules; final electron acceptor in many energy-yielding reactions
Nitrogen	N	7	14.0	5	3	Found in proteins, nucleic acids, and many other biological molecules
Calcium	Ca	20	40.1	2	1.5	Essential component of bones and teeth; important in muscle contraction; controls many cellular processes
Phosphorus	P	15	31.0	5	1	Component of nucleic acids and energy-carrying molecules such as ATP; found in many lipids
Sulfur	S	16	32.0	6	<1	Component of many proteins and other important biological molecules
Iron	Fe	26	55.8	2	<1	Component of electron carriers and oxygen carriers
Potassium	K	19	39.1	1	<1	Important in conduction of nerve signals
Sodium	Na	11	23.0	1	<1	Ion in solutions; important in conduction of nerve signals and transport mechanisms
Chlorine	Cl	17	35.4	7	<1	Ion in solutions; synthesis of HCl
Magnesium	Mg	12	24.3	2	<1	Important in enzyme-catalyzed reactions in most cells; important in photosynthesis in plants
Copper	Cu	29	63.6	1	T*	Important in some energy yielding reactions; important in photosynthesis in plants
Iodine	I	53	126.9	7	T	Essential part of thyroid hormone molecules
Fluorine	Fl	9	19.0	7	T	Prevents microbial growth
Manganese	Mn	25	54.9	2	T	Found in enzymes or important in activating enzymes
Zinc	Zn	30	65.4	2	T	Important in activating some enzymes
Selenium	Se	34	79.0	6	T	Part of an antioxidant enzyme
Molybdenum	Mo	42	95.9	1	T	Part of several enzymes

*T = Trace amount found in human body

Ionic bonds form between oppositely charged ions that are attracted to each other. For example, sodium ions having a positive charge combine with chloride ions having a negative charge (Figure 1.2). Ionic bonds are relatively weak bonds. When molecules with such bonds are put in water, the bonds easily break and the molecule **ionizes,** or forms ions. Therefore sodium chloride in body fluids exists as sodium and chloride ions.

Carbon, hydrogen, oxygen, and nitrogen atoms can be held together in molecules by **covalent bonds,** in which electrons are shared instead of being gained or lost (Figure 1.3). A carbon atom with 4 electrons in its outer shell can share an electron with each of 4 hydrogen atoms. At the same time, each of the 4 hydrogen atoms shares an electron with the carbon atom. Four pairs of electrons are shared, with each pair having an electron from carbon and an electron from hydrogen, and the outer shells of both atoms are filled. When forming covalent bonds, oxygen shares 2 electrons and nitrogen usually shares 3 electrons. Sometimes a carbon atom and another atom such as oxygen share two pairs of electrons to form a **double bond.** When writing structural formulas, chemists use a single line for a single pair of shared electrons and a double line for two pairs of shared electrons (Figure 1.3). Covalent bonds, which contain more energy than ionic bonds, hold molecules together more tightly than ionic bonds. Molecules with covalent bonds are more stable in solutions because they tend not to ionize. Many molecules in living things are stable because they contain covalent bonds.

Weak covalent bonds, called **hydrogen bonds,** are particularly important in biological structures, where they typically bind hydrogen atoms to oxygen or nitrogen atoms. In such bonds, oxygen or nitrogen atoms attract electrons. Shared electrons are pulled toward the atomic nucleus of oxygen or nitrogen and pulled away from hydrogen. Oxygen or nitrogen atoms have a partial negative charge, and hydrogen atoms have a partial positive charge.

Hydrogen bonds contribute significantly to the structure and properties of large molecules such as proteins and nucleic acids, which consist of long chains of atoms. Hydrogen bonds help to maintain each molecule in its characteristic three-dimensional shape; they also help to account for polar and nonpolar regions within large molecules. **Polar regions** have partial charges because of uneven distribution of positive and negative components of a molecule. Such regions attract other partially charged molecules, especially water. **Nonpolar regions** are uncharged and usually found below the surface of such molecules. Polar regions are **hydrophilic** (hi-dro-fil′ik), or water-loving, and nonpolar regions are **hydrophobic** (hi-dro-fo′bik), or water-fearing. Large biological molecules mix with water because of surface polar regions but do not dissolve in it because of internal nonpolar regions.

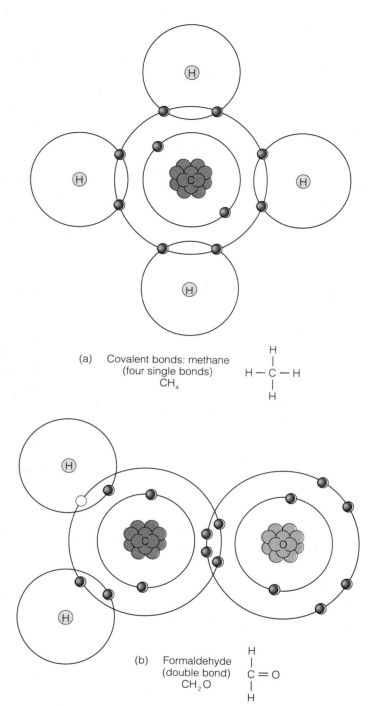

(a) Covalent bonds: methane (four single bonds) CH_4

(b) Formaldehyde (double bond) CH_2O

Figure 1.3

The sharing of electrons in covalent bonds: (*a*) In the gas methane, one carbon atom shares four electrons, one with each of four hydrogen atoms. The hydrogen atoms likewise share an electron with the carbon atom. (*b*) In formaldehyde, the carbon and hydrogen atoms share electrons as in methane, but carbon and oxygen each share two electrons. The sharing of two pairs of electrons forms a double bond.

Water mixes with biological molecules because of hydrogen bonds and polarity. Water molecules have a positive (hydrogen) pole and a negative (oxygen) pole because shared electrons stay closer to the oxygen than to the hydrogen atoms (Figure 1.4). The weak, partial charges allow the hydrogen side of one water molecule to form a hydrogen bond with the oxygen side of another water molecule. Water or any other compound with positive and negative regions can form hydrogen bonds and is called a **polar compound.**

See Questions—Objective 1

Mixtures, Solutions, and Colloidal Dispersions

Unlike a chemical compound the molecules of which contain atoms in specific proportions, a **mixture** consists of two or more compounds combined in any proportion but not chemically bound. A mixture retains the properties of its substances. For example, a sugar-salt mixture of any proportions will taste both sweet and salty, but the degree of each taste depends on the relative amount of each substance in the mixture.

A **solution** is a homogeneous mixture of two or more substances in which molecules are evenly distributed and usually will not separate upon standing. In a solution, the **solvent** is the medium in which one or more substances are dissolved, and the **solute** is any dissolved substance—atoms, ions, or molecules. In the human body, water is the solvent in nearly all solutions. Typical solutes include the sugar glucose, small protein molecules, the gases carbon dioxide and oxygen, and ions.

Water, the solvent in body fluids, is so essential to life that humans live only a few days without it. As noted earlier, it comprises 55 to 60 percent of the total human body weight—more in infants and less in the elderly and more in brain tissue and less in bone and fat. Several properties of water contribute to its importance in humans and other living things (Table 1.4). Because water is a polar compound, it acts as a good solvent and forms hydrogen bonds with other molecules of both water and other substances.

Water is an especially good solvent for ions because the polar water molecules orient around ions, forming a **hydration shell.** The positive regions of the water molecules surround negative ions and the negative regions surround positive ions. Ions thereby become evenly distributed through water with water molecules interspersed between them (Figure 1.5). Ions and other substances that mix with water are easily transported through blood and other body fluids.

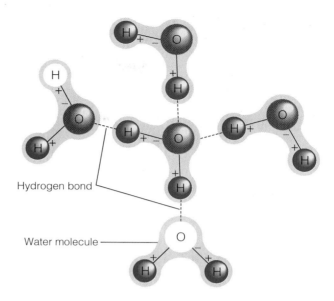

Hydrogen bond

Water molecule

Figure 1.4

In hydrogen bonding between water molecules, the slightly negative oxygen region of one molecule is attracted to the slightly positive hydrogen region of another molecule.

Table 1.4
Properties of Water

1. Good solvent ability helps to dissolve substances for transport in body fluids.

2. Ability to form layers (because of high surface tension) helps to keep membranes moist.

3. Ability to store or release large quantities of heat (high specific heat) helps to regulate body temperature.

4. Distribution throughout the body provides a medium for chemical reactions.

5. Chemical reactivity allows components of water (H^+ and OH^-) to participate in many chemical reactions.

Water forms thin layers because of its high **surface tension,** that is, it forms a thin, invisible, elastic layer on membrane surfaces. Water molecules beneath the surface are attracted to each other in all directions, but no such attraction exists between water molecules and gas molecules in the air. Attractions between water molecules cause any water layer to contract until it occupies a minimum area. Surface water molecules crowd with their hydrogen bonds projecting below the surface, thus creating sufficient tension to support waterstriders walking on a pond surface. On living membranes, high surface tension

Questions

Objective 1

(a) Define:

atom

element

molecule

compound

proton

electron

neutron

atomic number

atomic weight

cation

anion

(b) Use chemical formulas to illustrate the differences between ionic, covalent, and hydrogen bonds.

(c) What are the important properties of polar compounds?

---

maintains the continuity of water layers, which allows gases such as oxygen and carbon dioxide to diffuse across membranes and other membrane functions to occur.

Water has a high **specific heat;** it absorbs large quantities of energy with little change in temperature. This allows water to gain heat in the daytime and release it at night. Similarly, the high specific heat of water allows relatively large amounts of heat to be lost by the evaporation of small amounts of sweat, thereby regulating body temperature while conserving body water.

Finally, water provides a medium for chemical reactions and serves as a participant in many of these reactions. In synthetic reactions such as **condensation, or dehydration synthesis,** water is removed as two molecules form a larger molecule. Sugars combine to form complex carbohydrates and amino acids combine to form proteins by this kind of reaction. In degradative (breakdown) reactions such as **hydrolysis** (hi-drol′i-sis), water is added as a large molecule is broken down into smaller molecules.

For example, during digestion large food molecules are broken down into simple sugars, amino acids, and other small molecules by hydrolysis.

Particles with diameters between 1 and 100 nanometers (billionths of a meter) are called **colloids** (kol′oidz). Though too large to form true solutions, colloids can form **colloidal** (kol-oid′al) **dispersions.** Such particles are suspended in a medium by opposing electrical charges, layers of water molecules around the particles, and other forces. Gelatin dessert is a colloidal dispersion with the protein gelatin dispersed in water. In the body, colloidal dispersions consist of large protein molecules dispersed in water. Much of the **cytosol** (fluid or semifluid substance around organelles in cells) is a complex colloidal system. Proteins in plasma, the fluid portion of blood, also form a colloidal dispersion. Some colloidal systems have the ability to change from a semisolid **gel,** gelatin that has "set," to a more fluid **sol** state, akin to gelatin that has melted.

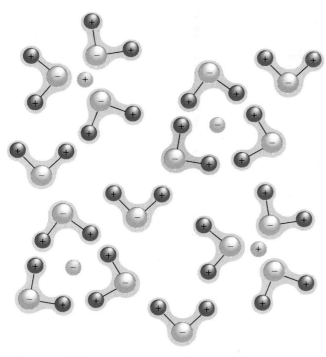

Figure 1.5

Water molecules surround positive and negative ions and help to hold these particles in solution.

The properties of chemical substances, including water, solutions, and colloidal suspensions, set limits on physiologic possibilities. The behavior of molecules in living organisms is limited by the properties of the molecules themselves.

See Questions—Objectives 2, 3, and 4

Acids, Bases, and pH

Except for the extreme acidity of the stomach, the body's external and internal environments are nearly chemically neutral—neither solidly acidic nor strongly basic (alkaline). A knowledge about acids and bases is needed to understand how the body maintains nearly neutral conditions in spite of acids that accumulate as cells use nutrients.

Acids and bases readily form ions in water. An **acid** releases or donates hydrogen ions (H^+) to a solution and is a hydrogen donor, or proton donor. (A hydrogen ion is a proton.) A **base** usually accepts hydrogen ions from a solution, but some bases release hydroxyl ions (OH^-) into solutions. A base is a proton acceptor or a hydroxyl ion donor. The acidity of a solution increases with the H^+ concentration and its alkalinity increases with the OH^- concentration. In body fluids, H^+ is often released by organic

Table 1.5 The pH of Various Substances	
Substance	**pH***
Hydrochloric acid (1 molar)	0.0
Stomach hydrochloric acid (0.1 molar)	1.0
Gastric juice	1.0–3.0
Lemon juice	2.5
Vinegar, beer, wine	3.0
Orange juice	3.5
Tomatoes, grapes	4.0
Coffee	5.0
Urine	5.0–7.0
Milk	6.5
Saliva	6.3–7.3
Pure water (at 25° C)	7.0
Blood	7.35–7.45
Eggs	7.5
Ocean water	7.8–8.2
Household bleach	9.5
Milk of magnesia	10.5
Household ammonia	10.5–11.8
Oven cleaner	13.5
Sodium hydroxide (1 molar)	14.0

*pH of body fluids measured at 37° C.

acid, or carboxyl ($-COOH$) groups (COOH ionizes to COO^- and H^+). H^+ is accepted by OH^- to form water or by amino ($-NH_2$) groups to form ammonia (NH_3).

To express acidity or alkalinity, physiologists use the concept of **pH,** which is the negative log of the hydrogen ion concentration in moles per liter.

$$pH = -\log [H^+]$$

(Brackets [ ] denote concentration.) A solution at pH 7 is neutral, neither acidic nor basic, and contains equal numbers of H^+ and OH^- ions. Such is the case in pure, distilled water because one ten-millionth (10^{-7}) part of each mole of water is ionized, or exists as H^+ and OH^- ions. The H^+ concentration is 10^{-7} moles per liter and the solution's pH is 7.

The pH scale (Figure 1.6) relates proton concentrations to pH and is logarithmic, that is, the proton concentration changes by a factor of ten for each unit of the scale. The usual range of the pH scale is from 0 to 14, but most tissues and body fluids have a pH between 5 and 8 except in the stomach. Table 1.5 shows the pH of some body fluids, foods, and other common substances.

Questions

Objective 2

(a) Define:

solvent

solute

solution

(b) How are solvents, solutes, and solutions related?

Objective 3

(a) In what ways is water important to living things?

(b) What disorders do you think are most likely to develop if humans are deprived of water?

Objective 4

(a) How do solutions and colloidal dispersions differ?

(b) How are solutions and colloidal dispersions alike?

Changes in the pH of body fluids have profound physiologic effects. For example, pneumonia and other diseases that interfere with gas exchange in the lungs allow carbon dioxide and H⁺ to accumulate in the blood by the following reactions:

$$CO_2 + H_2O \rightarrow H_2CO_3 \rightarrow H^+ + HCO_3^-$$

carbon dioxide   water   carbonic acid   hydrogen ion   bicarbonate

An overdose of aspirin stimulates respiratory centers in the brain and increases gas exchange, causing too much carbon dioxide and H⁺ to be removed from the blood as the reactions are reversed.

Buffers, substances that resist pH change, help to prevent such changes. Carbonic acid, which can donate H⁺, and bicarbonate, which can accept H⁺, act as natural blood buffers helping to keep the blood pH within a narrow, tolerable range.

See Questions—Objective 5

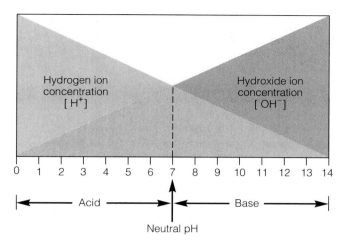

Figure 1.6

Solutions at pH 7 are neutral. Increasingly acid solutions have higher concentrations of H^+ and lower pH values. Increasingly basic solutions have lower concentrations of H^+ and higher pH values.

Complex Molecules

The study of the fundamentals of general chemistry has paved the way for considering **organic chemistry,** the study of most carbon-containing compounds. Such compounds occur in living things and their products and remains. The ability of carbon atoms to share electrons and to form long chains and rings makes the number of possible organic compounds almost infinite. Molecules that consist of chains are said to be **aliphatic** (al-eh-fat'ik) whereas those that contain rings are said to be **aromatic** (ar-o-mat'ik).

The simplest carbon compounds are hydrocarbons—chains of carbon atoms with associated hydrogen atoms. The simplest hydrocarbon, methane, has a single carbon atom, but gasoline and other petroleum products contain several carbons.

In addition to hydrogen, other atoms such as oxygen and nitrogen can bond to carbon chains, where they often form functional groups. A **functional group** is a molecular part that participates in chemical reactions and gives the molecule some of its properties. Functional groups demonstrate that structure and function are related even at the chemical level. Four categories of organic molecules with functional groups containing oxygen are alcohols, aldehydes, ketones, and organic acids (Figure 1.7). The functional groups of alcohols, called **hydroxyl** (hidrox'il) groups, are found almost anywhere in a molecule. **Carbonyl** (kar'bon-el) groups form aldehydes at the ends of chains and ketones within chains. The functional groups of organic acids are called **carboxyl** (kar-box'il) **groups.** An **amino** (ah-me'no) **group** ($-NH_2$) contains nitrogen but no oxygen. Amino groups, as in amino acids of proteins, account for most of the body's nitrogen.

Questions

Objective 5

(a) Define:

acid

base

pH

(b) How might altering the pH of a body fluid affect body functions?

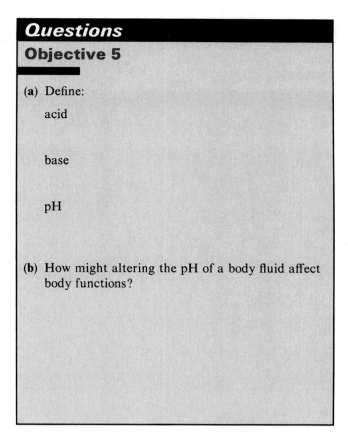

Figure 1.7

Oxygen-containing functional groups of organic molecules. The shaded portion of the molecule is the functional group. Molecules are arranged from most reduced (having the most hydrogen) to most oxidized (having the most oxygen).

The amount of oxygen in functional groups is related to the energy they contain. Alcohols, which have lots of hydrogen atoms and few oxygen atoms, contain more energy than organic acids, which have less hydrogen and more oxygen. Molecules with large amounts of hydrogen are said to be reduced and energy can be extracted from them as they are oxidized. **Reduction** is the addition of

Questions

Objective 6

(a) Define organic chemistry.

(b) Name four kinds of complex biological molecules.

(c) What functional groups are commonly found in biological molecules?

Table 1.6
Properties of Carbohydrates

Class of Carbohydrates	Examples	Description
Monosaccharides	Glucose	Six carbons, aldehyde functional group
	Fructose	Six carbons, ketone functional group
	Galactose	Six carbons, aldehyde functional group arranged differently than glucose
	Ribose	Five carbons, aldehyde functional group
	Deoxyribose	Five carbons, aldehyde functional group, one carbon has no oxygen
Disaccharides	Sucrose	Glucose and fructose
	Lactose	Glucose and galactose
	Maltose	Two glucose units
Polysaccharides	Starch	Polymer of glucose found in plants, digestible by humans
	Glycogen	Polymer of glucose stored in liver and skeletal muscles
	Cellulose	Polymer of glucose found in plants, not digestible by humans

hydrogen or the removal of oxygen. **Oxidation** is the addition of oxygen or the removal of hydrogen. Because gasoline and other hydrocarbons are extremely reduced they make good fuels. Glucose and other sugars are relatively reduced and are an important body fuel. Oxidized molecules such as CO_2, a product of glucose oxidation, contain little energy.

Biological molecules, or organic molecules found in living organisms, fall into four major classes: carbohydrates, lipids, proteins, and nucleic acids. Enzymes, a group of proteins, are of sufficient importance to warrant special consideration. We will look at the properties, categories, and physiologic importance of each class of molecules, and we will consider how they participate in chemical reactions in chapter 3.

See Questions—Objective 6

Carbohydrates

Foods such as bread, potatoes, table sugar, and honey contain large amounts of carbohydrate. **Carbohydrates** (kar-bo-hi'dratz) serve as the main source of energy for most living things (Table 1.6). All carbohydrates contain carbon, hydrogen, and oxygen, generally in the proportion of two hydrogen atoms for each carbon and oxygen atom (CH_2O).

Monosaccharides (mon-o-sak'ar-idz), the simplest of carbohydrates, consist of aldehydes or ketones with several hydroxyl groups. Glucose, a 6-carbon-sugar found in human blood, can be represented as a chain, a ring, or a three-dimensional structure (Figure 1.8). Regardless of how they are drawn on flat paper, all molecules are three-dimensional. Other common monosaccharides include fructose and galactose, which are found in many carbohydrate foods, and ribose and deoxyribose, both of which occur in nucleic acids (Figure 1.9).

Glucose, fructose, and galactose are **isomers** (is'om-erz). They contain the same numbers of each kind of atom ($C_6H_{12}O_6$) but the atoms are arranged differently. The aldehyde glucose and the ketone fructose are **structural isomers**—their structures differ significantly. Glucose and galactose, both of which are aldehydes, are **stereoisomers** (sta-re-o-i'-som-erz) and differ only by the positions of the hydrogen atom and hydroxyl group on carbon 4. Certain isomers are mirror images like left and right hands and are designated L (for *levo* or left-handed)

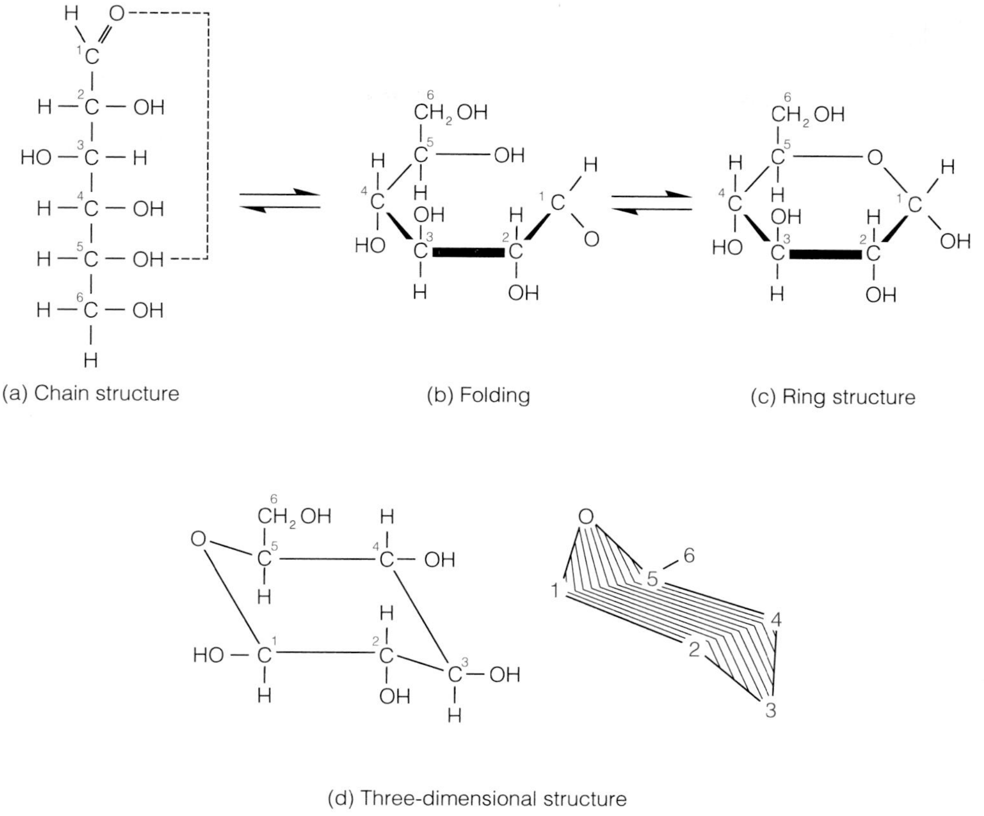

(a) Chain structure   (b) Folding   (c) Ring structure

(d) Three-dimensional structure

Figure 1.8

The structure of a glucose molecule can be represented in
several ways: (*a*) a chain structure, (*b*) the folding of a chain
to form a ring, (*c*) a ring structure, and (*d*) a three-
dimensional structure.

Fructose   Galactose   Ribose   Deoxyribose

6-carbon sugars   5-carbon sugars

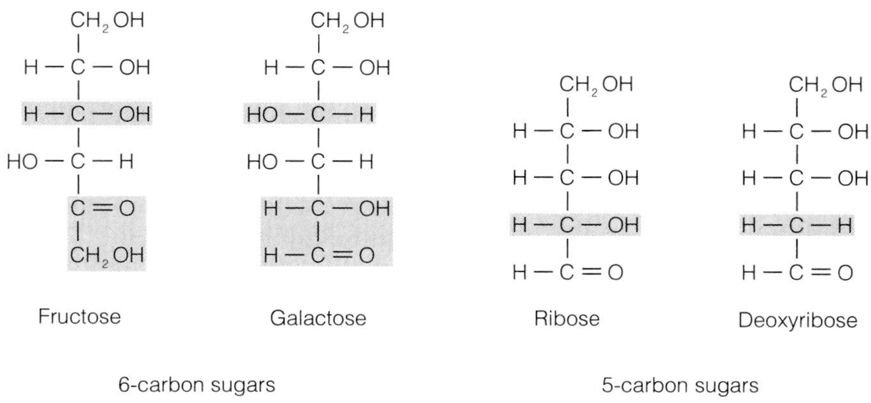

Figure 1.9

In addition to glucose, fructose and galactose also are
6-carbon-sugars. Ribose and deoxyribose are 5-carbon-sugars.
The sites where these molecules differ are shown in color.

Questions
Objective 7

(a) What are the distinguishing properties of carbohydrates?

(b) How do mono-, di-, and polysaccharides differ?

(c) In what ways are carbohydrates important in body function?

Table 1.7
Characteristics of Lipids

Class of Lipid	Characteristics
Simple lipids (fat)	Glycerol and three fatty acids form a triacylglycerl; mono- and diacylglycerols contain one and two fatty acids, respectively; fatty acids can be saturated (contain all the hydrogen they can) or unsaturated (have double bonds between carbons and lack some hydrogen)
Compound lipids	Contain other components in addition to glycerol and fatty acids; phospholipids contain a phosphate group instead of one of the fatty acids; found in cell membranes
Steroids	Contain a characteristic four-ring structure; include cholesterol, bile salts, vitamin D, and hormones from the adrenal and sex glands
Other	A variety of substances having some properties of lipids; include porphyrins (components of hemoglobin), vitamins A, E, and K, and prostaglandins (important regulatory molecules in cells)

and D (for *dextro* or right-handed). Naturally occurring sugars are D-sugars, but mirror-image L-sugars can be made in the laboratory. Because of their molecular shape L-sugars cannot be broken down in the body. It has been suggested that L-sugars be used to make synthetic foods for people on weight-loss diets, but such foods would be extremely expensive.

Two monosaccharide molecules joined by dehydration synthesis form a **disaccharide** (di-sak′ar-id), as in Figure 1.10a. Dietary disaccharides include sucrose, lactose, and maltose. Sucrose (table sugar) consists of glucose and fructose. Lactose (milk sugar) consists of glucose and galactose. Maltose contains two glucose molecules and is derived from the digestion of starch.

Polysaccharides (pol-e-sak′ar-idz) are polymers of monosaccharides (Figure 1.10b). A **polymer** (pol′imer) is a long chain of repeating molecular units. Common polysaccharide polymers include starch, glycogen, and cellulose, all of which are glucose polymers. They can be distinguished by the arrangement of the chemical bonds that hold the units together and by which enzymes digest them. Human enzymes break down starch and glycogen,

but cellulose passes through the digestive tract undigested as insoluble fiber or roughage.

Plants and animals use polysaccharides differently. Plants store the excess carbohydrate as starch, so potatoes, rice, grain products, and many other foods derived from plants contain starch. Animals use starch and sugars to make glycogen, which they store mainly in the liver and skeletal muscle. Plants use cellulose in cell walls and other structures. Many fruits and vegetables contain large amounts of cellulose. Celery, which is mainly cellulose and provides human cells with less energy than it takes to chew it, is more suitable for a weight-loss diet than L-sugars—and is considerably cheaper. Arthropods, such as lobsters, shrimp, and insects, make a complex polysaccharide called chitin, which serves as an external skeleton.

See Questions—Objective 7

Lipids

Butter and salad oil consist entirely of lipids; fatty meats and other fatty foods contain large quantities of lipids. **Lipids** are a chemically diverse group of substances—fats, steroids, vitamins A, D, E, and K—that are relatively insoluble in water but soluble in solvents such as ether and

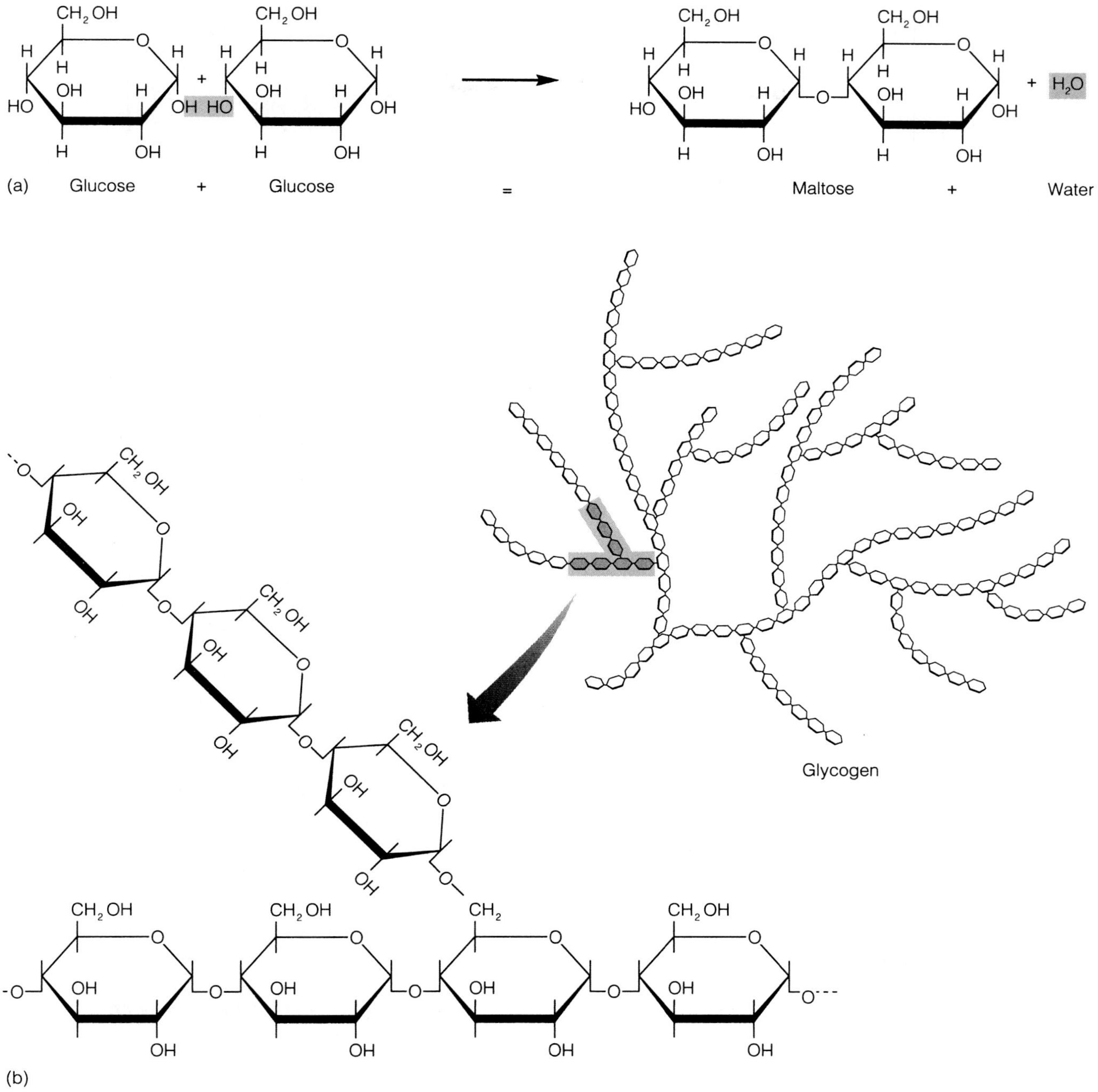

(a) Glucose + Glucose = Maltose + Water

Glycogen

(b)

Figure 1.10

(*a*) Two molecules of the monosaccharide glucose combine to form one molecule of the disaccharide maltose. The removal of water from hydroxyl groups of these molecules creates a glycosidic bond between the glucose units, and forms maltose. (*b*) The polysaccharide glycogen is composed of many glucose units linked together by glycosidic bonds.

benzene (Table 1.7). Lipids easily mix with these solvents because both the lipids and the solvents are nonpolar. Lipids serve several functions in the body. Some form part of cell membranes and others store energy. The vitamins participate indirectly in various chemical reactions. Cholesterol is used to make bile salts that help digest lipids.

Steroid hormones from the ovaries, testes, and adrenal glands regulate various body processes. Generally, lipids contain relatively more hydrogen and less oxygen, and therefore, more energy than carbohydrates. We should pay careful attention to the amount of lipids in our diet because it is easy to eat more than we need. As we shall see

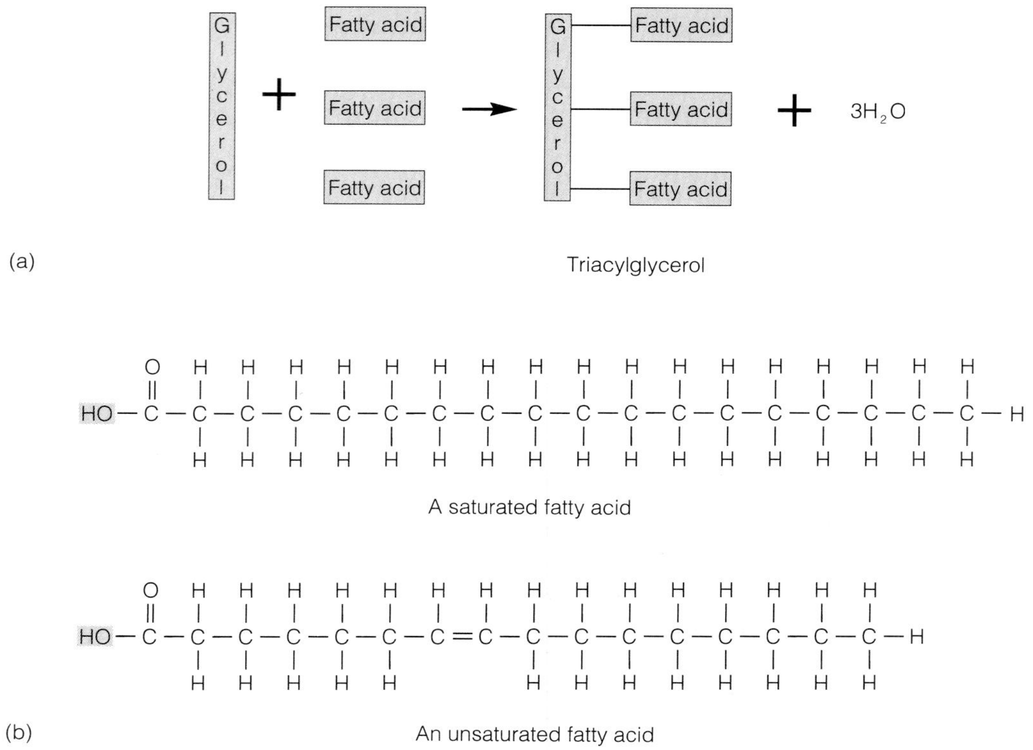

(a)   Triacylglycerol

A saturated fatty acid

(b)   An unsaturated fatty acid

Figure 1.11

Glycerol and fatty acids, either saturated or unsaturated, are the building blocks of a simple lipid. A molecule of water is removed as each fatty acid is attached to glycerol by an ester bond.

later, this can cause unwanted weight gain and increase the risk of a "heart attack," or heart muscle damage because of reduced blood flow.

Simple lipids, common in the diet, contain glycerol (a 3-carbon alcohol) and fatty acids (Figure 1.11). A **fatty acid** is a long chain of carbon atoms with attached hydrogen atoms and a carboxyl group at one end. Simple lipids are made by condensation reactions in which one or more fatty acids are joined to glycerol. Most are **triacylglycerols** (tri-as″il-gli′ser-olz), formerly called triglycerides, having 3 fatty acids connected to glycerol. A **monoacylglycerol** (monoglyceride) has 1 fatty acid and a **diacylglycerol** (diglyceride) has 2 fatty acids. Most mono- and diacylglycerols are formed by hydrolysis of triacylglycerols.

Fatty acids can be saturated or unsaturated. A **saturated fatty acid** is saturated with hydrogen—all its carbon atoms have the maximum number of hydrogen atoms bonded to them. An **unsaturated fatty acid** has lost one or more pairs of hydrogen atoms and a double bond is present between pairs of carbons that have lost hydrogen. Stearic acid is a saturated fatty acid, and palmitoleic acid an unsaturated fatty acid. Vegetable oils are high in unsaturated fatty acids; whereas animal fats contain more saturated fatty acids. Because diets high in fat, especially saturated fats, are associated with heart disease, limiting total dietary fat and consuming mainly vegetable fats may lower the risk of such disease.

Phospholipids (fos-fo-lip′idz), which are found in all cell membranes, are like simple lipids except that they have a phosphate group ($—HPO_4^-$) substituted for a fatty acid (Figure 1.12). The phosphate forms a polar hydrophilic region that enables the molecule to mix with water. On the surfaces of cell membranes, polar regions allow water to surround the membranes. On the surfaces of fat droplets being digested in the small intestine, they help water-soluble enzymes to mix with and digest fats.

Steroids have a characteristic 4-ring structure (Figure 1.13). Cholesterol, the main body steroid, is found in all cell membranes and is used to make steroid hormones, bile salts, and vitamin D. A few of the more important steroid hormones are testosterone from the testes, estradiol and progesterone from the ovaries, and cortisol from the cortex of the adrenal glands (small glands located above the kidneys). Bile salts help to mix fats with watery digestive juices and vitamin D makes calcium

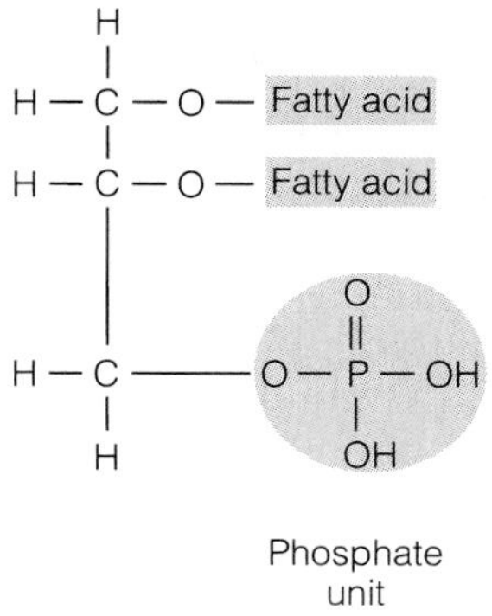

Figure 1.12

A phospholipid differs from a simple lipid by the substitution of a unit of phosphate for one of the fatty acids.

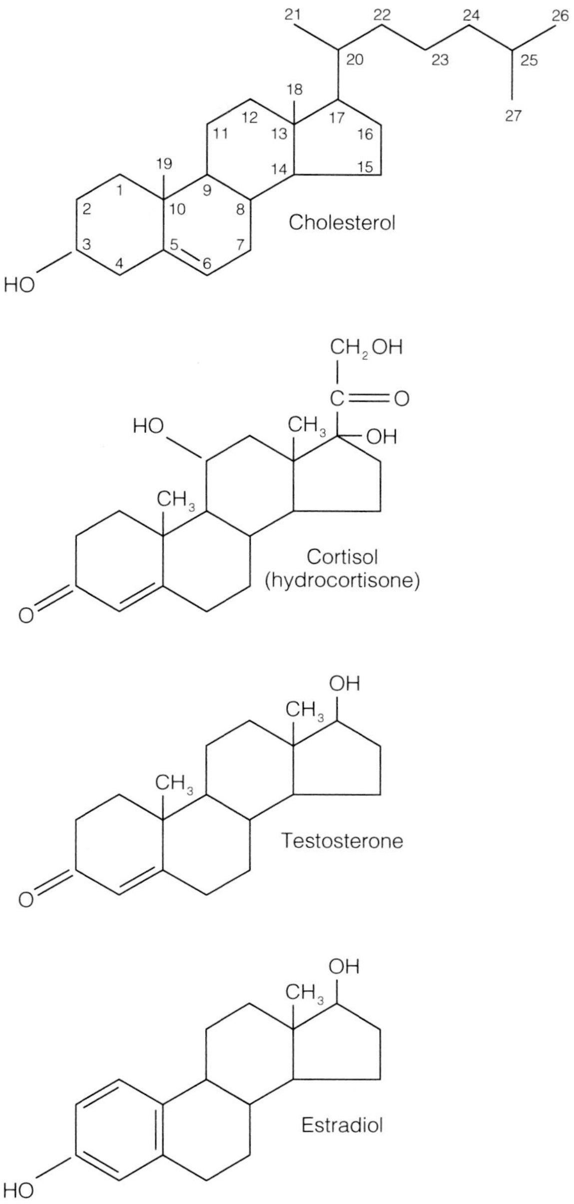

available for building and maintaining strong bones. In spite of the body's many uses of cholesterol, this fatty substance often becomes deposited in heart blood vessels increasing the risk of a heart attack (chapter 7).

The human body contains other lipids such as vitamins A, E, and K and prostaglandins. Vitamin A is required for normal vision, vitamin E prevents saturation of unsaturated fatty acids in cell membranes, and vitamin K is essential for blood clotting. Prostaglandins, derived from a 20-carbon fatty acid, arachidonic acid, are important regulators of functions such as contraction and relaxation of muscle cells within organs.

See Questions—Objective 8

Proteins

Foods such as meats, eggs, dairy products, legumes (peas and beans), nuts, cereal grains, and seeds contain significant amounts of protein (Table 1.8). Protein is essential for maintenance of cell membranes and organelles, synthesis of enzymes, cell division, and other processes. Children need protein for growth and normal brain development. Young children deprived of protein grow slowly and can become mentally retarded.

Proteins are composed of building blocks called **amino acids.** In addition to carbon, hydrogen, and oxygen, amino acids contain nitrogen and some also contain sulfur. Each amino acid has at least one basic amino (—NH$_2$) group and one acidic carboxyl (—COOH) group. Naturally occurring amino acids are L-isomers. The body can synthesize about half of the 20 amino acids found in proteins, some of which are shown in Figure 1.14. Amino acids the body cannot synthesize, called the **essential amino acids,** must be supplied in the diet (chapter 8).

A protein is a polymer of amino acids linked together by **peptide** (pep′tid) **bonds**—bonds between the ni-

Figure 1.13

The steroid cholesterol is typical of all steroids, which have a particular arrangement of the carbon rings and the long nonpolar side chain. Cholesterol is the precursor molecule from which all other steroids, such as sex hormones, are made.

trogen of one acid's amino group and the carbon of another acid's carboxyl group (Figure 1.15). Two amino acids linked together make a **dipeptide,** 3 make a **tripeptide,** and many (4 to 100) make a **polypeptide.** Some amino acids have sulfhydryl (—SH) groups—functional groups that combine to form disulfide (−S−S−) linkages between amino acid chains and maintain the three-dimensional shape of proteins.

Questions

Objective 8

(a) What are the distinguishing properties of different kinds of lipids?

(b) In what ways are lipids important in body function?

(c) How do saturated and unsaturated fatty acids differ?

Table 1.8
Classification of Proteins

Type of Protein	Characteristics
Structural	Form the structure of cell parts or cell products, including portions of cell membranes and organelles; keratin in skin, hair, and nails; collagen in bone, cartilage, and tendons
Motile	Fibrous proteins involved in muscle contraction and movement of microtubules within cells
Enzymes	Control the rate of chemical reactions in cells
Regulatory	Form receptor sites on cell membranes; act as hormones
Other	Antibodies that protect against infection, proteins that transport substances in the blood; albumins that help to maintain blood osmotic pressure

Proteins have increasingly complex levels of structure (Figure 1.16). The **primary structure** is made up of a specific sequence of amino acids in a polypeptide chain held together by peptide bonds. The **secondary structure** is formed by coiling or folding a polypeptide chain. Coiled polypeptides form a helix that has a shape resembling a bedspring, and folded polypeptides form a pleated sheet that resembles a Japanese fan in shape. Further folding and twisting of proteins into globular (irregular, spherical) shapes or fibrous threadlike strands produces the **tertiary structure.** All proteins have these three structural levels, but certain very large proteins, such as hemoglobin, contain several tertiary units that aggregate to form a **quaternary structure.**

Secondary and higher levels of protein structure are maintained by forces such as disulfide linkages and hydrogen bonds. Some proteins change shape as these weak forces are altered. Alterations in shape can alter molecular function. Such alterations can be caused by temperature or pH changes; thus physical changes can control some physiologic processes.

Acidic or basic conditions and above-normal temperatures disrupt the forces that maintain protein structure, causing **denaturation** (de-nat-ur-a'shun). The severity and permanency of denaturation depends on the harshness of conditions. Slight acidity or a moderate rise in temperature can temporarily separate the units of a quaternary protein or "unwind" the tertiary arrangement of a globular protein. The protein usually can function again when conditions return to normal and its structure is restored. Harsh conditions, such as boiling in acid, permanently disrupt all structural levels—even peptide bonds.

In addition to specific amino acid sequences and particular three-dimensional shapes, protein molecules have charged, polar hydrophilic (water-loving) regions and uncharged, nonpolar hydrophobic (water-fearing) regions. The presence of such regions causes molecular folding in which hydrophilic regions remain near the surface and hydrophobic ones get tucked inside the molecule. Water aversion of hydrophobic regions holds them close together and stabilizes the shape of the molecule.

OH
|
C
HC   CH
HC   CH
C
|
CH₃ CH₃   CH₂

H   R   H   O   H   O
\ | \   CH   //   \   CH₂   //
N — C — C=O   N — C — C   N — C — C
/ | \   / | \   / | \
H | OH   H | OH   H | OH
           H   H   H

Amino group Carboxylic acid group   Valine   Tyrosine

Amino acid structure   **Nonpolar amino acids**

H₂N — C = NH
|
NH   SH   O   OH
| | \\ /
(CH₂)₃   CH₂   C
| | |
H | O   H | O   CH₂
\ | //   \ | // |
N — C — C   N — C — C   H | O
/ | \   / | \   \ | //
H | OH   H | OH   N — C — C
         H   H   / | \
                                                    H | OH
                                                           H

Arginine   Cysteine   Aspartic acid
(basic)   (sulfur-containing)   (acidic)

Polar amino acids

Figure 1.14

All amino acids have at least one amino group and one carboxyl group (shown in color). Nonpolar amino acids such as valine and tyrosine contain one each of amino and carboxylic acid groups. Polar amino acids can have an extra amino group and be basic as in arginine, a sulfhydril group as in cysteine, or an extra acid group as in aspartic acid.

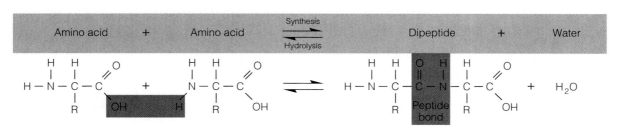

Figure 1.15

Amino acids are attached to each other by the removal of water—OH from the carboxyl group of one amino acid and H from the amino group of another amino acid—in dehydration synthesis. A peptide bond forms between two amino acids from the carbon of a carboxyl group to the nitrogen of an amino group.

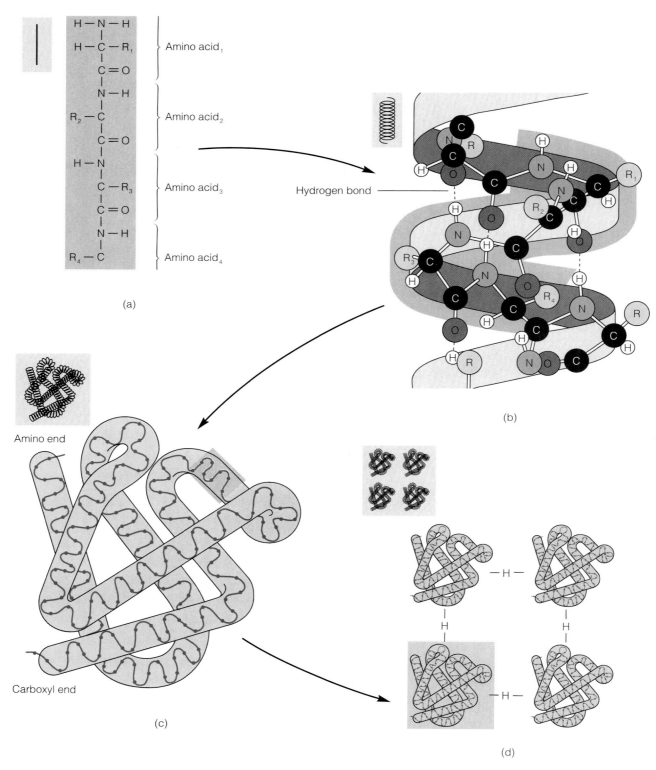

(a)

(b)

(c)

Amino end

Carboxyl end

(d)

Hydrogen bond

Figure 1.16

Proteins have at least three and sometimes four structural levels: (*a*) The primary structure consists of a chain of amino acids connected by peptide bonds. This is analogous to an uncoiled "Slinky®." (*b*) The secondary structure often consists of a helix held together by hydrogen bonds and disulfide linkages. This is analogous to a coiled "Slinky." (*c*) The tertiary structure can consist of complex folding of the helical structure. This is analogous to a tangled "Slinky." (*d*) Large proteins sometimes have a quaternary structure, which consists of several tertiary structures bound together. This is analogous to three or four "Slinkies" tangled together. Tertiary and quaternary structures are held together by the same kinds of forces that maintain the secondary structure.

Slinky® is manufactured in the United States by James Industry, Inc., Hollidaysburg, PA 16648.

Certain configurations of surface shape and charge on protein molecules form **receptors,** or **binding sites** (Figure 1.17), at which a particular molecule, atom, or ion can bind, or ligate, to the protein. A substance that binds to a receptor is called a **ligand** (li′gand). Some receptors are highly specific—they only bind to one ligand. Others are only relatively specific—they bind to any of a class of ligands with shapes and charges that complement those of the receptor. The affinity between receptors and ligands also depends on how tightly they fit. Different ligands that bind to the same site can compete for receptors, and receptors can become saturated. Which ligands bind to receptors depends on affinities, ligand concentrations, availability of receptors, and random collisions of ligands with receptors. These effects are demonstrated later with the discussion of enzymes and their inhibition.

Binding of ligands to protein receptors is very important in physiology: It allows communication at the molecular level, and the information communicated, in turn, contributes to the control of a living system. Such control occurs in the actions of enzymes and hormones, reactions of the immune system, in neural functions, and in many other processes. The presence of products of reactions can sometimes provide negative feedback to the control, or ligand-binding, component of the system.

Proteins can be classified by their functions in the body as structural proteins, motile proteins, enzymes, or regulatory proteins. **Structural proteins** form the structure of body parts. Various proteins form portions of membranes and organelles of all cells. The protein keratin contributes to the structure of the skin, hair, and nails. The protein collagen contributes to the structure of bone, cartilage, tendons, ligaments, and other connective tissues. Keratin and collagen, which form long, strong fibers, are classified as fibrous proteins.

Motile proteins contribute to movement from the cellular to the whole body level. They include actin, myosin, and other proteins that are abundant in muscle cells. These proteins also appear in limited quantity in microtubules and other components of the internal skeleton of cells.

Enzymes (en′zimz) control the rate of chemical reactions in the body. Often globular, enzymes can change shape as they control a reaction. Some proteins in membranes can function as enzymes and also contribute to the structure of cells.

In addition to enzymes, **regulatory proteins** include some hormones and proteins in cell membranes that exert control over cellular functions. Binding of a hormone ligand to a receptor on a membrane protein can often regulate cell function. Other proteins are harder to classify. Antibodies protect against infections; fibrinogen and

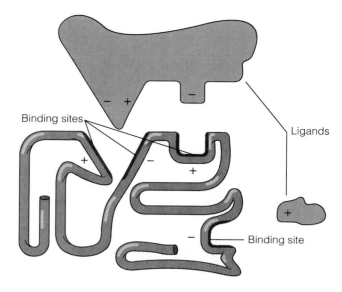

(a) Protein molecule

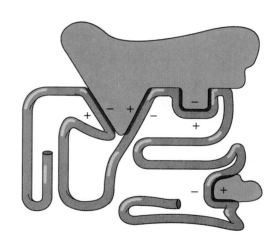

Protein molecule with ligands attached to binding sites

(b)

Figure 1.17

(*a*) Protein molecule and unbound ligands. (*b*) The attachment of ligands to binding sites of a protein molecule.

certain other proteins cause blood to clot; albumins help to maintain blood osmotic pressure; and some proteins transport lipids and other substances in the blood.

See Questions—Objective 9

Questions

Objective 9

(a) What are the distinguishing properties of proteins?

(b) Why are receptors and ligands important?

(c) How are proteins classified?

(d) Define:

essential amino acid

peptide bond

denaturation

(e) What are the four levels of protein structure?

Nucleotides and Nucleic Acids

Nucleotides are found in high-energy compounds such as ATP, molecules called coenzymes that assist certain enzymes, and the nucleic acids DNA and RNA. A **nucleotide** (nu′kle-o-tĭd) has three parts: (1) a nitrogenous base, so-called because it contains nitrogen and has alkaline properties, (2) a 5-carbon-sugar, and (3) one or more phosphate groups (Figure 1.18a).

The nucleotide **adenosine** (ad-en′o-sen) **triphosphate** (ATP) is a cell's main energy currency molecule because it captures energy from food and releases it in a form cells can use (Figure 1.18b). **Guanidine** (gwan′id-en) **triphosphate** (GTP) and **uridine** (ur′id-en) **triphosphate**

(UTP) also can capture and release energy. Enzymes control the hydrolysis of bonds between phosphates so as to make energy available at appropriate sites in cells as needed.

Nucleic (nu-kle′ik) **acids,** long polymers of nucleotides, are information molecules that control heredity and protein synthesis. **Ribonucleic acid** (RNA) consists of one strand of nucleotides and **deoxyribonucleic acid** (DNA) consists of two strands arranged as a double helix. The helix is about 2 nm wide, makes a full turn every 3.4 nm, and has bases that are separated by a distance of .34 nm. DNA and RNA have somewhat different building blocks (Table 1.9). DNA contains four nitrogenous bases: adenine, cytosine, guanine, and thymine. RNA contains

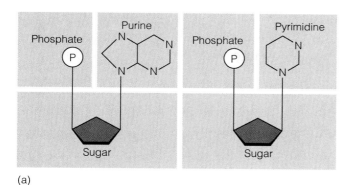
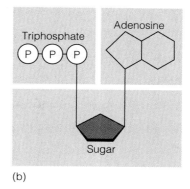

(a)   (b)

Figure 1.18

(*a*) A nucleotide consists of a nitrogenous base, a five-carbon-sugar, and at least one phosphate. (*b*) The nucleotide ATP consists of adenine, ribose, and three phosphate units.

Table 1.9
Components of DNA and RNA

Component	Found in DNA	Found in RNA
Adenine	X	X
Guanine	X	X
Cytosine	X	X
Thymine	X	
Uracil		X
Ribose		X
Deoxyribose	X	
Phosphoric acid	X	X

the same bases except that uracil is substituted for thymine. Of these bases, adenine and guanine are purines, molecules with a double-ring structure; thymine, cytosine, and uracil are pyrimidines, molecules with a single-ring structure. RNA contains ribose and DNA contains deoxyribose, which has one less oxygen atom than ribose, and both contain phosphate.

In both DNA and RNA, sugar and phosphate molecules form the "backbone" of the strand from which the nitrogenous bases protrude (Figure 1.19). The two strands of DNA are held together mainly by hydrogen bonds between pairs of bases on adjacent strands, in which adenine always pairs with thymine and cytosine always pairs with guanine. This linking of bases is called **complementary base pairing.** The same kind of complementary base pairing also occurs as information is transmitted from DNA to RNA in protein synthesis (chapter 2). Complementary base pairing is a kind of communication—*it provides the sole means by which genetic information directs cellular activities.*

Any particular DNA or RNA strand contains several hundred nucleotides with bases in a particular sequence. This sequence of nucleotides, like the sequence of letters in words and sentences, conveys information that specifies which particular proteins an organism will have. Like changing a letter in a word, changing a nucleotide in a sequence can change the information it carries. The large numbers of nucleotides in nucleic acids makes possible a nearly infinite number of different base sequences—and a great many different pieces of information—in DNA and RNA.

The functions of DNA and RNA are to convey and express genetic information. DNA, located in cell nuclei and transmitted from one generation to the next, determines the heritable characteristics of the new individual by supplying the information for the synthesis of its proteins. Except for identical twins, each human being has a unique set of DNA molecules. RNA, which functions outside the nucleus, uses information from DNA to direct protein assembly. More will be said about these processes in chapter 2.

See Questions—Objective 10

Bioenergetics

Bioenergetics (bī′o-en-er-jet′iks), the study of the flow of energy in living systems, is a branch of science that deals with energy changes in living organisms. It especially includes laws governing chemical changes and coupled reactions.

Three laws—the first and second laws of thermodynamics and the law of mass action—are especially important in understanding energy flow in living systems. Because life cannot exist without energy, we need at least a simple introduction to these laws. Building a perpetual motion machine, a machine that once started would run

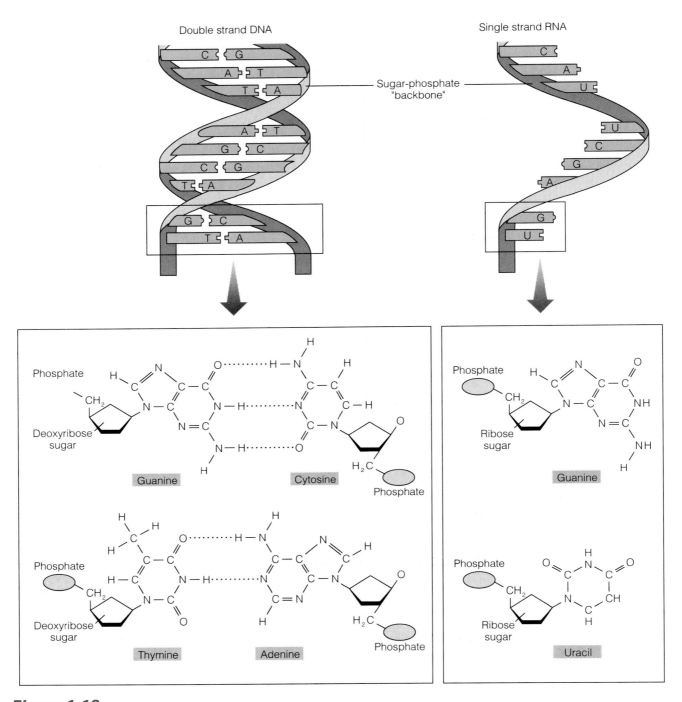

Figure 1.19

Nucleotide polymers are arranged in strands in DNA and RNA. DNA contains two strands held together by complementary base pairing—A with T and C with G. RNA contains a single strand of nucleotides. Bases are connected to sugars in the formation of nucleotides and to each other in the nucleic acid DNA. The bases guanine and cytosine always pair with each other as do thymine and adenine. Uracil replaces thymine in the nucleic acid RNA.

Questions

Objective 10

(a) Briefly describe the structure and function of:

ATP

nucleotide coenzymes

DNA

RNA

(b) What properties of nucleic acids contribute to their ability to store and transmit information?

a hill. At the top of the hill the rock has a quantity of **potential energy,** or stored energy. As it rolls down the hill it releases **kinetic energy,** or energy in action, into the surroundings as heat and contains less potential energy when it reaches the bottom of the hill. Potential energy at the hill top equals the potential energy remaining at the bottom plus the energy released as heat. Some energy was transformed into heat but none was lost, so the system obeys the first law of thermodynamics.

Like the rock rolling down a hill, some chemical reactions also release energy. The energy in the **reactants** (substances that enter a reaction) equals the energy in the **products** (substances remaining at the end of the reaction) plus energy released as heat. Such reactions are **exergonic** (ex-er-gon'ik). They include reactions that occur during **catabolism** (kat-ab'o-lizm), the **degradation** (breakdown) of large molecules into smaller ones, in which energy is released from chemical bonds.

In contrast, some systems gain energy as a rock would if someone carried it up a hill. Some chemical reactions also gain energy, and the energy in the reactants plus energy added from the surroundings equals the energy in the products. Such reactions are **endergonic** (end-er-gon'ik). They include reactions that occur during **anabolism** (an-ab'ol-izm), the **synthesis** (building) of larger molecules from smaller ones with energy incorporated in chemical bonds.

Endergonic and exergonic reactions will share energy with one usually providing for the other, and both reactions will obey the first law of thermodynamics. Energy is transferred from the exergonic reaction to the endergonic and the total energy in the system of coupled reactions remains constant. In general, exergonic reactions are spontaneous whereas endergonic reactions are not.

The melting of ice is a spontaneous endergonic reaction—an exception to above statement. In surroundings warmer than 0° C, ice spontaneously takes up energy from the surroundings. Fluid water that forms as ice melts contains more energy than solid water in the ice. The **second law of thermodynamics** dictates that in any process involving spontaneous change from the initial to the final state, the randomness, or disorder, of the system and its surroundings will increase. Such randomness is called **entropy** (en'tro-pe). Melting ice uses energy and water molecules become more randomly distributed. (An irregular puddle is certainly less orderly than a block of ice.) Any spontaneous change is accompanied by an increase in entropy. Living organisms fight a constant battle against entropy—eventually losing the battle in death.

Many chemical reactions are reversible, that is, they can go either direction from A to B or from B back to A. If more A than B is present, the reaction will go toward B. If more B than A is present, the reaction will go toward A. The ability of reversible reactions to be driven by the substance present in greatest concentration is called the **law of mass action.**

forever without an energy supply, has captured the attention of many inventors. No matter how ingeniously designed, such machines eventually run down because of friction between the mechanical parts. Careful measurements on such machines show that heat energy from friction exactly equals the energy used to start the machine—an example of the first law of thermodynamics.

According to the **first law of thermodynamics,** energy can be neither created nor destroyed but can be transferred in various ways, including rolling a rock down

Endergonic reactions usually are coupled to, or occur simultaneously with, exergonic reactions. Frequently coupled reactions in living organisms include the capture and release of energy in ATP and oxidation-reduction reactions.

The formation of ATP, when a cell captures energy from food

$$ADP + P_i + energy \rightarrow ATP$$

(where P_i is an inorganic phosphate), is coupled with the release of energy from glucose or other foods.

$$glucose + 6\ O_2 \rightarrow 6\ CO_2 + 6\ H_2O + energy$$

Similarly, hydrolysis of ATP, when a cell needs energy,

$$ATP \rightarrow ADP + P_i + energy$$

is coupled with chemical reactions such as those that transport substances across cell membranes or cause muscles to contract.

In oxidation-reduction reactions, oxidation can occur by the addition of oxygen or the removal of hydrogen atoms or electrons, and reduction can occur by the removal of oxygen or the addition of hydrogen atoms or electrons. For example, during oxidation of a reduced organic molecule, hydrogen atoms are removed and used in reduction of a coenzyme, such as FAD.

$$succinate\text{-}H_2 + FAD \rightarrow fumarate + FAD\text{-}H_2$$

Similarly, certain coenzymes called cytochromes contain an iron ion. The iron is oxidized when it gains an electron

$$Fe^{2+} + e^- \rightarrow Fe^{3+}$$

and reduced when it loses an electron.

$$Fe^{3+} \rightarrow Fe^{2+} + e^-$$

In the final reaction of cellular oxidation-reduction, 2-hydrogen atoms combine with an oxygen atom to form water. One can think of such a reaction as the simultaneous oxidization of hydrogen and reduction of oxygen.

See Questions—Objectives 11 and 12

Enzymes

The laws governing chemical changes apply to all chemical reactions, including those in living organisms. Most reactions are not feasible at physiologic temperatures, so **enzymes,** or **biological catalysts** (kat′al-istz), are needed to make them occur at rates fast enough to support life. Enzymes increase the rate of reactions. Most can catalyze a given reaction in either direction, and the relative concentrations of reactants and products determine the direction of reactions. To see how enzymes increase reaction

rates, we need to know more about the properties of enzymes and the factors that affect enzyme reactions.

In general, spontaneous reactions occur without input of energy from the surroundings but often at unmeasurably low rates. Though oxidation of glucose is thermodynamically a spontaneous reaction, it does not occur unless energy to start the reaction is available. The energy required to start such a reaction is called **activation energy.** Activation energy can be thought of as a hurdle over which molecules must be raised to get a reaction started (Figure 1.20). By analogy, a rock resting behind a hump at the top of a hill would easily roll down the hill if pushed over the hump. Activation energy is the energy required to lift the rock over the hump. Another way to activate a reaction is to heat the reactants, thus increasing their molecular movement. The quantity of heat needed to activate many reactions would denature proteins and evaporate liquids in living organisms. Enzymes lower the activation energy needed to start a reaction at body temperature and keep the reaction going at a sufficiently high rate to maintain life.

Enzymes, like catalysts in inorganic reactions, are not totally consumed in the reactions they initiate. Though enzyme molecules eventually "wear out" probably by losing the molecular shape necessary for their catalytic properties, they can be recycled and used over again. Also like inorganic catalysts, enzymes are not permanently affected by the reaction they catalyze.

Many enzymes are named by adding the suffix -ase to the name of the **substrate,** or substance upon which they act. For example, phosphatases act on phosphates, lipases act on lipids, sucrase digests the sugar sucrose, and peptidases break peptide bonds.

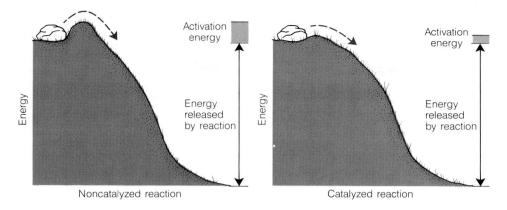

Figure 1.20

Enzymes catalyze reactions in living organisms by lowering the activation energy needed for a reaction to occur. This is analogous to a rock behind a small hump near the top of a hill. The rock will not move down the hill until enough "activation energy" has been added to raise it over the hump.

Enzymes increase the rate of spontaneous reactions by reducing the activation energy required to start them. This allows biological reactions to proceed rapidly at temperatures that living organisms can tolerate.

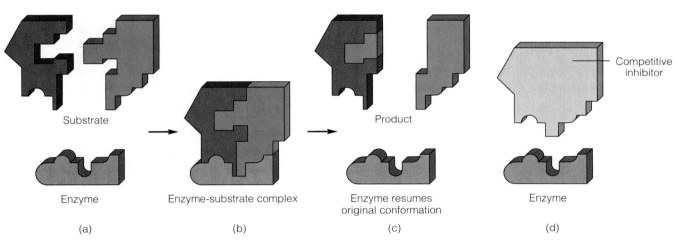

Figure 1.21

(*a*) The surface of an enzyme has an active site on which its normal substrate fits. (*b*) The substrate and enzyme form a complex. (*c*) The substrate that fits the active site is converted to the product(s) of the specific reaction catalyzed by the enzyme. (*d*) A nonsubstrate molecule that fits the active site and prevents the substrate molecules from occupying it can inhibit the action of the enzyme. Because it competes with the normal substrate, it is called a competitive inhibitor.

Enzymes provide a surface on which reactions occur. Each enzyme has an **active site,** a surface binding site at which it forms a loose association with its substrate, or ligand (Figure 1.21a). After binding to the enzyme, the substrate undergoes chemical change and the product or products are formed.

Sometimes a ligand similar to a substrate binds to the active site but fails to react. If the ligand forms a reversible attachment with the active site **competitive inhibition** occurs (Figure 1.21d). Such inhibition is concentration dependent. Having the inhibitor occupy a few active sites part of the time slightly slows the reaction rate.

Having it occupy many sites most of the time greatly slows the reaction rate.

Enzymes also can be inhibited by **noncompetitive inhibitors** such as mercury, lead, and other heavy metals. Such inhibitors permanently attach to an enzyme at a site other than the active site and distort the enzyme so the shape of the active site is altered (Figure 1.22). Affected enzyme molecules are permanently inactivated and the reaction rate is greatly reduced regardless of the concentration of the substrate. Mercurochrome is used to disinfect skin and mercuric chloride is used to disinfect inanimate objects, but substances containing mercury are

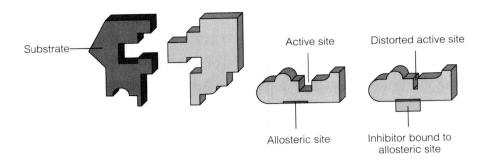

Figure 1.22

In noncompetitive inhibition, an inhibitor binds to a site other than the active site. It distorts the molecular shape so the substrate cannot bind to the enzyme's active site.

highly toxic if ingested. Lead, which is found in exhaust fumes from leaded gasoline and some paints, replaces calcium in bones, interferes with hemoglobin synthesis, and causes brain edema (fluid accumulation). Lead also causes gout by allowing the waste product uric acid to accumulate in the body. The high incidence of gout among port wine drinkers in the nineteenth century has been attributed to lead solder in wine-making equipment.

Another extremely important characteristic of enzymes is **specificity.** All enzymes catalyze a specific reaction and most act on a particular substrate according to the shape and charges at their active sites. Enzymes that act on more than one substrate usually act on a group of substances with a particular functional group or chemical bond. For example, a peptidase can break down many proteins, but it acts specifically on peptide bonds. From a functional point of view, the specificity of enzymes subjects them to control. A single enzyme can be turned on or off without affecting other enzymes and particular enzymes can be synthesized in accordance with the needs of cells.

Many enzymes require coenzymes or cofactors to catalyze a reaction. Coenzymes, such as NAD and FAD, are large molecules bound to or loosely associated with an enzyme. Many coenzymes are synthesized from vitamins—NAD from niacin and FAD from riboflavin—found in whole grains and some other foods. Vitamins are essential in the diet because they are needed to make necessary coenzymes. As we have seen, coenzymes often act as carriers of hydrogen atoms or electrons in oxidative reactions from which cells derive energy. **Cofactors** usually are inorganic ions such as copper, magnesium, and zinc, that are found in trace amounts of many foods. These ions

often improve the fit of an enzyme's active site with its substrate; their presence is essential to allow the reaction to proceed.

Factors that affect enzyme activity, or the rate at which enzymes catalyze reactions, include temperature, pH, and concentrations of substrate, product, and enzyme. Like other proteins, enzymes are subject to alteration by heat and extremes of pH that can affect net charges on protein molecules, thereby altering enzyme activity.

Most human enzymes have an **optimum temperature,** the temperature at which they have the greatest activity—near normal body temperature (Figure 1.23a). Enzyme activity decreases gradually as the temperature drops below normal, but activity decreases rapidly as the temperature rises above normal, especially above 42° C. Nearly all human enzymes are denatured at 50° C.

Similarly, many human enzymes have an **optimum pH** near neutral (Figure 1.23b). Their activity decreases gradually as the pH becomes either more acidic or more basic. Some human enzymes have an acidic or basic optimum pH. Pepsin, a peptidase that breaks down proteins in the stomach, has an optimum pH of 2. Two kinds of phosphatases, enzymes that remove phosphate groups from molecules, occur in human cells. Acid phosphatase has an optimum pH of 5.5 and alkaline phosphatase has an optimum pH of 9.0.

The rate of enzyme catalyzed reactions also is affected by the concentrations of enzymes, substrates, and products. The quantity of enzyme determines how fast the product accumulates if enough substrate is available. With 100 enzyme molecules acting on a substrate, the product will accumulate ten times faster than with 10 enzyme molecules acting on a substrate. The reaction rate in-

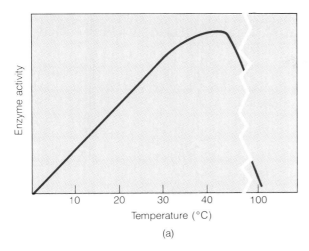

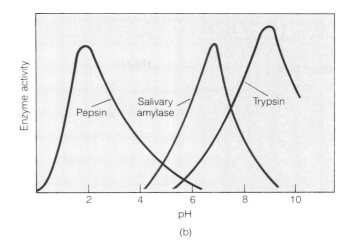

Figure 1.23

(*a*) Most enzymes found in human cells have an optimum temperature of 37° C. Lowering the temperature gradually reduces enzyme activity, but raising it above the optimum temperature rapidly reduces activity because it denatures the enzyme. (*b*) Though most enzymes have an optimum pH near neutral, some work best in acidic or alkaline conditions. Altering the pH from the optimum level also reduces enzyme activity, and sufficiently large changes also denature the enzyme.

creases in proportion to the number of enzyme molecules present and reaches a maximum when all enzyme molecules catalyze a maximum number of reactions per minute. If the concentration of a substrate is too low to keep all the enzyme molecules working at maximum capacity, the substrate concentration will determine the reaction rate.

Substrates and products obey the law of mass action. Large quantities of substrate drive the reaction toward formation of product. Large quantities of product would drive the reaction toward reformation of substrate, but, in most cellular reactions, the product of one reaction quickly becomes the substrate of another reaction in a metabolic pathway.

A **metabolic pathway** is a sequence of enzyme controlled reactions that accomplishes a particular task that usually takes place within a cell. As we shall see in chapter 4, different pathways exist to break down glucose, fatty acids, or proteins for energy and to synthesize glycogen, lipids, proteins, or the waste product urea. Cells have various ways to regulate enzyme activity, and regulating the rate of one enzyme reaction often controls the rate of an entire pathway. For example, suppose a pathway begins with substrate A and goes through **intermediates** B, C, and D, ending with product E. In **feedback inhibition,** the product of a reaction several steps along a metabolic pathway (E) acts to inhibit an early reaction in the pathway (A→B). Feedback inhibition is more common in microbes than in humans, but in one instance in humans

accumulation of the amino acid alanine determines whether pyruvic acid (a product of glucose metabolism) is used for energy or to make more alanine.

See Questions—Objectives 13 and 14

Radioactivity

Radioactivity, the capacity of isotopes of some elements to emit radiation and energy, was first observed in 1896 by French scientist Marie Curie. It was noticed when photographic plates that had inadvertently been laid next to a piece of uranium ore were developed and found to be clouded. Study showed that the uranium must have shed radiation similar to light, but that was invisible and capable of penetrating material light could not. This quality of uranium (and many other elements) is now called radioactivity.

Radioactive emissions—alpha and beta particles and gamma rays—are emitted from unstable nuclei of atoms of radioactive elements. **Alpha particles** consist of 2 protons and 2 neutrons. **Beta particles** are electrons from the disintegration of a neutron into a proton and an electron. **Gamma rays** are short wavelength electromagnetic radiation. Emissions occur at a constant rate—the element's **half-life** ($t_{1/2}$), the time for half the atoms in a sample to emit radiation and disintegrate.

Questions

Objective 12

(a) What two processes are coupled in energy transfer reactions?

(b) What two processes are coupled in oxidation-reduction reactions?

Objective 13

(a) Define:

catalyst

active site

specificity

(b) What properties of enzymes contribute to their ability to catalyze reactions?

(c) In what ways are enzymes physiologically important?

(d) How do competitive and noncompetitive inhibition of enzymes differ?

Objective 14

(a) How do temperature and pH affect the rate of enzyme controlled reactions?

(b) How do the concentrations of enzyme, substrate, and product affect reaction rates?

Radioactive emission causes several kinds of cellular damage. **Ionizing radiation** forms highly reactive ions often by splitting water molecules into H^+ and OH^- ions. The OH^- ions can combine to form hydrogen peroxide, H_2O_2, an unstable product that releases charged oxygen, O_2^-. Both charged oxygen and free H^+ damage enzymes and cellular molecules. **Cosmic radiation,** which passes through the earth's atmosphere from outer space, can cause mutations (changes in DNA) that kill cells, make them cancerous, or alter genetic information in egg or sperm cells.

Different emissions vary in their effects. Alpha particles, though large and potentially harmful, move slowly and do not penetrate the skin. Smaller, faster beta particles penetrate only a few millimeters into tissues and are relatively harmless. Fast-moving gamma rays, such as X rays, easily enter and damage tissues. Elements with long half-lives have great potential for damage if incorporated into tissues. For example, strontium (^{90}Sr), which is sometimes found in milk, has a half-life of 28 years. (The superscript indicates the atomic weight of the isotope.) Being similar to calcium, this isotope concentrates in bones and emits radiation for years. Nuclear power plant wastes have half-lives longer than any available containers.

See Questions—Objective 15

Questions
Objective 15

(a) What is radiation, and how was it discovered?

(b) What properties distinguish each kind of emission from radioactive material?

Study and Review

Did you get the essentials?

The following summary contains the basic concepts from chapter 1 except for the key terms that have been omitted. Try to fill in the blanks from memory. If you cannot fill in all the blanks, review the chapter paying particular attention to boldface terms.

Matter is composed of _____ _____ _____ , which can be combined to make different substances. The sum total of chemical reactions in the body is called _____ . Matter composed of one kind of _____ is an _____ . Two or more atoms combined make a _____ and atoms of different elements make a _____ . The nucleus of an atom contains positively charged _____ and uncharged _____ . Negatively charged _____ are found in an _____ _____ around the nucleus.

Chemical bonds are _____ when they form between ions and _____ when they share electrons. _____ _____ are associated with unequal sharing of _____ ,

Continued on next page

Study and Review

causing molecules to have _____ polar regions and _____ nonpolar regions.

Water, a common _____ (medium in which things dissolve), absorbs large quantities of _____ , has high _____ , and participates in _____ and _____ reactions.

An _____ donates H$^+$ and a _____ accepts H$^+$. On the _____ scale, 7 is _____ , below 7 is _____ , and above 7 is _____ . A _____ is a substance that resists pH change.

_____ chemistry deals with carbon-containing compounds. Such compounds usually have _____ _____ that participate in reactions.

The general name for compounds with hydroxyl groups and an aldehyde or ketone group is _____ . This group includes simple _____ two-unit molecules called _____ , and long polymers called _____ .

_____ , which dissolve in nonpolar solvents, include _____ made of glycerol and fatty acids, _____ that also contain a phosphate, and _____ that have a complex 4-ring structure. Fats are _____ when they contain a maximum amount of hydrogen and _____ when they contain one or more double bonds in the carbon chain.

Nitrogen containing _____ _____ are connected together by _____ bonds to make _____ . The body cannot make _____ _____ _____ . Heat and acidic or basic conditions cause _____ , an alteration in the shape of proteins. Binding of _____ to protein _____ on the surface of cells provides a means of communication at the molecular level. _____ proteins form body parts, _____ proteins contribute to movement, and _____ control the rate of chemical reactions.

A _____ consists of a nitrogenous base, a 5-carbon sugar, and one or more phosphates. ATP contains _____ _____ and _____ . Dinucleotides such as NAD and FAD help enzymes by transporting _____ . DNA provides _____ information and contains two _____ of _____ with bases _____ , _____ , _____ , and _____ ; the sugar _____ , and phosphate. In _____ _____ pairing, adenine pairs with _____ and cytosine pairs with _____ . RNA directs _____ synthesis and contains one _____ of _____ with the same bases except that _____ replaces _____ and the sugar is _____ .

According to the _____ law of thermodynamics, _____ can be _____ but neither created nor destroyed. In an _____ reaction, reactants release _____ as they form products. Such reactions occur during _____ . In an _____ reaction, reactants require _____ to form products. Such reactions occur during _____ . According to the _____ law of thermodynamics, spontaneous reactions proceed toward randomness or _____ . According to the law of _____ _____ , reversible reactions are driven by the substance present in the greatest concentration. Examples of coupled reactions include the transfer of _____ and _____ -_____ reactions.

Enzymes act by lowering the _____ _____ needed for a reaction to occur. They have _____ , that is, they act on a particular substrate after it binds to the enzyme's _____ _____ . Inhibition of an enzyme reaction is _____ when the _____ competes with the substrate and _____ when the _____ binds permanently and distorts the enzyme. _____ assist enzymes. Factors that affect enzyme reactions include _____ , _____ , and the concentrations of _____ , _____ , and _____ .

Basic Cellular Chemistry

2

OBJECTIVES

Cells and Cell Membranes

1. Briefly describe the chemical properties of cells and use the fluid-mosaic model to explain membrane properties.

Passage of Materials Across Membranes

2. Distinguish between active and passive transport across cell membranes.
3. Describe the processes of simple diffusion, facilitated diffusion, and osmosis and explain their significance.
4. Use the sodium-potassium pump to explain the process of active transport and its significance.

Organelles

5. Briefly describe the chemical functions of each organelle.
6. Explain how cellular functions are integrated.

DNA Replication

7. Summarize the steps in DNA replication and explain its significance.

Protein Synthesis

8. Summarize the steps in protein synthesis and explain its significance.

Genetics

9. Explain the basic principles of genetics and the nature of mutations.

Study and Review

Cells and Cell Membranes

In general, cells consist of internal structures called **organelles** (or-gan-elz') that are suspended in a semifluid **cytoplasm** and surrounded by a **cell membrane** (Figure 2.1). Each organelle and each enzyme in the cytoplasm perform a specific function; the cell membrane defines the boundaries of a cell and regulates the movement of substances into and out of the cell.

Cell functions involve chemical processes in several ways. Cells use oxygen and molecules derived from food to capture energy in ATP. They use that energy to enable muscles to exert force and the nervous system to relay signals. Certain kidney cells remove wastes and adjust concentrations of substances in the blood, and endocrine cells regulate body functions by synthesizing and secreting chemical messengers called hormones. Nearly all chemical reactions in the body are controlled by cellular enzymes.

Biochemical and biophysical studies of cells have complemented the microscopic studies, and many new techniques have been developed to study cell function. Cells can be centrifuged at different speeds in different media separating their organelles for independent study. Enzyme functions in one organelle can be distinguished from those of other organelles by this technique. Molecules in living cells can be made radioactive or fluorescent and followed as they go through metabolic processes.

The **plasma membrane,** or **cell membrane,** defines the boundaries of a cell and membranes within cells define the boundaries of certain organelles. These membranes all have the same structural components (Figure 2.2). According to the **fluid-mosaic model,** a membrane consists of a semifluid lipid matrix in which proteins are imbedded to form a mosaic. (A mosaic is a composite of many small pieces.) The lipid matrix is made of **phospholipids** arranged in 2 layers. The polar hydrophilic phosphate regions of the molecules are oriented toward the extracellular and intracellular surfaces, which are exposed to aqueous fluids. The nonpolar hydrocarbon ends point toward each other in the interior of the membrane. The lipid matrix also contains cholesterol, which contributes to the fluidity of the membrane at body temperature. Membrane proteins are interspersed among the lipids. **Integral proteins** extend partially or completely through the lipid layers, and **peripheral proteins** are loosely attached to membrane surfaces.

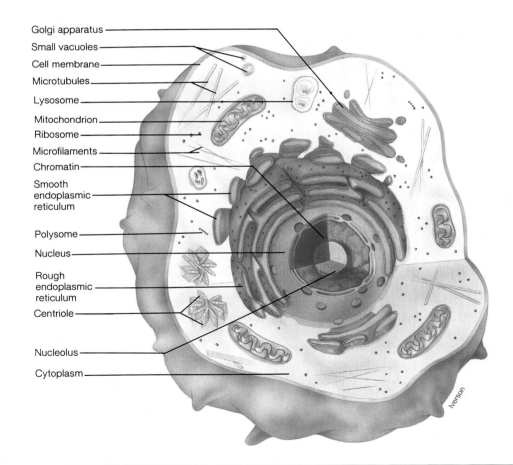

Golgi apparatus
Small vacuoles
Cell membrane
Microtubules
Lysosome
Mitochondrion
Ribosome
Microfilaments
Chromatin
Smooth endoplasmic reticulum
Polysome
Nucleus
Rough endoplasmic reticulum
Centriole
Nucleolus
Cytoplasm

Figure 2.1

An animal cell and its major organelles.

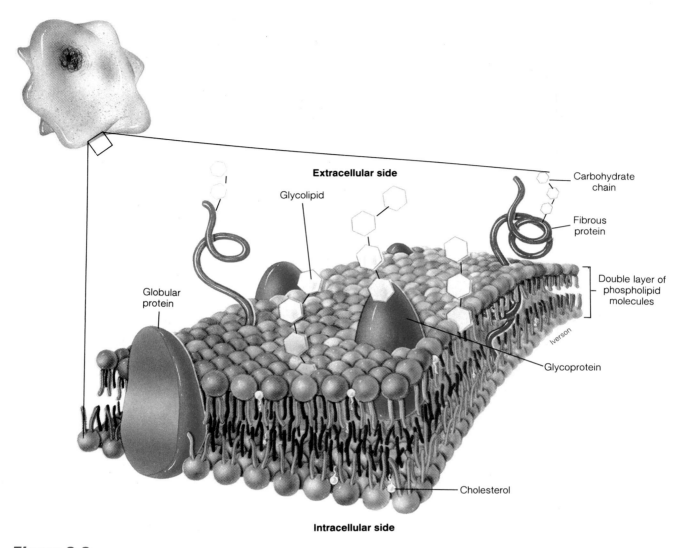

Extracellular side

Glycolipid

Carbohydrate chain

Fibrous protein

Double layer of phospholipid molecules

Globular protein

Glycoprotein

Cholesterol

Intracellular side

Figure 2.2

Each cell is surrounded by a plasma membrane (cell membrane) that separates it from neighboring cells. The widely accepted Singer-Nicholson fluid-mosaic model shows the plasma membrane to consist of two layers of lipids with protein molecules interspersed through it in a mosaic pattern.

Integral proteins are nearly surrounded by lipids, whereas peripheral proteins extend from the outer or inner surfaces of the membrane. The glycocalyx, or cell coat, contains glycoproteins and glycolipids.

Sugars combine with some membrane proteins to form **glycoproteins** (gli″ko-pro′te-inz). When sugars combine with some membrane lipids they form **glycolipids** (gli″ko-lip′idz). Glycoproteins and glycolipids cover about 7 percent of a cell's extracellular surface, or **glycocalyx** (gli′ko-kal′iks). These sugary molecules act as cell recognition sites, which are important in cell-to-cell interactions. For example, molecules called antigens on the surface of red blood cells identify the cells—and the person's blood type—as A, B, or AB. They also form binding sites that allow various chemical substances such as hormones to attach to the cell's surface and influence cell function.

The primary function of the plasma membrane is to control the passage of substances into or out of the cell and both its lipids and it proteins participate in this process. The plasma membrane regulates the environment within the cell and helps to maintain intracellular homeostasis.

The lipid portion of the membrane allows lipid-soluble substances to move across it but passively prevents water-soluble substances from doing so. Many substances both inside cells and in the interstitial fluids around them are water soluble, so membrane lipids, which are insoluble in water, provide an ideal boundary between the cell and its environment. Phospholipids have polar and nonpolar

Questions

Objective 1

(a) What is meant by this statement, "Many body functions take place at the cellular level"?

(b) Describe the structure of a cell membrane as proposed by the fluid-mosaic model.

(c) What would be the effect on cell membrane function if any of the following components were absent or nonfunctioning?

phospholipids

proteins

glycoproteins

glycolipids

---

ends as explained in chapter 1. Their hydrophilic polar ends are exposed to aqueous solutions, but their hydrophobic nonpolar portions prevent water and substances dissolved in it from crossing the membrane.

Some integral proteins form channels (already shown in Figure 2.2) just large enough to allow ions and water molecules to pass through the membrane. Living membranes are highly permeable to water. The proteins of channels are free to change shape, to open and close channels, and to move laterally in the semifluid lipid matrix. They regulate the movement of substances into and out of cells by several mechanisms to be described later in this chapter and in subsequent chapters.

In addition to regulating movements of substances, proteins of the plasma membrane perform other functions. They give the membrane structural stability. Some proteins are enzymes that regulate chemical reactions taking place in the membrane or on the membrane's intracellular surface. Membranes associated with organelles also control movements of substances into and out of the organelles, but they perform other functions, as will be explained, for each organelle.

See Questions—Objective 1

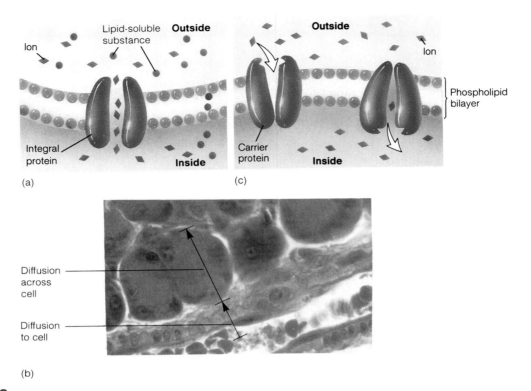

Figure 2.3

(*a*) Lipid-soluble substances diffuse through the phospholipid bilayer. Water, ions, and small water-soluble molecules diffuse through small pores formed by some of the integral proteins. (*b*) Substances easily diffuse from blood to nearby cells and to interior of cells if they have a high enough surface-to-volume ratio. (*c*) A possible model to explain facilitated diffusion. The carrier molecule, which is imbedded in the membrane, may have several binding sites. The carrier moves in the membrane so as to "push" the diffusing substance across the membrane. The same mechanism could explain movement of substances in either direction across the membrane.

Passage of Materials Across Membranes

A living cell is a dynamic entity with substances constantly moving in and out of it across the plasma membrane. Understanding how these movements occur is essential to understanding how a cell functions. Polar substances, such as water and small ions, are believed to move across membranes by passing though channels composed of membrane proteins. Nonpolar substances, such as lipids and other uncharged particles, dissolve in the membrane lipids and diffuse through them (Figure 2.3a). Still other substances are moved through the membrane by carrier molecules.

The manner of movement of an ion or molecule across a membrane is determined by a combination of the following factors: particle size; electrical charge; relative concentrations of the substance on both sides of the membrane; lipid solubility; and the availability of carrier molecules in the membrane. Substances move across membranes by passive processes that require no cellular energy and by active processes that require it.

Passive transport processes involve movements of substances down a **concentration gradient,** that is, from a region of higher concentration to a region of lower concentration, without requiring cells to supply energy. Passive processes include simple diffusion, facilitated diffusion, and osmosis. In contrast, **active transport processes** require energy from ATP to move substances across the cell membrane.

Molecules are in constant motion and have kinetic energy proportional to the temperature. **Simple diffusion** is the net movement of molecules from their region of higher concentration to their region of lower concentration along a concentration gradient. Diffusion occurs because of the random motion of molecules in a liquid or gas. The net movement of molecules is down the concentration gradient. At equilibrium, no gradient exists, random movement continues but no net movement occurs.

The time required to reach equilibrium by diffusion increases with molecule size and the distance between the regions of high and low concentration. For ions and small molecules, equilibrium is reached across a distance of 1 micrometer (μm) in 0.06 seconds, across 10 μm in 6 seconds, across 100 μm in 11 minutes, and across

1,000 μm (1 mm) in 18 hours. Cells lying within 10 μm to 20 μm of a blood vessel receive small molecules in a matter of seconds, but cells located farther from blood vessels receive nutrients more slowly.

Cells with diameters no greater than about 20 μm typically have a high **surface-to-volume ratio**—the amount of surface area relative to the volume. A sufficiently high surface-to-volume ratio allows small molecules and ions to enter and diffuse throughout the interior of a cell in a matter of seconds. The cells receive adequate nutrients and rid themselves of wastes (Figure 2.3b). Large cells sometimes achieve an acceptable surface-to-volume ratio with unusual cell shapes. For example, muscle cells (fibers) can be quite long (up to 0.5 m), but they have a relatively small diameter. Substances diffuse rapidly over the distance from the cell surface to its center.

When diffusion occurs across a living plasma membrane, the rate of diffusion is determined, not only by the concentrations of substances on the two sides of the membrane, but also by **membrane permeability.** The membrane itself severely limits diffusion. Another limiting factor is the **unstirred water layer,** a 100 μm to 400 μm layer of relatively stationary water molecules that coat membrane surfaces. Because living membranes allow some substances to pass through and prevent others from doing so they are said to be **selectively permeable.**

Substances can move through plasma membranes either by diffusing through the lipid layer or by passing through the small channels provided by integral proteins. The rate of diffusion through the lipid layer is affected by the solubility of the diffusing substance in lipid, temperature, and the concentration gradient. Nonpolar substances such as fatty acids, steroid hormones, and gases cross the membrane rapidly by dissolving in the nonpolar fatty acids in the membrane.

The rate of diffusion through membrane channels is affected by the size and charge of the diffusing particles and by the size of the channels and the charges on their surfaces. Studies of small particle movement through channels indicate that the channels are less than 0.8 nm in diameter. Only water, small water-soluble molecules, and ions such as H^+, K^+, Na^+, Ca^{2+}, and Cl^- pass through these channels, and special channels with a certain size, configuration, and charge exist for each ion type.

Facilitated diffusion is the diffusion of a substance across a membrane with the assistance of a carrier molecule. Glucose, some amino acids, and vitamins enter cells by facilitated diffusion at rates faster than is possible by simple diffusion. Like simple diffusion, facilitated diffusion involves movement down a concentration gradient without expenditure of ATP. Unlike simple diffusion, facilitated diffusion requires a carrier molecule, a membrane-bound protein that binds to one or a few specific molecules and assists in their movement across the membrane. A proposed mechanism of facilitated diffusion

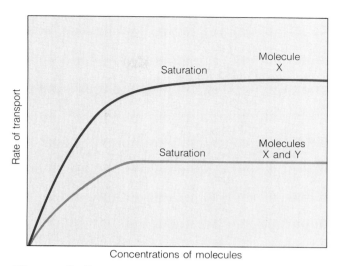

Figure 2.4

In facilitated diffusion, the maximum flow of a substance into a cell occurs when the carrier molecules become saturated, that is, when they are carrying the diffusing substance as fast as they can. When two substances transported by the same carrier are present in equal concentrations, at any one moment, half of the carrier molecules will be transporting one substance and half the other substance. Assuming equal affinities of the substances for the carrier, the total number of molecules transported will remain the same.

is that the carrier acts like a revolving door, providing a channel for the movement of substances across a membrane (Figure 2.3c).

Because a membrane contains only a limited number of carrier molecules, facilitated diffusion displays carrier saturation and competition between structurally similar molecules for a carrier (Figure 2.4). **Carrier saturation** occurs when all the available carrier molecules are transporting their specific substance at as rapid a rate as possible. For example, if a cell had 10 glucose carrier molecules and each could transport 10 molecules per second, the carriers would be saturated when they move 100 glucose molecules per second. Because of the limited number of carrier molecules, no further rate increase is possible.

Competition occurs between the substance normally transported and another substance with similar shape, charge, and affinity for the carrier. Suppose that another sugar has equal affinity for the glucose carrier molecules on certain cells. When both sugars are present in the same concentration, the carrier molecules would transport about the same number of molecules of each sugar. Using the above example again, the membrane would transport 50 glucose molecules and 50 molecules of the other sugar per second but never more than 100 molecules per second. Even when different proportions of transported substances or substances with different affinities are used, the total number of molecules transported at one time remains constant.

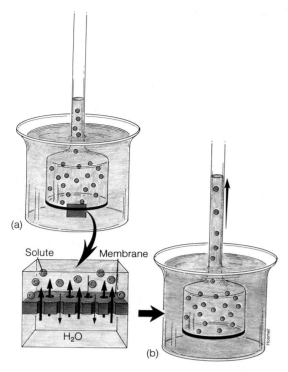

(a)

Solute   Membrane

H_2O

(b)

Figure 2.5

A laboratory demonstration of osmosis: (*a*) at the beginning of experiment and (*b*) after osmosis has occurred. Water molecules move by osmosis from their region of higher concentration to their region of lower concentration in much greater numbers than they move in the opposite direction. The membrane in the diagram, like a living membrane, is permeable to water but not to protein molecules.

Water passes through cell membranes rapidly in both directions by **osmosis**—the diffusion of water across a selectively permeable membrane from a region of high water concentration to one of lower water concentration. To explain osmosis, we will use a hypothetical situation in which a membrane permeable only to *water* separates two compartments. One compartment contains pure water and the other contains relatively less water and some protein molecules that cannot cross the membrane (Figure 2.5). Osmotic pressure exists because of differences between the two solutions. Water is free to move and it diffuses into the compartment containing the protein. The pressure necessary to prevent such movement is called **osmotic pressure.**

The property of solutions that leads to osmotic pressure is called **osmolarity** and is measured in osmols (osm). The number of osmols per liter of a solution depends on the number of dissolved particles in the solution. In solutions where every molecule and ion contributes to osmolarity, molar and osmolar concentrations are directly related. A 1-molar solution of a nonionizing substance such as glucose has osmolarity of 1 osm. A 1-molar solution of

a salt such as NaCl, which ionizes nearly completely into 2 ions per molecule, has osmolarity near 2 osm.

Osmotic pressures in living systems, which are much smaller than osmols, are expressed in **milliosmols** (mOsm). A milliosmol is 1/1,000 of an osmol. In body fluids, some molecules ionize partly or not at all and others bind to protein molecules, so the actual osmotic pressure in body fluids is less than would be predicted from the concentrations of the substances in them. Most body fluids have an osmolarity of about 300 mOsm/l.

Physiologists often use intracellular fluid as a standard and determine pressure differences between it and other fluids. Solutions with no pressure difference when compared with intracellular fluids are **isosmotic** (i-sos-mot′ik), those where pressure is higher outside the cell are **hyperosmotic** (hi″per-os-mot′ik), and those where pressure is lower outside the cell are **hyposmotic** (hip-os-mot′ik).

Tonicity (to-nis′ĭ-te), which is related to, and sometimes confused with, osmolarity, is determined by observing the behavior of cells in a solution. A solution surrounding a cell is **isotonic** (i-so-ton′ik) when it causes no change in cell volume. The solution is **hypertonic** (hi-per-ton′ik) to the cell if the cell shrinks as water moves out of it into the solution; it is **hypotonic** (hi-po-ton′ik) to the cell if the cell swells or bursts as water moves from the solution into the cell (Figure 2.6). Where a selectively permeable membrane allows only water and *no* solutes to cross the membrane, osmolarity and tonicity are the same. In biological systems where some solutes *do* cross the membrane, osmolarity and tonicity are not the same.

In contrast to the passive processes described above, **active transport** moves substances against concentration gradients from regions of lower to higher concentration. Active transport is analogous to rolling something uphill and requires the cell to use energy from ATP. In fact, cells can spend as much as half the energy they use in a resting state for active transport. Active transport requires membrane proteins that are both enzymes and carriers. Carriers have specificity in that each binds to and transports a single substance or a few closely related ones.

Active transport is essential for normal animal cell function because passive processes such as diffusion and osmosis cannot concentrate substances against a gradient. Active transport is the main means by which cells establish such gradients and several other processes depend in one way or another on maintaining gradients. For example, active transport of sodium ions contributes to polarization of nerve and muscle cell membranes. Active transport of chloride ions in the kidneys creates a gradient that is essential for concentrating urine and conserving body fluids. Active transport of iodide ions in the thyroid gland concentrates those ions, so they are available for synthesis of thyroid hormones. Active transport of calcium ions is important in several processes—muscle contraction, transmission of signals from one neuron to the

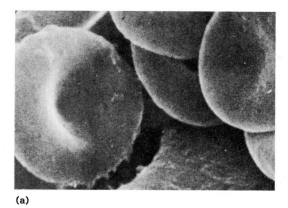

(a)

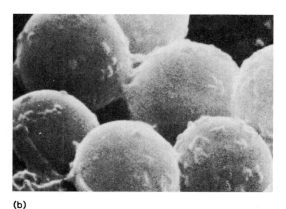

(b)

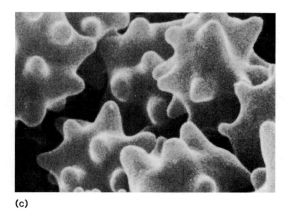

(c)

Figure 2.6

Scanning electron micrographs showing the effects of solutions on red blood cells. A cell placed in (*a*) isotonic solution will have no net movement of water because the same amount of water moves in each direction. A cell placed in (*b*) hypotonic solution will gain more water than it loses; the cell will swell and eventually burst. A cell placed in (*c*) hypertonic solution will lose more water than it gains; the cell will shrink and the membrane will shrivel. This is called crenation. (Cells in (*b*) are less highly magnified than the cells in the other micrographs.)

Courtesy S. J. Slinger

next, and in a variety of cellular control mechanisms. By concentrating substances against a gradient, active transport plays an important role in homeostatic mechanisms in many body systems.

We will use the **sodium-potassium pump** to illustrate active transport (Figure 2.7). Most animal cells maintain concentration gradients of Na^+ and K^+, where the Na^+ concentration outside the cell is about 150 mmol/l and that inside is about 10 mmol/l, and in which the K^+ concentration inside the cell is about 140 mmol/l and that outside is about 4 mmol/l. Under these conditions, Na^+ constantly diffuses down its concentration gradient into cells and K^+ constantly diffuses down its concentration gradient out of cells. Active transport counteracts diffusion by pumping Na^+ out of cells and K^+ back into them. As much as half of the body's resting energy may be used to operate the Na^+-K^+ pump.

The sodium-potassium pump operates as follows: The membrane protein binds to a Na^+ ion at the inner membrane surface (Figure 2.7a). This binding activates the enzyme ATPase. In the presence of Mg^{2+} ions, ATP is hydrolyzed thus releasing sufficient energy to power the pump and a phosphate group that binds to the protein (Figure 2.7b). Binding of phosphate causes the protein to change shape placing the Na^+ binding site at the outer membrane surface and releasing Na^+ (Figure 2.7c). The K^+ binding site is exposed on the outer membrane surface, and K^+ from the interstitial fluid binds to it (Figure 2.7d). The protein returns to its original shape carrying the K^+ ion to the inner membrane surface (Figure 2.7e), and K^+ is released into the cell (Figure 2.7f). One each of the Na^+ and K^+ ions is used in the example, but three Na^+ and two K^+ ions are transported each time the membrane protein goes through the above cycle.

All processes that move substances across membranes (Table 2.1) affect concentrations on both sides of the membrane. Sometimes the movements directly help to maintain conditions within the cell or its immediate environment in a narrow and tolerable range. Other movements, such as movements of ions that cause a neuron to send a signal for a muscle to contract, are necessary for cells to contribute to the normal function of a tissue, organ, system, or the whole body.

See Questions—Objectives 2, 3, and 4

Organelles

Scientists who first observed cells called their substance **protoplasm** and identified the **nucleus** and **cytoplasm,** the substance surrounding the nucleus. We now know that cytoplasm contains many discrete structures called **organelles** (including the nucleus), which carry out specific intracellular functions. Organelles are the functional units of a cell (refer back to Figure 2.1). Cytoplasm also contains small particles suspended in a thick fluid, **cytosol.**

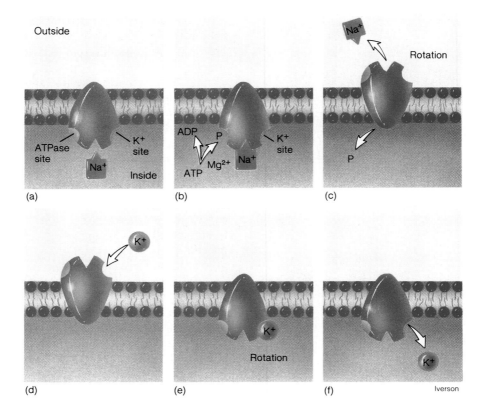

Outside

(a) ATPase site · K+ site · Na+ · Inside

(b) ADP · P · Mg²⁺ · ATP · Na+ · K+ site

(c) Na+ · Rotation · P

(d) K+

(e) K+ · Rotation

(f) K+

Iverson

Figure 2.7

A model of the sodium-potassium active transport system. (*a*) A protein complex (a carrier and the enzyme ATPase) in the membrane binds Na⁺ from inside the cell. (*b*) Na⁺ binding activates ATPase in the presence of Mg²⁺, ATP is hydrolyzed and the cleaved phosphate group binds to the protein complex. (*c*) Phosphate binding causes a change in the shape of the complex so the Na⁺ binding site rotates to the outside surface of the membrane and releases Na⁺. (*d*) The now exposed K⁺ binding site binds K⁺. (*e*) Having released Na⁺ and bound K⁺, the protein complex carries K⁺ into the cell as it returns to its original shape. (*f*) The K⁺ binding site then releases K⁺ into the cell.

Table 2.1
Mechanisms of Movement of Substances Across Cell Membranes

Process	Characteristics
Simple diffusion	Net movement down a gradient with or without a membrane, requires no energy
Facilitated diffusion	Net movement of molecules down a gradient, requires a carrier molecule but no energy
Osmosis	Movement of water along its own gradient; occurs across a selectively permeable membrane that prevents some molecules from diffusing
Active transport	Movement of substances across a membrane, usually against a gradient, requires energy expenditure and a carrier molecule

Cytosol consists of dissolved substances, enzymes, and several kinds of granules mostly containing glycogen or fat. The cytosol and dissolved substances, which are usually ions, gases, and nutrient molecules, comprise the intracellular fluid. Though many cellular enzymes are confined to organelles, those that control glycolysis, the first sequence of reactions in the metabolism of glucose, are found in the cytosol.

As the control center of the cell, the **nucleus** is essential at some point in a cell's life cycle. It is surrounded by a folded, double layer of membrane called the **nuclear envelope,** which has many large pores that can be closed by a protein associated with them. Opening and closing pores allows RNA to leave the nucleus, and certain complex molecules to enter it. The **nucleoplasm** (nu′kle-o-plazm) in the nucleus includes the nucleolus (or two or more nucleoli) and the chromosomes. The **nucleolus** (nu-kle′o-lus), which lacks a membrane and changes shape as a cell grows and divides, is the site of assembly of organelles called ribosomes.

Questions

Objective 2

(a) What properties distinguish active transport?

(b) What properties distinguish passive transport?

Objective 3

(a) How do each of the following occur?

simple diffusion

facilitated diffusion

osmosis

(b) What is the significance of each of these processes?

(c) Why are these processes said to be passive?

Objective 4

(a) Describe the mechanism of active transport.

(b) Why is it said to be an active process?

(c) What is the significance of active transport?

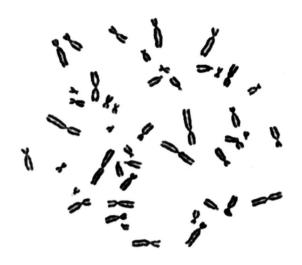

Figure 2.8

Chromosomes of a normal human cell. This is the historic photograph by which Drs. Joe Hin Tjio and Albert Levan demonstrated that human cells contain 46 chromosomes.

Courtesy Kleberg Cytogenetics Lab Inst. for Molecular Genetics

The **chromosomes** (Figure 2.8), which condense into short rods during cell division, exist as more diffuse, elongated fibers called **chromatin** at other times. Chromatin exists in two forms: **euchromatin** consists of active DNA used by the cell to make proteins and more condensed **heterochromatin** consists of inactive DNA. Human cells contain 46 chromosomes—22 pairs of autosomal chromosomes and 2 sex chromosomes (XX in females and XY in males). Chromosomes are known to consist of DNA wrapped around a protein core with surface proteins in certain locations.

DNA is important in two cellular processes. First, it dictates the nature of RNAs and proteins that a cell can synthesize and thereby controls the functions of the cell. Second, it transmits information for cellular control from one generation to the next—from parents to offspring in eggs and sperm, for example.

The **endoplasmic reticulum** (en-do-plaz′mik re-tik′u-lum), or ER, is an extensive network of interconnected flattened vesicles and tubules bounded by membranes of the same basic structure as the plasma membrane. Tubules of ER serve as a transport system for proteins and possibly for other large molecules synthesized by a cell. ER can be rough or smooth. Rough ER is coated with ribosomes, where protein synthesis occurs. Lipid and cholesterol synthesis and some lipid breakdown occurs in smooth ER. Cells vary in the relative amounts of rough and smooth ER they contain. Pancreatic cells have large amounts of rough ER, because they produce large quantities of protein enzymes for secretion. In contrast, liver and fat cells have large amounts of smooth ER, because they are active in lipid metabolism.

Ribosomes (ri-bo-sōmz), small bodies lacking membranes, are found on the surface of the rough ER or in the cytosol where they serve as sites for protein synthesis. Chemically, ribosomes consist of ribonucleic acid (RNA) and protein. Ribosomes attached to the ER synthesize proteins to be secreted, such as some hormones and digestive enzymes. Researchers believe that as such proteins are synthesized, they move through the membrane, to the lumen (inner cavity) of the ER, to the Golgi apparatus, and are eventually secreted. Ribosomes found free in the cytosol make proteins for the cell's own use.

The **Golgi** (gōl′je) **apparatus,** usually located near the nucleus, is a stack of vesicles called **cisternae** (sister′ne), which processes newly synthesized proteins. Processing can include removal of a segment of a protein, the addition of carbohydrate or lipid to the protein, and the wrapping of the product in a membranous envelope. Products are extruded from the cell when a vesicle fuses with the plasma membrane.

The Golgi apparatus also makes organelles called **lysosomes** (li′so-sōmz), which contain hydrolytic enzymes. When cells engulf foreign substances in vacuoles, as white blood cells do in infection and injury, lysosomes fuse with the vacuoles and pour their enzymes into the vacuoles. The enzymes digest the foreign matter. Even cells that do not engulf foreign substances contain lysosomes. As long as such cells remain healthy, the lysosomal membranes apparently prevent the release of enzymes. When a cell is damaged or dies, it accumulates acid, which causes lysosomes to rupture and their enzymes to digest the remains of the cell.

Frequently called the powerhouses of a cell, **mitochondria** (mi″to-kon′dre-ah) are the sites of most of the oxidative reactions that transform energy into a form usable by cells. Cells contain hundreds to thousands of these rod-shaped or ellipsoid organelles with the most metabolically active cells containing the largest numbers of mitochondria. Structurally, a mitochondrion has an outer smooth membrane and an inner folded membrane. **Cristae** (kris′te), folds in the inner membrane, contain enzymes and other molecules that capture and store energy in ATP. The **matrix** (center) of a mitochondrion contains enzymes that break down fatty acids and pyruvic acid (from glucose), so enzymes in the cristae can capture energy from these nutrients.

Recognized only since the early 1950s, **peroxisomes** (per-ox′eh-sōmz) are membrane-covered organelles containing oxidative enzymes that destroy toxins. Some reactions release hydrogen peroxide, which is broken down to oxygen and water by the peroxisome enzyme catalase. Certain human genetic defects involving absent or defective peroxisomes cause severe impairments including seizures, sight and hearing loss, uncoordinated movements, and liver and kidney disease. Thus defects at the cellular level can have profound effects on whole body function.

Cells have an internal **cytoskeleton** that give them rigidity and that allow movement of whole cells and particles within cells. The cytoskeleton consists of a complex network of microtubules and microfilaments, both of which consist of a protein devoid of any membrane covering. **Microtubules,** which are hollow fibers of the protein **tubulin** and smaller **microfilaments,** which consist mainly of the protein **actin,** form a network. That network holds the nucleus in place and to provide binding sites anchoring protein molecules, organelles, and other cellular particles once thought to be randomly distributed in the cytosol. The cytoskeleton also contributes to cellular movements.

Microtubules form the spindle along which chromosomes move as they separate during cell division and the tracks along which various particles move within cells. The moving particles include mitochondria, large vesicles of surplus cell membrane accumulated from engulfing particles, and small vesicles containing neurotransmitters—a chemical substance that relays signals from one neuron to the next. Attached to microtubules are molecules of a motor protein with enzyme activity that seems to use energy from ATP to move particles around a microtubule in a path similar to the belt on a "caterpillar" wheel.

The roles of organelles in cellular functions are summarized in Table 2.2. Many organelle functions are closely associated with the enzymes they contain. Such functions are closely integrated, so that organelles work together to perform cell functions and maintain cellular homeostasis.

Many mechanisms for regulating cell function exist, and most involve protein binding sites, or receptors. These receptors usually are located on outer membrane surfaces, but some, such as those concerned with the actions of steroid hormones, are located inside the nucleus. Certain proteins have several specific receptors, each of which bind a different ligand. Binding of ligands to receptors is a reversible process, so binding and unbinding can function as a control process much like turning on and off a switch. Receptors also provide a means by which cells can recognize, and thereby respond to, specific ligands that act as stimuli, so ligand-receptor binding also is a communication process. Receptors that bind ligands are a fundamental control and communication process at the cellular level.

The ability for spontaneous self-assembly of cell components is important in living systems. For example, bicycle parts would never assemble themselves into a bicycle no matter how the parts were laid out or how long one waited. However, many complex structures within cells can, in fact, assemble themselves. In other words, they produce order from entropy. The significance of self-assembly of cell components is that once the sequence of amino acids in proteins has been established from information in DNA, no further information is required. The

proteins take on particular shapes and acquire binding sites that lead to spontaneous self-assembly of many cell components.

See Questions—Objectives 5 and 6

Table 2.2
Characteristics of Organelles

Organelle	Characteristics
Nucleus	Large body surrounded by a nuclear envelope that consists of a double membrane with large pores; contains nucleoplasm, chromosomes, and one or more nucleoli; functions as the control center of the cell
Chromosomes	Long strandlike bodies within the nucleus consisting of DNA and protein; contain genetic information that controls the activities of cells
Nucleolus	Irregularly shaped body in the nucleus, site of RNA synthesis
Endoplasmic reticulum	Network of interconnected vesicles and tubules bound by membranes; serves as transport system for molecules synthesized by the cell; rough ER has ribosomes on the surface, smooth ER lacks ribosomes and is the site of lipid metabolism
Ribosomes	Small round bodies lacking membranes found on ER or in cytosol, consist of RNA and protein, function in protein synthesis
Golgi apparatus	Large membrane-bound structure consisting of a stack of vesicles; serves as a storage and processing organ for cell products, may remove a segment of a protein, add carbohydrate to a protein, wrap protein in a membrane; extrudes secretions from cell; forms lysosomes and secretory granules
Lysosomes	Membrane-bound bodies that contain hydrolytic enzymes, enzymes are released into vacuoles or into dead or injured cells
Mitochondria	Rod-shaped or spherical bodies that have a smooth outer membrane and a folded inner membrane, contain enzymes for reactions that capture energy in ATP
Microtubules	Hollow fibers composed of the protein tubulin, provide motility and internal support for cells; form centrioles, cilia, and flagella
Microfilaments	Very thin, solid structures composed of the protein actin, account for motility and contractility of cells

Questions

Objective 5

(a) How does cytoplasm differ from cytosol?

(b) List the organelles and other internal components of cells and give the function of each.

ribosomes

Golgi apparatus

cytoskeleton

(c) What deficiencies in function would a cell lacking one of the following display?

nucleus

mitochondria

Objective 6

(a) How are cellular functions integrated?

(b) Why is integrated function important?

DNA Replication

The ability of cells to divide and produce new cells is an essential property of living things. Cell division requires **replication** (duplication) of DNA, which is a chemical process (Figure 2.9). It begins as segments of the nucleotide strands of each double helix of DNA separate and expose strands of unpaired bases. Each strand is replicated separately and a small segment of RNA primer starts, or primes, the process. Each nucleotide strand serves as a **template,** or pattern, for the synthesis of a partner strand. The enzyme **DNA polymerase** (pol-im′er-āz) binds to the primer and its paired base in the template. Phosphorylated nucleotides are plentiful in the surrounding nucleoplasm, and a nucleotide containing a complementary base combines with the exposed base adjacent to the enzyme-primer complex. Energy from a nucleotide allows the enzyme to join the nucleotide to the primer, thereby starting a new strand of DNA. The enzyme moves along the template; another nucleotide binds to the complex and is added to the new strand of DNA. The process continues until a new DNA strand that matches the original partner of the template strand is made.

Each new chromosome consists of one old and one new strand of DNA, that is, it undergoes **semiconservative replication.** This was demonstrated by providing radioactively labeled nucleotides to cells while they were replicating DNA. In such experiments, half the radioactive material appears in each chromosome, so each chromosome must contain half the newly synthesized DNA. One strand of a DNA molecule is the **sense strand,** its information is used to direct protein synthesis. The other is the **missense strand,** which functions only during replication. Each sense molecule serves as a template for the missense molecule and each missense molecule serves as a template for the sense molecule.

> **See Questions—Objective 7**

Protein Synthesis

Proteins are important as part of the structure of cells, enzymes, carriers in transport systems, antigens and antibodies, and regulatory molecules. Protein synthesis is essential for growth and for replacement of cell components. How cells make proteins is a complex process that has been elucidated by painstaking research within the last few decades.

As we have seen, DNA conveys genetic information to new cells and directs the synthesis of all kinds of proteins, including enzymes. Because enzymes control the synthesis of nonprotein cell components, DNA directly or indirectly determines the nature of all cell com-

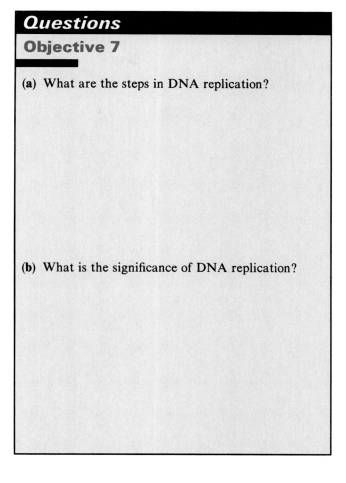

Questions

Objective 7

(a) What are the steps in DNA replication?

(b) What is the significance of DNA replication?

ponents. DNA acts in protein synthesis by relaying information to several kinds of RNA that work together to assemble proteins.

Protein synthesis is initiated by factors such as a steroid hormone entering the nucleus and binding to receptors that activate a segment of DNA. The helix of activated DNA unwinds, exposing unpaired bases, and the sense strand serves as a template for RNA synthesis. Protein synthesis involves two processes—**transcription,** in which RNA is synthesized from DNA, and **translation,** in which protein is synthesized according to information in RNA (Figure 2.10).

The terms transcription and translation have the same meaning in describing protein synthesis as in their less technical usage. For example, a stenographer transcribes shorthand notes into typewritten words; both are in the same language. Likewise, the cell transcribes information in DNA to RNA; both are in "nucleotide" language. A linguist translates one language to another—English to French, Spanish to English, for example. Likewise, the cell translates information in mRNA to an amino acid sequence, that is, from the language of nucleotides to the language of amino acids.

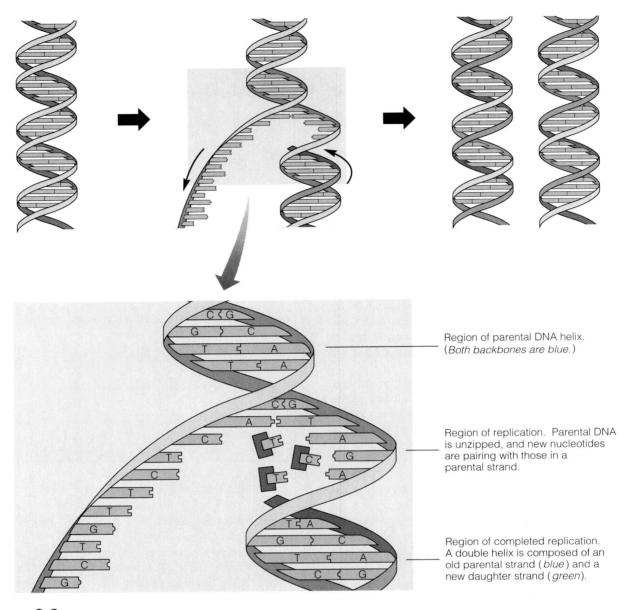

Region of parental DNA helix. (*Both backbones are blue.*)

Region of replication. Parental DNA is unzipped, and new nucleotides are pairing with those in a parental strand.

Region of completed replication. A double helix is composed of an old parental strand (*blue*) and a new daughter strand (*green*).

Figure 2.9

The replication of DNA. After separation of parent strands, replication of only one strand is shown. The other strand also is replicated but in the opposite direction. In each new molecule of DNA one strand is "old" DNA from the parent cell and one strand is recently synthesized "new" DNA. This is called semiconservative replication.

Transcription (Figure 2.11) occurs in the nucleus where phosphorylated RNA nucleotides are plentiful. The enzyme RNA polymerase binds to the exposed template DNA, the appropriate phosphorylated nucleotide joins the DNA base-enzyme complex. For example, the nitrogenous base uridine attaches by base pairing to the adenine of DNA and pyrophosphate (P-P) is released. The enzyme moves to the next DNA base and the appropriate phosphorylated nucleotide joins the complex. The phosphate of the second nucleotide is linked to the ribose of the first nucleotide, forming the first link in the "backbone" of a new molecule of RNA. Energy to form this link comes from the hydrolysis of ATP. The enzyme continues to move along the DNA and this process is repeated until the end of a DNA template is reached and the RNA molecule is completed. Punctuation exists in DNA to indicate the end of a sequence that will perform a particular function.

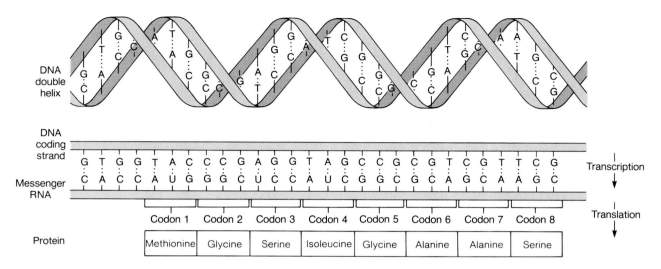

Figure 2.10

An overview of the processes of transcription and translation.

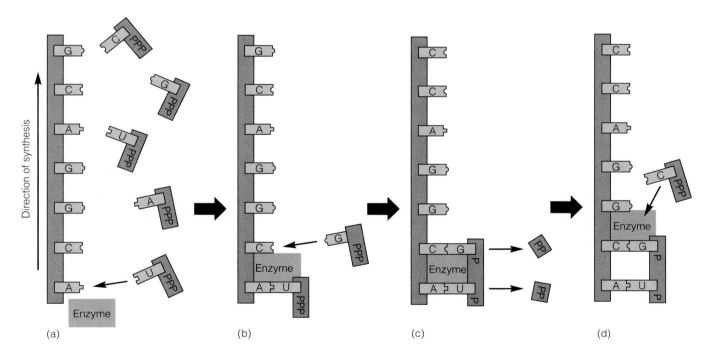

Figure 2.11

The transcription for RNA from template DNA: (*a*) An enzyme binds to the first exposed base and the appropriate phosphorylated nucleotide attaches to the DNA by base pairing. (*b*) The enzyme moves to the next base and a second nucleotide attaches. (*c*) The nucleotides bond together, releasing pyrophosphate, to form the first two units of RNA. (*d*) The process is repeated until a code in DNA indicates the end of a molecule of RNA.

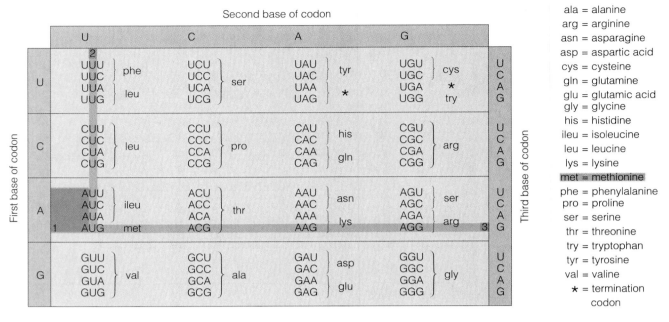

Figure 2.12

The genetic code consists of codons of messenger RNA. Most codons specify amino acids. The methionine codon always signifies the start of a protein, and one of three termination codons signifies the end of a protein. To find the codon for methionine, follow the shaded blocks numbered 1, 2, and 3.

The RNA base sequence is determined by base pairing with the DNA template. Pairing occurs as in DNA replication, except that the RNA uracil, and not thymine, pairs with adenine. Base pairing is critical in protein synthesis because it transfers information from DNA to RNA and between RNAs. Three kinds of RNA are synthesized in the nucleus from information in DNA—messenger RNA, ribosomal RNA, and transfer RNA. Each RNA consists of a single strand of nucleotides and each is formed by transcription.

Messenger RNA (mRNA) is synthesized in specific length segments. Ultimately, an mRNA segment includes the exact number of bases to convey genetic information from DNA for the synthesis of a polypeptide. Many original mRNA molecules contain sequences of nucleotides not used to code amino acids in a protein. Such molecules are clipped into segments, unneeded segments are discarded, and coding segments are reunited. This is done in the nucleus by enzymes that apparently can recognize which segments to delete and which to retain. Once synthesized, mRNA moves through a nuclear membrane pore and binds to a ribosome. Such an mRNA is ready to code the sequence of amino acids in a protein.

Ribosomal RNA (rRNA) combines with ribosomal structural proteins to form ribosomal subunits. To form a subunit, a previously made protein enters the nucleus and combines with nucleolar rRNA. Ribosomal subunits in animal cells are of two sizes: a large 60S subunit

and a small 40S subunit. (The S refers to the sedimentation coefficient of the subunit, which is determined by its size, shape, and density.) The subunits pass through nuclear membrane pores and enter the cytosol. Any mRNA present in the cytosol binds to a 40S subunit and then binds to a 60S subunit to form a complete ribosome. The 40S subunit facilitates amino acid placement as a protein is made, and enzymes of the 60S subunit catalyze peptide bond formation between amino acids. Some ribosomes bind to endoplasmic reticulum and others cluster as **polyribosomes** in the cytosol.

In translation, information in mRNA is read in 3-base sequences. These base triplets called **codons** specify particular amino acids or act as punctuation marks to indicate the beginning and the end of a polypeptide. At least 1 codon exists for each of the 20 amino acids found in proteins, and several codons exist for some amino acids. The full set of codons is called the **genetic code** (Figure 2.12). This code is the same, or nearly the same, in all organisms, and its universality allows us to apply findings from other organisms to our understanding of information transmission in human cells.

Bringing amino acids to the protein synthesis site requires **transfer RNA** (tRNA). Numerous tRNAs are synthesized in the nucleus and move to the cytosol. A tRNA typically consists of 75 to 80 nucleotides arranged in a cloverleaf shape with an amino acid binding site at one end and a 3-base sequence called an **anticodon** at the

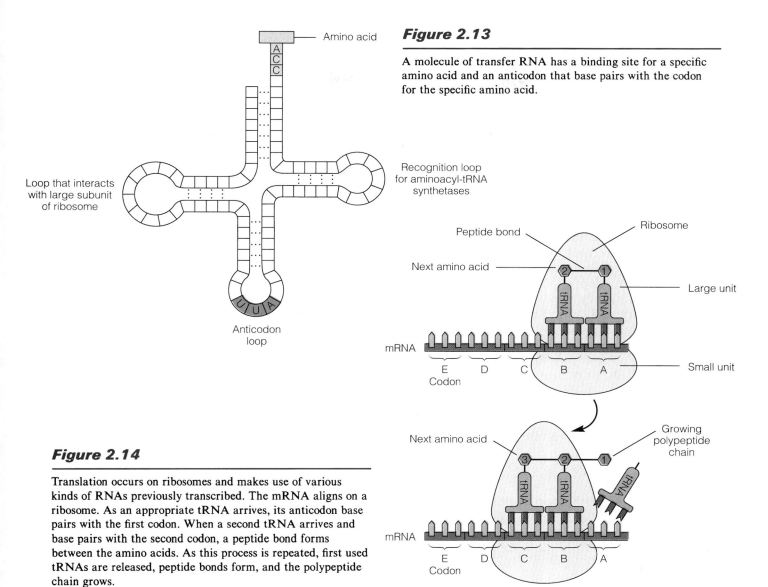

Figure 2.13

A molecule of transfer RNA has a binding site for a specific amino acid and an anticodon that base pairs with the codon for the specific amino acid.

Figure 2.14

Translation occurs on ribosomes and makes use of various kinds of RNAs previously transcribed. The mRNA aligns on a ribosome. As an appropriate tRNA arrives, its anticodon base pairs with the first codon. When a second tRNA arrives and base pairs with the second codon, a peptide bond forms between the amino acids. As this process is repeated, first used tRNAs are released, peptide bonds form, and the polypeptide chain grows.

other end (Figure 2.13). Each tRNA transports a specific amino acid, which must be activated by phosphorylation from ATP hydrolysis before it can bind to tRNA. At a ribosome, the tRNA anticodon attaches to an mRNA codon by base pairing. Any specific tRNA transports only the amino acid specified by the codon to which the tRNA's anticodon attaches.

All RNAs are reusable, but a precise quantity of mRNA is made according to the cell's needs for a particular enzyme or other protein. Control processes apparently assure that cells make appropriate quantities of particular proteins, so higher-level functions occur in an orderly fashion. Once an mRNA has joined a ribosome, the ribosome initiates translation and provides a protein assembly site. Translation (Figure 2.14) occurs in the following steps: Once mRNA is aligned on the ribosome (a),

the first tRNA arrives (b). The small ribosomal unit orients tRNA, so its anticodon can base pair with the mRNA codon. Such base pairing allows mRNA codons to specify the sequence of amino acids in the protein by specifying the sequence of tRNA anticodons, which carry the amino acids. When the second amino acid arrives (c), an enzyme in the large ribosomal subunit forms a peptide bond between the two amino acids (d). ATP hydrolysis provides energy to form the bond. As the ribosome moves along the mRNA, a third amino acid arrives and the first tRNA is released into the cytosol (e). This tRNA can transport another amino acid molecule. A second peptide bond is then formed (f), and the process continues until a terminator codon is recognized by the ribosome.

When the finished protein is released, the mRNA can synthesize another molecule of the same protein. In fact, several ribosomes can "read" different parts of a molecule of mRNA at the same time. Protein synthesis is summarized in Table 2.3.

Once a polypeptide chain has been synthesized according to the information in mRNA, it can be modified in a variety of ways. Methionine, the first amino acid in any polypeptide, can be removed or modified. Other amino acids in the chain can be modified by the addition of methyl, phosphate, hydroxyl, or other groups. Proteins destined to be secreted are often transported in the endoplasmic reticulum to the Golgi apparatus for processing. A portion of the polypeptide chain can be removed, as occurs in the processing of insulin. Carbohydrates or lipids can combine with the protein to form glycoproteins or lipoproteins, respectively. Proteins secreted in an inactive form are activated as needed. Pancreatic proteolytic enzymes are activated when they reach the small intestine. Blood clotting factors are activated by substances released from injured tissues.

Cells have genetic information to synthesize thousands of proteins, but they make only the proteins they need and only a few at a time. For example, though pancreatic cells contain information to make any protein, different cells make insulin, glucagon, and digestive enzymes. Even cells that make digestive enzymes probably do not make large quantities of several different enzymes at the same time. For the proteins a cell normally makes, regulation of when and how much protein is made can be imposed at any point in the process of protein synthesis. Regulatory mechanisms work together to cause the right amounts of the right proteins to be made at the appropriate times for optimal cell function, and thus for whole organism function.

The first mechanism proposed to regulate transcription, the operon mechanism, was described in bacteria in 1961 by Francois Jacob and Jacques Monod, who received a Nobel Prize for their work. Briefly, an operon is a set of genes that turns on and off the synthesis of one or more proteins, usually enzymes that perform a metabolic process. Finding operons in bacteria suggested they might be present in higher organisms. Though no operons have been found, other mechanisms have been identified. They include condensation of chromatin and the actions of histones and nonhistone proteins. When chromatin becomes greatly condensed **heterochromatin** the information in the DNA is no longer accessible. In cell-free extracts of DNA, histones bind to DNA and prevent factors such as steroids from activating genes. Nonhistone proteins appear to selectively activate genes, possibly by counteracting histones.

The rate of synthesis of any particular protein has been thought to be influenced mainly by how fast the cell synthesized the proper mRNA. Recent evidence suggests that how fast the cell degrades the mRNA may be more important. It is already known that the mRNAs for a cell's

Table 2.3
Summary of the Process of Protein Synthesis

Transcription

1. Bonds between strands of DNA break and expose a portion of the DNA template.
2. An enzyme attaches to the first exposed base in DNA while phosphorylated nucleotides are available in the nucleoplasm.
3. The phosphorylated nucleotides are paired with appropriate bases in DNA and bound together using energy from attached phosphates.
4. The process continues until a complete molecule of mRNA is produced.

Translation

5. Messenger RNA moves from the nucleus to the cytosol and one end binds to the smaller unit of a ribosome; ribosomal units combine.
6. First and then second tRNAs carrying their respective amino acids combine with mRNA by base pairing of codons and anticodons.
7. The first two amino acids are linked by a peptide bond; energy from hydrolysis of ATP is used to form the bond.
8. The third tRNA combines with mRNA and the first tRNA is released.
9. Another peptide bond is formed.
10. Amino acids are carried to their proper positions and peptide bonds are formed until an entire polypeptide is synthesized.
11. When the ribosome recognizes a terminator codon, the polypeptide is released from the ribosome.

most abundant proteins are stable, long-lasting molecules, whereas the mRNAs for scarce proteins are unstable and short-lived.

See Questions—Objective 8

Genetics

Each of a cell's proteins, including its enzymes, is synthesized according to information in a specific segment of DNA called a **gene** (jēn). Together all genes on all chromosomes constitute the information a person inherits from his or her parents—information that sets limits on body structure, function, and behavior. Most genes occur in two or more forms called **alleles** (al-ēlz). For example, a gene *S* directs synthesis of a normal polypeptide of hemoglobin. Another gene *s* directs synthesis of an abnormal polypeptide of hemoglobin in people with sickle-cell anemia. Cells have two copies of most genes, one on each member of a pair of chromosomes. If these two copies carry

Questions

Objective 8

(a) What cell components are involved in protein synthesis?

(b) Distinguish between transcription and translation.

(c) Summarize the process of protein synthesis from DNA to the finished protein.

(d) What happens in posttranslational processing?

(e) How is protein synthesis controlled?

the same information (*SS* or *ss*), the alleles are said to be **homozygous** (ho-mo-zi'gus). If the copies are different (*Ss*), the alleles are said to be **heterozygous** (het-er-o-zi'gus).

The combination of alleles is the **genotype** (jēn'o-tīp); the characteristic expressed is the **phenotype** (fēn'o-tīp). For example, genotype *SS* produces normal erythrocytes; genotype *ss* produces erythrocytes that can take on a sickle shape under certain conditions. Genotype *Ss* produces normal erythrocytes because the *S* allele is a

dominant allele and is expressed in the phenotype. The *s* allele is a **recessive** (re-ses'iv) allele and is expressed only in the absence of dominant alleles. Blood tests can determine both genotype and phenotype for these alleles.

In human cells, the 22 pairs of homologous chromosomes called the **autosomal** (aw-to-so'mal) **chromosomes** control most inherited characteristics, and the nonhomologous **sex chromosomes** control sex-linked inheritance. Most genes in sex-linked inheritance are on the X chromosome, but a Y chromosome is necessary for normal male development. Because males have only 1 X chromosome, any recessive alleles on that chromosome will be expressed. When males have a colorblindness allele (X^c), they are colorblind (genotype $Y\ X^c$). Because females have 2 X chromosomes, they are colorblind only if they have 2 copies of the recessive gene. Women who have a normal allele (X^C) and a colorblindness allele (X^c) have normal color vision (genotype $X^c\ X^c$) but can transmit the colorblindness allele to their offspring. Genetic terminology is summarized in Table 2.4.

Many genetic diseases are due to alleles that lead to an absent or defective enzyme or other protein (Table 2.5). Prospective parents' genotypes for a particular characteristic often can be determined by laboratory tests or by studying family members. Such information is used to calculate the probability of their having a child with a certain genetic defect. For example, if both parents carry the sickle cell anemia gene (genotype *Ss*), they have a probability of 0.5 (a 50–50 chance) of transmitting allele *S* to each child. They also have a probability of 0.5 of transmitting allele *s*. Allelic combinations and probabilities are shown in Figure 2.15. *These probabilities remain the same for each child regardless of how many children the couple has.*

Any change in the base sequences in DNA constitutes a **mutation** (mu-ta′shun), and mutations provide a source of genetic variability. To understand the effects of mutations, recall that protein synthesis is directed by codons consisting of base triplets of mRNA. We can represent a small segment of such mRNA with a sentence consisting of three-letter words:

OUR BIG BAD DOG BIT OUR OLD FAT CAT.

A **point mutation** is the substitution of a single base as a segment of DNA is replicated:

OUR BIG BAF DOG BIT OUR OLD FAT CAT.

Except for the word "baf," the sentence still has meaning. As shown in Figure 2.16a, such a change may or may not specify a different amino acid. If it does, the protein will differ from normal by 1 amino acid. This is exactly the kind of error that produces sickle cell anemia, as valine is substituted for glutamic acid in a polypeptide of hemoglobin. If in spite of the change, the RNA codon still specifies the normal amino acid, the protein is made without error.

A **frameshift mutation** consists of adding or deleting one or more bases in DNA. Adding a letter to our sentence produces:

OUR XBI GBA DDO GBI TOU ROL DFA TCA T.

Deleting a letter yields:

OUR (B) IGB ADD OGB ITO URO LDF ATC AT.

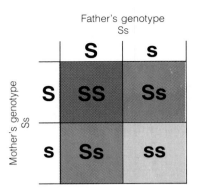

Father's genotype
Ss

Probability ¼ SS: ½ Ss: ¼ ss

Figure 2.15

Probability of each genotype among offspring of parents who are carriers of the sickle-cell anemia gene.

Table 2.4
Basic Terminology of Genetics

Term	Definition
Allele	One of two or more alternate forms of a gene that occupies a particular locus (site) on a particular chromosome
Homozygous	A condition in which the alleles of a gene on the two homologous chromosomes are alike
Heterozygous	A condition in which the alleles of a gene on the two homologous chromosomes are not alike
Genotype	The combination of the alleles at a given locus carried by an individual; the genetic makeup of the individual for all genes
Phenotype	The outward appearance of the individual with respect to a particular characteristic (or all characteristics) determined by the genotype
Dominant	An allele that is expressed in the phenotype when it is carried either homozygously or heterozygously
Recessive	An allele that is expressed in the phenotype only in the absence of a dominant gene
Carrier	A heterozygous individual of normal phenotype who has in his or her genotype a hidden recessive allele and can, thus, transmit the recessive allele to offspring
Autosomal inheritance	Inheritance of characteristics that are carried on one of the 22 pairs (in humans) of autosomal, not sex, chromosomes
Sex-linked inheritance	Inheritance of characteristics carried on the X chromosome (or rarely on the Y chromosome)

Table 2.5
Selected Human Genetic Diseases

Disease and Inheritance Pattern*	Description
Achondroplasia (AD)	Dwarfism, with especially short limbs and adult height of about 4 ft. Defect has a high mutation rate and can appear in families never before affected.
Albinism (AR)	Lack of an enzyme to convert amino acid tyrosine to the pigment melanin. Individuals have white hair and very fair skin; the iris of the eye appears pink due to blood vessels unobscured by normal pigmentation.
Brachydactyly (AD)	Individuals have very short, stubby fingers.
Bruton agammaglobulinemia (XR)	Almost complete inability to form antibodies to infectious agents. Affected individuals require lifelong gamma globulin treatment and remain susceptible to infections.
Cystic fibrosis (AR)	Excess mucus blocks respiratory passages and impairs release of digestive enzymes. Affected individuals who receive intensive therapy sometimes live to adulthood.
Duchenne muscular dystrophy (XR)	Muscle wasting usually begins by age 6 and leads to paralysis in childhood. High rate of mutation leads to frequent appearance of disease in previously unaffected families.
Huntington's disease (AD)	Onset from 10 to 60 years of age, but often not until after the affected individual has had children. Progressive deterioration of nervous system leads to uncontrollable movements, personality changes, mental impairment, and, ultimately, death. Carriers of the gene can be identified.
Lesch-Nyhan syndrome (XR)	Enzyme defect causes overproduction of uric acid. The affected individual is mentally retarded and engages in self-mutilation. Uric acid accumulation leads to kidney damage, but this effect can now be minimized by drug treatment. Tests are available to detect carriers.
Marfan's syndrome (AD)	Produces long spidery appendages, eye defects, and aortic disease. It has been speculated that Abraham Lincoln was mildly affected by this disease.
Phenylketonuria (AR)	Enzyme defect causes inability to metabolize the amino acid phenylalanine. Mental retardation results unless the patient is placed on a low-phenylalanine diet in the first few weeks of life. Tests are available to detect this defect in neonates.
Sickle-cell anemia (AR)	Enzyme defect causes production of abnormal hemoglobin. Effects are chronic anemia and crises with extreme pain in joints, blockage of blood vessels by sickle-shaped red blood cells, and need for transfusions. Heterozygous individuals are resistant to the disease malaria.
Tay-Sachs disease (AR)	Defect in lipid metabolism. Results in progressive deterioration of nerve tissue in an apparently normal infant until death at about age 3 or 4. Tests are available to identify carriers and to detect affected fetuses prenatally.

*Inheritance patterns of genetic diseases: AD = autosomal dominant, AR = autosomal recessive, XR = sex-linked recessive

In both examples the sentence has become nonsense. Such changes alter all base triplets beyond the mutation, making nonsense of the code. As shown in Figure 2.16b, the resulting protein will be vastly different from the normal protein. Many frameshift mutations insert a termination codon in mRNA, so that only a short nonfunctional polypeptide, and not a whole protein, is made.

Various environmental agents called **mutagens** (mu'tă-jenz) can cause mutations. Chemical mutagens include nitrous acid, mustard gases, and base analogs such as bromouracil. Other mutagens include X ray and ultraviolet radiation. The ways mutations alter body structure and function are varied, but most modify structural proteins or enzymes. A mutation in an egg or a sperm is passed to all the cells of the new individual. This process accounts for the initial appearance of genetic diseases in families with no previous history of carrying the disease. A mutation in body cells affects only the progeny of the mutated cell in a particular individual and is not transmitted from one generation to the next. Cancer can result from such mutations.

See Questions—Objective 9

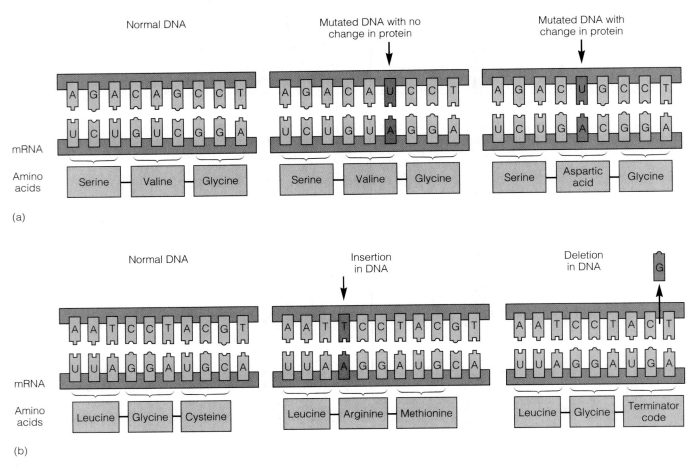

(a)

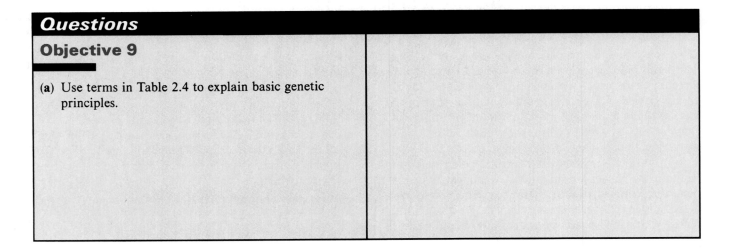

(b)

Figure 2.16

(*a*) The effects of a point mutation. Arrow indicates the substituted base in DNA. (*b*) The effects of frameshift mutations. Insertions and deletions are indicated by arrows.

The deletion mutation happens to create a DNA sequence for which mRNA contains a terminator code at the third codon. Protein synthesis stops after the formation of a dipeptide.

Questions

Objective 9

(a) Use terms in Table 2.4 to explain basic genetic principles.

Questions Continued

Objective 9 Continued

(b) How are some common genetic disorders inherited?

a point mutation that would affect the protein

(c) What is a mutation and how do frameshift and point mutations differ?

(f) explain the effects of adding or deleting a base at the second base in the DNA nucleotide sequence.

(d) Determine at least one possible sequence of bases in DNA, mRNA, and tRNA that would produce the following protein segment: alanine, glutamic acid, serine, valine, arginine, leucine, tryptophan.

(g) What are some causes of mutations?

(e) For the DNA in question (d) above, write the sequence for:
a point mutation that would not affect the protein structure

Study and Review

Did you get the essentials?

The following summary contains the basic concepts from chapter 2 except for the key terms that have been omitted. Read the chapter. Then try to fill in the blanks from memory. If you cannot fill in all the blanks, review the chapter paying particular attention to boldface terms.

In general, cells consist of _____ suspended in a semifluid _____ surrounded by a _____ _____ , or _____ _____ . According to the _____ -_____ model, all membranes in cells have a semifluid lipid matrix made of _____ arranged in two layers. Interspered in a mosaic pattern are _____ _____ that extend partially or completely through the layers and _____ _____ that are loosely attached to the surface.

_____ transport processes do not require the cells to expend energy. They include _____ _____ in which small molecules and ions move down their _____ _____ , _____ in which water moves down its gradient, and _____ _____ in which substances are moved down their gradient by a carrier molecule. Carriers are subject to _____ (all working as fast as possible) and _____ (alternate transportable substances).

Extracellular fluids compared to intracellular fluids are _____ when osmotic pressure is the same in both and _____ when there is no fluid movement. When extracellular pressure is _____ , cells shrink and when it is _____ , cells swell.

_____ _____ is a process whereby cells use energy from ATP and carriers to move substances against a concentration gradient. The _____ -_____ _____ is an example.

What early cell biologists called _____ is now known to contain discrete structures called _____ and the fluid around them is called _____ .

The cell's control center is the _____ . It is surrounded by a double membrane, the _____ _____ , and contains one or more irregularly shaped _____ , 46 _____ , parts of which are active _____ and parts of which are inactive _____ .

Portions of the _____ _____ , a network of vesicles and tubules that are smooth, make and break down lipids. Portions that are rough have attached _____ that serve as a site for protein synthesis. The _____ _____ processes newly synthesized proteins for secretion and enzymes for packing into _____ . The powerhouses of the cell called _____ capture energy in a form cells can use.

Cells have an internal _____ , which consists of a complex network of hollow _____ made of _____ and smaller _____ made of _____ , a protein found in abundance in muscle cells. DNA replication (or _____) is _____ because each new chromosome contains half-old and half-new DNA. The _____ strand directs protein synthesis, and the _____ strand is a template for replication of the _____ strand.

In protein synthesis, _____ refers to the synthesis of mRNA and _____ refers to the placement of _____ _____ in a new protein molecule. The order of bases in mRNA is determined by _____ _____ _____ with bases in the sense strand of _____ . When mRNA reaches a _____ sets of 3 bases in it act as _____ that pair with _____ in _____ , thereby placing particular _____ _____ .

_____ is the study of heredity and the _____ is the basic unit of heredity. _____ are different kinds of information that can be found in a gene. For genes on _____ (not sex) chromosomes, each person has two copies of each gene. They are _____ if alike and _____ if different. The genetic makeup is the _____ and the appearance resulting from it is the _____ . Genes that are expressed when present are _____ and those that are expressed only in the absence of a dominant gene are _____ .

A _____ is a change in DNA. If the change is the substitution of one base for another it is a _____ _____ , but if it is the addition or deletion of 1 or more bases it is a _____ _____ . Environmental agents that can alter DNA are _____ .

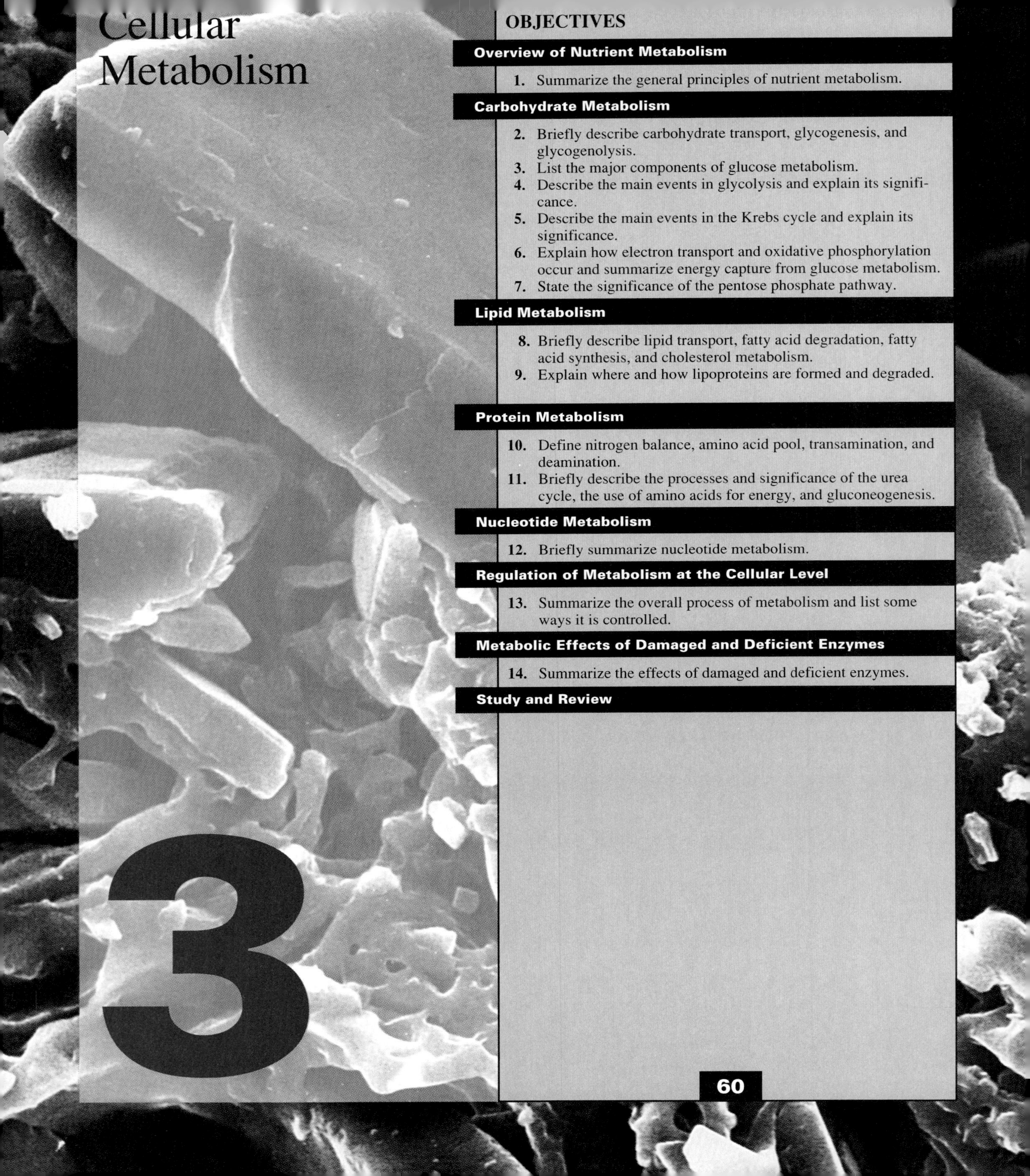

Cellular Metabolism

OBJECTIVES

Overview of Nutrient Metabolism

1. Summarize the general principles of nutrient metabolism.

Carbohydrate Metabolism

2. Briefly describe carbohydrate transport, glycogenesis, and glycogenolysis.
3. List the major components of glucose metabolism.
4. Describe the main events in glycolysis and explain its significance.
5. Describe the main events in the Krebs cycle and explain its significance.
6. Explain how electron transport and oxidative phosphorylation occur and summarize energy capture from glucose metabolism.
7. State the significance of the pentose phosphate pathway.

Lipid Metabolism

8. Briefly describe lipid transport, fatty acid degradation, fatty acid synthesis, and cholesterol metabolism.
9. Explain where and how lipoproteins are formed and degraded.

Protein Metabolism

10. Define nitrogen balance, amino acid pool, transamination, and deamination.
11. Briefly describe the processes and significance of the urea cycle, the use of amino acids for energy, and gluconeogenesis.

Nucleotide Metabolism

12. Briefly summarize nucleotide metabolism.

Regulation of Metabolism at the Cellular Level

13. Summarize the overall process of metabolism and list some ways it is controlled.

Metabolic Effects of Damaged and Deficient Enzymes

14. Summarize the effects of damaged and deficient enzymes.

Study and Review

3

Overview of Nutrient Metabolism

Metabolism refers to the sum total of chemical reactions in the body. It includes **catabolic** (kat-ah-bol′ik), or **degradative,** reactions that use nutrients to provide energy for cells and **anabolic** (an-ah-bol′ik), or **synthetic,** reactions that produce the cell's substance. Compared to reactions in the digestive tract, reactions in cells are also catalyzed by enzymes. Three kinds of nutrients—carbohydrates, lipids, and proteins—are the substrates, or reactants, of most catabolic reactions. Their breakdown products serve as building blocks for anabolic reactions. In many metabolic pathways, several molecules called **intermediates** appear between the initial reactant and the final product (Figure 3.1). Water participates in reactions directly or as a reaction medium. Vitamins are used to make coenzymes and minerals act as cofactors in many enzyme reactions.

Glucose provides energy by the following summary reaction:

glucose + oxygen → carbon dioxide + water + energy.

Outside the body this reaction could occur by simple burning, but inside cells, it takes place in a series of specific enzyme-controlled reactions. In certain reactions, energy is captured in a form that can be released later in a controlled manner for cells to use in their many activities.

Certain coenzymes transport electrons through oxidation reactions that ultimately capture energy in ATP. (Recall from chapter 1 that oxidation is the addition of oxygen or the removal of electrons or hydrogen.) Coenzymes work with mitochondrial enzymes to accept electrons or whole hydrogen atoms as they are removed from substrates. FAD (flavin adenine dinucleotide) and NAD (nicotinamide adenine dinucleotide) illustrate this:

$$FAD + 2H \rightarrow FADH_2$$
$$NAD^+ + 2H \rightarrow NADH + H^+$$

oxidized   reduced

See Questions—Objective 1

Carbohydrate Metabolism

Carbohydrates comprise a large part of a normal diet, mostly in the form of glucose. We will see how carbohydrates are transported, how glycogen is stored and released, and how glucose is broken down for energy.

The simple sugars glucose, fructose, and galactose are absorbed and carried in the blood to the liver where most of the fructose and galactose is converted to glucose. The liver releases glucose to maintain a normal blood level

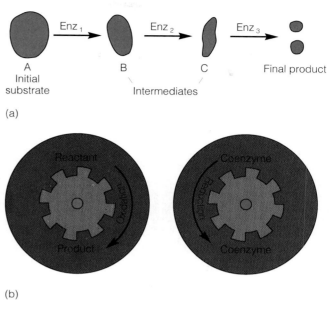

(a)

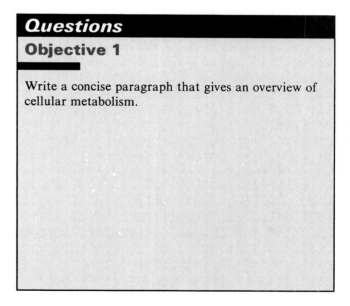

(b)

Figure 3.1

(*a*) Metabolic pathways often involve a series of enzyme-controlled reactions in which the product of one reaction becomes the reactant of the next. (*b*) Many metabolic reactions require coenzymes.

Questions

Objective 1

Write a concise paragraph that gives an overview of cellular metabolism.

and stores any excess as glycogen. Insulin facilitates the entry of glucose into many cells; without insulin cells can starve while surrounded by fluids rich in glucose.

Once in a cell, glucose undergoes **phosphorylation** (fos-for-il-a′shun) which uses ATP to add a phosphate group to its end (6th) carbon, and, thus, forming glucose-6-phosphate (G-6-P). Fructose and galactose are phosphorylated to form F-6-P and gal-6-P, and these sugars are interconvertible. Phosphorylation keeps sugars

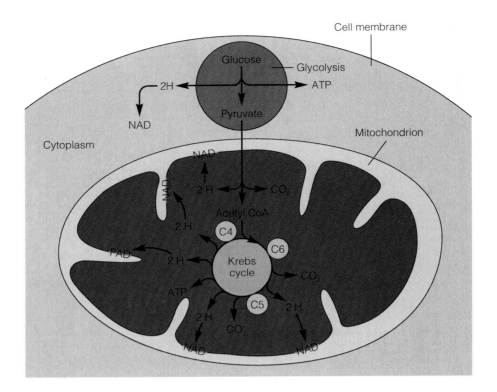

Figure 3.2

An overview of glucose metabolism.

from crossing the membrane again and enables them to undergo other reactions. Because sugars are phosphorylated as they enter cells, unphosphorylated sugars move down concentration gradients to enter cells.

Synthesis of glycogen from glucose, or **glycogenesis** (gli-ko-jen′es-is), occurs when blood glucose is above normal. Breakdown of glycogen to glucose, **glycogenolysis** (gli-ko-jen-ol′is-is), occurs when blood glucose is below normal. Glycogen can be made from phosphorylated 6-carbon-sugars, using energy from uridine triphosphate (UTP). Removing glucose from glycogen requires no energy but an inorganic phosphate group is added to form G-1-P. When the phosphate is removed, free glucose becomes available. These processes, catalyzed by different sets of enzymes, occur mainly in the liver.

Insulin fosters glycogenesis when abundant supplies of glucose are present, whereas epinephrine and glucagon foster glycogenolysis when cells need more glucose. These and other hormones precisely regulate the amount of glucose available to cells. The physiological advantage of such precise regulation is to provide all body cells with a source of energy as they need it.

Catabolism of glucose for energy requires three processes: glycolysis, which occurs in the cytosol, the Krebs cycle, and electron transport, which occurs in mitochondria (Figure 3.2). Glycolysis breaks glucose down to pyruvate and does not require oxygen. The Krebs cycle and electron transport break pyruvate down to carbon dioxide and water and do require oxygen. All these pathways involve a series of chemical reactions, and each reaction requires a specific enzyme. Many intermediates are acids that release H^+ and become ions in the aqueous reaction medium. Such ions have names ending in "ate." For example, pyruvic acid becomes pyruvate and acetic acid becomes acetate.

As soon as glucose enters a cell it undergoes **glycolysis** (gli-kol′is-is), in which a molecule of glucose is split into 2 molecules of pyruvate (Figure 3.3). Without oxygen, glycolysis proceeds from pyruvate to lactate. The steps in glycolysis are as follows:

1. Glucose is phosphorylated to glucose-6-phosphate (G-6-P), using energy and phosphate from ATP. (This is the same reaction that keeps glucose inside cells.) The G-6-P molecule rearranges to form fructose-6-phosphate (F-6-P). (Fructose and galactose can be converted to F-6-P and catabolized by glycolysis.)
2. F-6-P is phosphorylated to form fructose-1,6-diphosphate (F-1,6-diP), using energy and phosphate from another ATP.
3. F-1,6-diP is split into two 3-carbon molecules, dihydroxyacetone phosphate (DHAP) and 3-phosphoglyceraldehyde (3-PGA). Each DHAP is

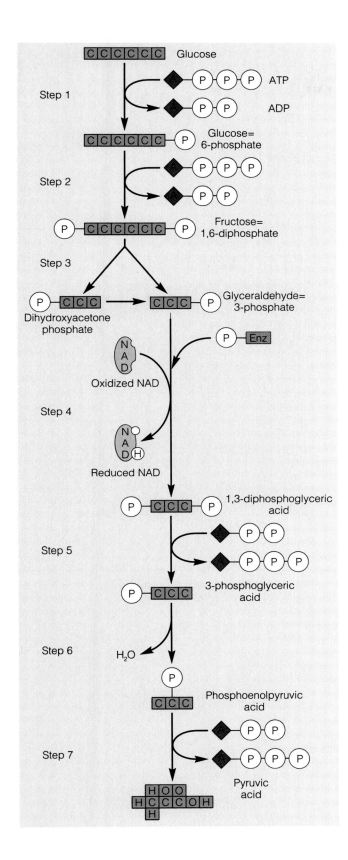

converted to 3-PGA. From this point, two 3-carbon molecules must be catabolized to account for 1 glucose molecule.

4. Inorganic phosphate from the cytosol is added to each 3-PGA molecule. Two hydrogen atoms are transferred from it to NAD^+ to form $NADH + H^+$ and 1,3-diphosphoglycerate (1,3-DPG).

5. Phosphate and energy from 1,3-DPG are added to ADP to make ATP and 3-phosphoglycerate, which becomes 2-phosphoglycerate.

6. Water is removed from 2-phosphoglycerate to form phosphoenolpyruvate (PEP).

7. PEP is converted to pyruvate and phosphate and energy is added to ADP to form ATP. The fate of pyruvate depends on whether oxygen is available.

In muscles during strenuous exercise when glycolysis is rapid and pyruvate is made faster than mitochondria can use it, some pyruvate is converted to lactate by the addition of hydrogen from NAD (Figure 3.4a). Lactic acid released from skeletal muscles goes to the liver where it is used to make glucose. The glucose goes to muscles where it is metabolized to lactic acid. This process is called the **Cori cycle** (Figure 3.4b).

Moving hydrogen from $NADH + H^+$ to pyruvate to make lactate frees NAD to remove more hydrogen from 3-PGA (step 4 above) and keeps glycolysis operating, so cells have some energy even when oxygen is not available. When oxygen becomes available hydrogen temporarily stored in lactate is transferred to NAD and on to oxygen. Each 3-carbon molecule formed from glucose generates 2 ATP molecules for a total of 4 ATP per 6-carbon glucose molecule. Because energy from 2 ATP was used in phosphorylation, the net energy yield from glycolysis in the absence of oxygen is 2 ATP per glucose molecule.

When oxygen is available, pyruvate enters mitochondria and is converted to **acetyl-coenzyme A** (as'ēt'al ko-en'zīm a), or acetyl-CoA, by a complex reaction (Figure 3.5) in which pyruvate loses carbon dioxide, releases 2 hydrogen atoms to NAD, and acquires energy as it combines with coenzyme A. Acetyl-CoA now contains the energy it needs to enter the Krebs cycle.

Figure 3.3

Glycolysis. The energy from two molecules of ATP is required to raise the energy level of glucose so that it can undergo the remainder of the reactions. The phosphorylated 6-carbon molecule is broken into two 3-carbon molecules. Each 3-carbon molecule loses hydrogen to NAD and releases sufficient energy to form two molecules of ATP. The end product of glycolysis is two molecules of pyruvate.

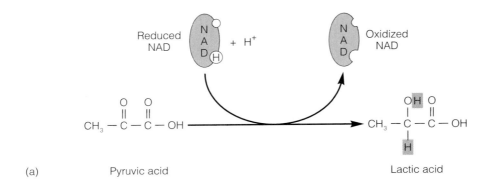

(a) Pyruvic acid — Lactic acid

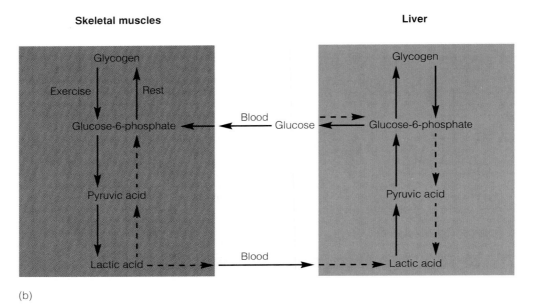

(b)

Figure 3.4

(a) The interconversion between pyruvate and lactate. The reaction goes to the right when oxygen is in short supply and to the left when the oxygen supply is adequate. (b) The Cori cycle.

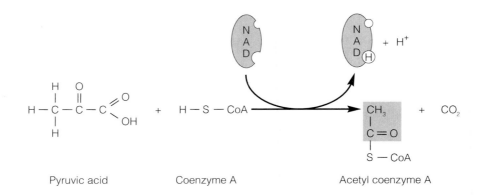

Figure 3.5

Pyruvate is converted to acetyl-CoA. In this complex reaction, CO_2 is released and hydrogen is transferred to NAD^+.

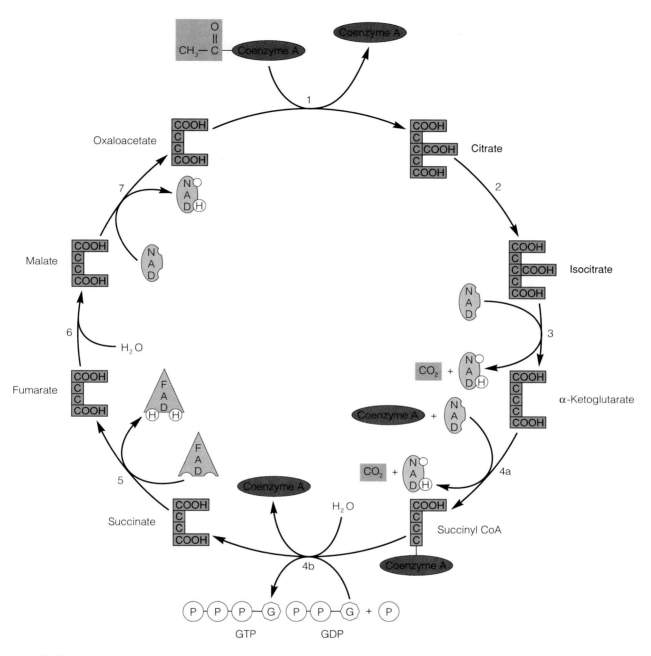

Figure 3.6

The Krebs cycle.

The **Krebs cycle** is a sequence of chemical reactions in which acetyl-CoA is oxidized to carbon dioxide and hydrogen atoms or hydrogen ions and electrons. It also is called the **citric** (sit'rik) **acid cycle** (its first product being citric acid) or the **tricarboxylic** (tri-kar-box-il'ik) **acid cycle** because citric and some other acids have 3 carboxyl groups. Hydrogen atoms and electrons removed from Krebs cycle intermediates undergo electron transport during which energy is captured in ATP. The Krebs cycle and electron transport occur simultaneously and only in the presence of oxygen.

Enzymes that control Krebs cycle reactions are located in the matrix of mitochondria, and molecules are passed from one enzyme to the next as they go through the cycle. The cycle begins with acetyl-CoA combining with oxaloacetate. When the cycle is completed oxaloacetate becomes available to combine with another acetyl-CoA. Oxaloacetate is continuously recycled.

The Krebs cycle (Figure 3.6) can be summarized as follows:

1. Acetyl-CoA combines with oxaloacetate, water is added, and CoA is removed to form citrate.
2. Citrate rearranges to form isocitrate.
3. Isocitrate loses carbon dioxide and 2 hydrogen atoms forming reduced NAD and alpha ketoglutarate.
4. Alpha-ketoglutarate loses carbon dioxide and hydrogen while gaining water and forming reduced NAD and succinate. During this reaction, energy is transferred to a guanosine triphosphate (GTP), a nucleotide similar to ATP.
5. Succinate loses 2 more hydrogen atoms, forming reduced FAD and fumarate.
6. Fumarate gains water to form malate.
7. Malate loses 2 hydrogen atoms, forming reduced NAD and oxaloacetate, which can combine with acetyl-CoA to start the cycle over again.

Significant events in the Krebs cycle include the following:

1. The release of two molecules of carbon dioxide, thereby disposing of the 2 carbons in an acetyl group.
2. The transfer of 4 pairs of electrons (or hydrogen atoms) to coenzymes for later oxidation.
3. The capture of energy in GTP directly from the substrate. Energy in GTP can be transferred to ATP:

$$GTP + ADP \rightarrow GDP + ATP.$$

So far, we have encountered six reactions in glucose metabolism that transfer electrons to coenzymes—from 3-PGA during glycolysis, from pyruvate, and from four intermediates of the Krebs cycle—citrate, alpha-ketoglutarate, succinate, and malate. To metabolize a whole glucose molecule each reaction takes place twice—once for each 3-carbon molecule into which glucose was split. Thus, 12 pairs of hydrogen atoms enter electron transport for each glucose molecule metabolized.

Electron transport (Figure 3.7) involves the transfer of electrons from a substrate through a series of carriers and the capture of energy in ATP at certain transfers. Electrons and hydrogen ions ultimately combine with oxygen to form water. During each transfer one substance is oxidized as the other is reduced. For example, a substrate with hydrogen (substrate-H_2) loses hydrogen (is oxidized) at the same time coenzyme NAD gains electrons (is reduced) and H^+ is released. Substrate-H_2 refers to any substrate that can be oxidized—3-PGA, pyruvate, and Krebs cycle intermediates.

Oxygen must be available for electron transport and the Krebs cycle to occur. Enzymes that catalyze electron transport are located on the cristae of mitochondria, adjacent to those in the matrix that catalyze Krebs cycle reactions. Electrons removed when a substrate in the matrix is oxidized are used in a coupled reaction to reduce coenzymes in the cristae. Each reaction requires a specific enzyme and most also require a coenzyme.

At three points, electron transport is accompanied by **oxidative phosphorylation.** It occurs where sufficient energy is available to add phosphate and energy to ADP to form ATP.

During electron transport, electrons from most substrates go to NAD and on to other coenzymes—FMN (flavin mononucleotide), coenzyme Q, and the cytochromes b, c_1 and c, a and a_3. Electrons that go to FAD enter the electron transport system, not at the beginning, but at coenzyme Q. All electron carriers are proteins and most contain an iron atom and some sulfur. Oxidized iron has a valence of 3^+ and reduced iron (after it accepts an electron) has a valence of 2^+ as shown below.

$$Fe^{3+} + e^- \rightarrow Fe^{2+}$$

At the end of the chain, two H^+ combine with two electrons and an atom of oxygen to form water as shown in the equation below.

$$2e^- + 2H^+ + 1/2O_2 \rightarrow H_2O$$

Thus, oxygen is the final hydrogen acceptor. All twelve pairs of hydrogen atoms from the metabolism of one molecule of glucose go through the electron transport process:

$$12H_2 + 6O_2 \rightarrow 12H_2O.$$

Oxidative phosphorylation occurs at three sites along the electron transport chain as electrons are transferred (1) from FMN to cytochrome b, (2) from cytochrome b to cytochromes c_1 and c, and (3) from cytochrome a to cytochrome a_3. Energy capture by oxidative phosphorylation during electron transport is analogous to water flow down a multitiered waterfall (Figure 3.7). Electron transfers that release little energy are represented as small cascades and those that provide sufficient energy to make ATP are represented as large cascades.

Oxidative phosphorylation is not fully understood, but the **chemiosmotic** (kem-e-os-mot'ik) **theory** for which Peter Mitchell received a Nobel Prize in 1978 offers this explanation. During oxidation, H^+ from the matrix concentrates in the intermembranous space of a mitochondrion, thus creating a **chemiosmotic gradient.** This gradient of both electrical charge and osmotic pressure drives phosphorylation reactions. The movement of a pair of protons down the gradient (back across the membrane to the matrix) makes energy available to form ATP.

Pairs of electrons transferred to NAD^+ go through the entire electron transport process and yield 3 molecules of ATP by oxidative phosphorylation. Pairs of electrons transferred to FAD enter electron transport after the first phosphorylation and yield only 2 molecules of ATP. In

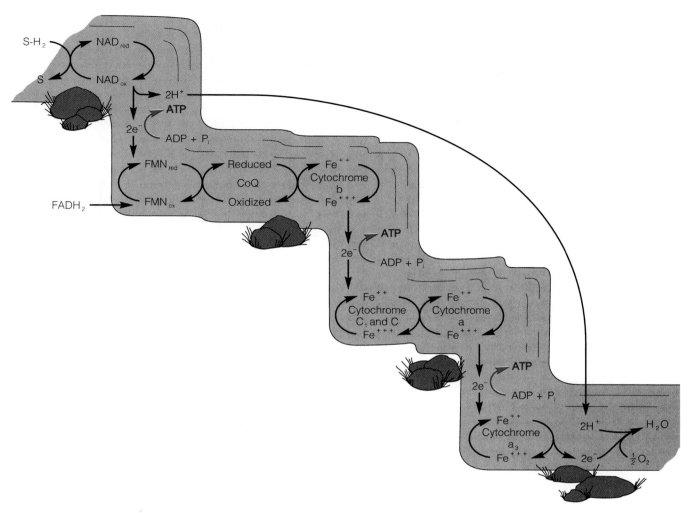

Figure 3.7

Electron transport and oxidative phosphorylation: coupled
reactions and analogy of a multitiered waterfall.

other words, electrons on NAD go down the entire waterfall, those on FAD go down the second and third largest cascades. Oxidative phosphorylation is significant because it makes energy available in a form cells can use. We can move because muscles exert force on bones as individual muscle cells expend energy from ATP in the contraction process.

How much energy is captured from a glucose molecule depends on whether it is metabolized with or without oxygen. Under anaerobic (without oxygen) conditions, glucose is metabolized to lactic acid in so-called **fast glycolysis** with a net yield of only 2 ATPs. Under aerobic (with oxygen) conditions, glucose is completely metabolized by **slow glycolysis,** the Krebs cycle, and electron transport to carbon dioxide and water, and yields 38 ATPs. Glycolysis yields a net of 2 substrate level ATPs and the Krebs cycle yields 2 more substrate level ATPs, 1 per cycle.

Most ATP comes from electron transport of 12 pairs of electrons—10 pairs entering at NAD produce 30 ATPs (3 per pair) and 2 pairs entering at FAD produce 4 ATPs (2 per pair). Anaerobic and aerobic metabolism of glucose is summarized in Table 3.1.

Though most carbohydrates are catabolized via glycolysis, the Krebs cycle, and electron transport, sometimes as much as 30 percent of the body's glucose is metabolized by the **pentose phosphate pathway** (Figure 3.8). This pathway does not require oxygen and operates in the cytosol of cells—mainly in the liver, fatty tissues, and red blood cells. Though it does not produce ATP directly, the pentose phosphate pathway is significant because it makes reduced NADP that is required for fatty acid synthesis, the 5-carbon-sugars required for synthesis of nucleic acids, and many structural proteins such as collagen.

Table 3.1
Energy Yields in ATPs for Glucose by Anaerobic and Aerobic Metabolism

Process	Anaerobic Metabolism	Aerobic Metabolism
Glycolysis (substrate level)	4	4
Less energy to start process	−2	−2
Krebs cycle—2 acetyl-CoAs (substrate level)		2
Oxidative phosphorylation		
Hydrogen to NAD		30
Hydrogen to FAD		4
Total	2	38

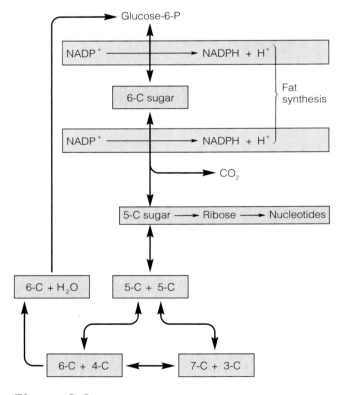

Figure 3.8

Some of the reactions of the pentose phosphate pathway.

See Questions—Objectives 2 through 7

Lipid Metabolism

Most human dietary lipids are triacylglycerols (formerly triglycerides), but cholesterol and fat-soluble vitamins also are plentiful. Half of the energy in foods Americans eat comes from lipids, and as much as half of the dietary carbohydrate is converted to lipids—later used for energy or stored as fat. Thus, lipid metabolism is an extremely important part of total metabolism.

After absorption, acylglycerols, phospholipids, cholesterol, and other lipids are packaged with protein in **chylomicrons** and are transported in lymph and blood to the liver. By the time chylomicrons enter the blood after a meal, blood glucose and insulin already have increased. Insulin activates **lipoprotein lipase,** an enzyme bound to the membranes of blood vessel lining cells. As chylomicrons pass through blood vessels, lipoprotein lipase digests their acylglycerols and phospholipids into smaller molecules, such as fatty acids and glycerol, that can enter cells.

Lipids are stored mainly in adipose (fat) cells where they can comprise 95 percent of the cell volume. When insulin causes glucose to enter fat cells, it is metabolized by glycolysis to 3-carbon molecules, which are converted to glycerol. Fatty acids that diffuse into fat cells combine with glycerol to form acylglycerols. Lipids are excellent energy storage molecules because they are insoluble in water (and do not dissolve in cell fluids) and contain much hydrogen (which yields large amounts of energy in electron transport). Lipids in subcutaneous tissues insulate the body and delay heat loss when the environment is cooler than body temperature. Fat deposits contain mostly the same fatty acids that occur in foods—stearic acid (a saturated 18-carbon molecule), oleic acid (an unsaturated 18-carbon molecule), and palmitic acid (a saturated 16-carbon molecule).

Triacyglycerols, the main lipids used for energy, are hydrolyzed to glycerol and 3 fatty acids (chapter 1). Glycerol is converted to 3-PGA and can be metabolized by glycolysis or the pentose phosphate pathway.

Fatty acids from fat hydrolysis are broken down into 2-carbon units by **beta oxidation** (Figure 3.9), so named because oxidation occurs at beta carbons (the second carbon from the carboxyl group). Beta oxidation takes place in the mitochondrial matrix, but fatty acids require processing before they can enter the mitochondria. In the cytosol, a fatty acid combines with coenzyme A using energy from ATP to form a molecule called fatty acid-CoA, which then combines with carnitine, a 7-carbon nitrogenous substance, to form fatty acid-carnitine and free coenzyme A. Carnitine ferries the fatty acid across the inner mitochondrial membrane. Coenzyme A replaces carnitine, and carnitine is released to transport more fatty acid molecules.

Questions

Objective 2

(a) How are components of carbohydrates transported to cells?

(b) What are the differences between glycogenesis and glycogenolysis?

Objective 3

(a) What are the major components of carbohydrate metabolism?

(b) Where do glycolysis, the Krebs cycle, and biological oxidation occur?

Objective 4

(a) What are the main steps in glycolysis?

(b) What are the products and signficance of glycolysis?

(c) How does the metabolism of glucose differ under aerobic and anaerobic conditions?

Continued on next page

Questions

Objective 5

(a) What are the main steps in the Krebs cycle?

(b) What are the products and significance of the Krebs cycle?

(b) What are the main steps in electron transport?

(c) What are the products and significance of oxidative phosphorylation?

Objective 6

(a) Distinguish between electron transport and oxidative phosphorylation.

Objective 7

(a) What is the significance of the pentose phosphate pathway?

Fatty acid-CoA undergoes beta oxidation in the mitochondrial matrix by the following steps:

1. Fatty acid-CoA loses 2 hydrogen atoms to FAD and a double bond forms between the alpha and beta carbon atoms.
2. Reduced fatty acid-CoA gains water at its double bond and immediately loses hydrogen atoms. Oxygen forms a ketone group at the beta carbon, and electrons are transferred to NAD.
3. The bond between the alpha and beta carbons breaks releasing acetyl-CoA, which goes to the Krebs cycle.
4. Coenzyme A bonds to the beta carbon, making a new fatty acid-CoA shorter by 2 carbons.

Each new shorter fatty acid-CoA undergoes beta oxidation until finally a 4-carbon fatty acid forms two molecules of acetyl-CoA. Thus, an 18-carbon fatty acid

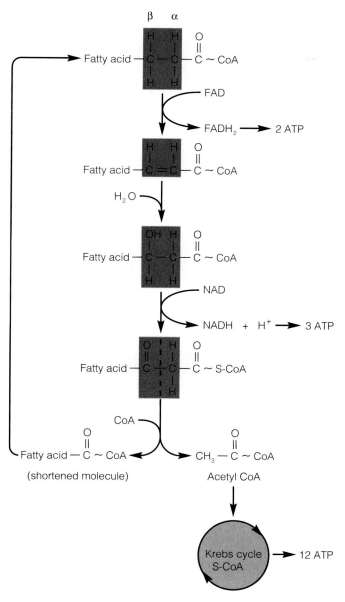

Figure 3.9

Beta oxidation.

Table 3.2
Energy Yields in ATPs for Saturated and Unsaturated Fatty Acids

Process	Saturated Fatty Acid (Stearic) (No Double Bonds)	Unsaturated Fatty Acid (Linoleic) (2 Double Bonds)
Beta oxidation	40	36
Less ATP used to start process	−1	−1
Krebs cycle and biological oxidation	108	108
Total	147	143

$CH_3 - C(=O) - CH_2 - C(=O) - OH \longrightarrow CH_3 - CH(OH) - CH_2 - C(=O) - OH$

Acetoacetic acid   Beta-hydroxybutyric acid

$CH_3 - C(=O) - CH_3$

Acetone

Figure 3.10

Ketone bodies are products of incomplete fatty acid metabolism.

goes through beta oxidation eight times and yields 9 acetyl-CoA molecules. (Eight, and not nine, trips through beta oxidation are required because the last trip produces 2 acetyl-CoAs.) Most fatty acids have an even number of carbons, so beta oxidation yields 2-carbon acetyl groups almost exclusively.

Fatty acids contain relatively more hydrogen and less oxygen (are more reduced) than sugars, so they yield more energy on oxidation. Each beta oxidation produces 5 ATPs by electron transport of hydrogens (2 ATPs from FAD and 3 ATPs from NAD$^+$). For an 18-carbon fatty acid, 8 beta oxidations yield 40 ATPs less 1 ATP required to start the process for a net of 39 ATPs. In addition, 9

acetyl-CoA molecules go through the Krebs cycle and their hydrogens through electron transport with 12 ATPs being captured from each acetyl-CoA. This yields another 108 ATPs. Of course, beta oxidation, like the Krebs cycle, is coupled to electron transport and requires oxygen, NAD, and FAD. In unsaturated fatty acids, double bonds already exist at some beta oxidation sites and hydrogens are not transferred to FAD. The energy yield from beta oxidation decreases by 2 ATPs per double bond. Energy yield from fatty acids is shown in Table 3.2.

Liver cells rapidly metabolize fatty acids and can make more acetyl-CoA than they need. They use excess acetyl-CoA to make **ketone** (ke′tōn) **bodies**—acetoacetic acid, beta-hydroxybutyric acid, and acetone (Figure 3.10). Sometimes cells make ketone bodies from incompletely metabolized fatty acids. Ketone bodies diffuse freely out of the liver, and acetone, being volatile, is exhaled in the lungs. Ketone bodies that enter cells are converted back to acetyl-CoA and used for energy.

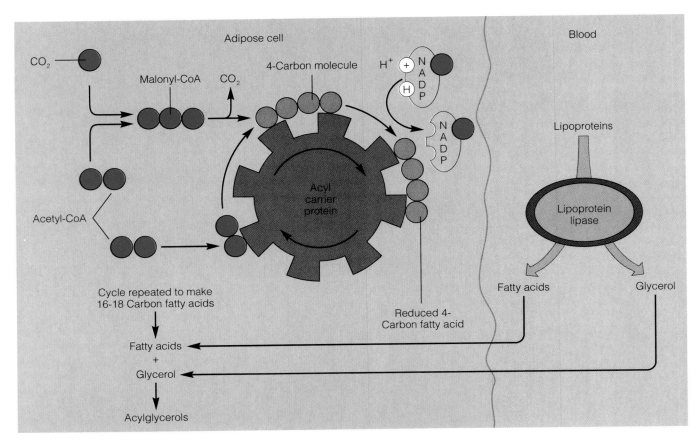

Figure 3.11

Synthesis of fatty acids and fats.

In addition to making ketone bodies, the fat cells (and the liver) can use acetyl-CoA to make fatty acids by **beta reduction.** Though beta reduction is the reverse of beta oxidation, it is controlled by different enzymes and occurs on a large cytoplasmic molecule called **acyl** (as'il) **carrier protein** instead of in mitochondria. In addition to acetyl-CoA, fatty acid synthesis requires reduced NADP from the pentose phosphate pathway and bicarbonate ions (HCO_3^-) from CO_2. As excess acetyl-CoA is used to make fatty acids or as lipoprotein lipase release them from chylomicrons, the fatty acids move to adipose cells and combine with glycerol to make acylglycerols (Figure 3.11).

Most cells synthesize fatty acids when they have more than enough glucose and acetyl-CoA and use them for energy when they lack glucose. When acetyl-CoA is plentiful, an intermediate in fatty acid synthesis, malonyl-CoA binds to carnitine and inhibits the fatty acid breakdown. Conversely, when glucose is in short supply fatty acid-carnitine inhibits the conversion of pyruvate to acetyl-CoA. Thus, a cell synthesizes or oxidizes fatty acids but not both at the same time (Figure 3.12).

In addition to providing energy, lipids form structural components of membranes and certain other important molecules. Cells can make all the lipids they need

provided **linoleic acid** is available. Linoleic acid is an **essential fatty acid,** that is, it is essential in the diet because human cells cannot make it. It is needed to synthesize membranes and prostaglandins and maybe for the transport and metabolism of cholesterol.

Cholesterol, present in any diet containing animal fats, is absorbed and packaged in chylomicrons like other lipids. Most cells make a little cholesterol and liver cells make a lot. Cholesterol gives membranes a fluidity at body temperature, and the more cholesterol a membrane contains the higher its fluidity. Cholesterol is completely insoluble in water—an advantage in keeping membranes from dissolving but a disadvantage in keeping it suspended in blood.

Cholesterol is needed to make and maintain cell membranes and to synthesize steroid hormones and bile salts. It helps to waterproof skin. The derivative 7-dehydrocholesterol is used to synthesize vitamin D. Though human cells can make cholesterol, they lack enzymes to break it down. The body's main means of getting rid of cholesterol is through excretion of bile salts in the feces. This process removes only about half a gram of cholesterol per day. Even when more cholesterol is lost than is taken in in foods, the liver easily replaces it.

Lipids, being insoluble in water, are transported in aqueous plasma as **lipoproteins,** which are combinations of lipid and protein. They include chylomicrons, very low density lipoproteins (VLDLs), low density lipoproteins (LDLs), and high density lipoproteins (HDLs). Each lipoprotein particle consists of a core of lipids such as acylglycerols and cholesteryl esters that are surrounded by a lipoprotein membrane made of a protein called apolipoprotein, phospholipids, and cholesterol. (Cholesteryl esters consist of cholesterol and a fatty acid attached by an ester bond.) Lipoprotein composition is summarized in Table 3.3.

Lipoproteins circulate in the blood and participate in the following processes (Figure 3.13).

1. Dietary fat leaves the intestine in chylomicrons.
2. In the blood, lipoprotein lipase partly degrades chylomicrons to fatty acids, monoacylglycerols, and glycerol, which can enter adipose and other tissues.
3. In the liver, chylomicron remnants are incorporated into HDLs.
4. HDLs, made in the liver and intestine, appear to activate lipoprotein lipase and to consume cholesterol. They transport cholesteryl esters and help to prevent cholesterol deposition in coronary and other blood vessels.
5. VLDLs are made mostly from acylglycerols and cholesterol that are synthesized in the liver.
6. Lipoprotein lipase releases fatty acids and glycerol from VLDLs and a little surface cholesterol moves to cell membranes leaving VLDL remnants.
7. Some VLDL remnants bind to liver cell receptors and the remainder become cholesterol-laden LDLs. LDLs also bind to receptors, enter cells by endocytosis, and are degraded by lysosomal enzymes. Receptors are returned to the cell membrane. Familial hypercholesterolemia is caused by defective LDL receptors.
8. HDLs incorporate some LDLs and VLDL remnants.

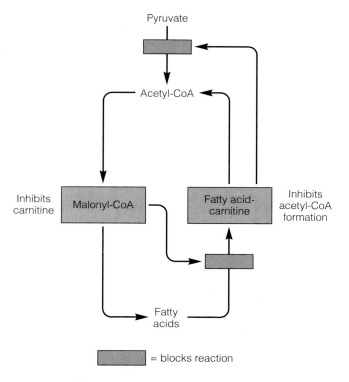

Figure 3.12

Regulation of synthesis and breakdown of fatty acids. When fatty acids are being synthesized, malonyl-CoA inhibits the formation of fatty acid-carnitine. When fatty acids are being oxidized, fatty acid-carnitine inhibits the formation of acetyl-CoA from pyruvate. Thus, a cell can synthesize or oxidize fatty acids, but it cannot do both at the same time.

As more is learned about lipoproteins, it is becoming increasingly clear that the risk of coronary artery blockage increases when total cholesterol and LDL cholesterol are too high, when HDL cholesterol is too low, or when the ratio of total cholesterol to HDL cholesterol is greater than 4.5. However, excessive lowering of blood cholesterol is not necessarily beneficial because a high incidence of cerebral hemorrhages (ruptured cerebral blood vessels) has been reported among people with total cholesterol under 150 mg/dl. The current view of blood cholesterol values is summarized in Table 3.4.

Table 3.3
Lipid and Protein Composition (in percent) of Lipoproteins

Lipoprotein	Protein	Acylglycerols	Cholesterol	Cholesteryl esters	Phospholipids
Chylomicrons	2%	84%	2%	5%	7%
VLDLs	9	54	7	12	18
LDLs	21	11	8	37	22
HDLs	50	4	2	20	24

Source: Data from R. Montgomery, et al., *Biochemistry: A Case-Oriented Approach,* 1983.

Table 3.4
Interpretation of Cholesterol Levels

Kind of Cholesterol	Concentrations (in mg/dl)			
	Low	*Normal*	*Borderline*	*High*
Total cholesterol	<150	150–200	200–240	>240
LDL cholesterol		<130	130–160	>160
HDL cholesterol	<35	35 or more		

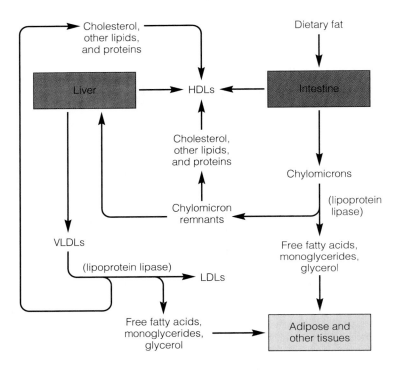

Figure 3.13

Lipoprotein metabolism.

See Questions—Objectives 8 and 9

Protein Metabolism

For several hours after a meal, amino acids from protein digestion are absorbed into the blood. Once in the blood, they are actively transported into cells in 5 to 10 minutes.

Unlike carbohydrates and lipids, proteins are not stored in cells. Furthermore, enzymes and other cell proteins are regularly degraded and replaced in **protein turnover,** a process that occurs continuously at an overall rate of 350 to 400 grams of protein per day. Rates in tissues vary according to metabolic activity from high in liver, moderate in muscle, to low in cartilage.

When proteins are turned over, many of their amino acids are reused in protein synthesis, but some are degraded and their nitrogen excreted. Nitrogen lost in this way is replaced from dietary proteins to maintain **nitrogen balance.** If nitrogen intake exceeds nitrogen loss, the body is in **positive nitrogen balance,** as occurs in growth, pregnancy, or tissue repair after injury. If nitrogen loss exceeds nitrogen intake, the body is in **negative nitrogen balance,** as occurs in starvation, certain wasting diseases, and weight-loss diets that contain too little protein.

Of 20 amino acids normally found in proteins, 8 **essential amino acids**—leucine, isoleucine, lysine, methionine, threonine, phenylalanine, tryptophan, and valine—

Questions

Objective 8

(a) How are lipids transported in blood, and how are they stored in fat cells?

(b) How do the processes by which fatty acids are degraded and synthesized differ?

(c) How are these processes regulated?

(d) How is cholesterol metabolism related to lipoproteins?

Objective 9

(a) Where and how are the different kinds of lipoproteins formed?

(b) How are lipoproteins degraded?

---

must be in the diet because human cells lack the enzymes to synthesize them. Human cells synthesize only small amounts of arginine and histidine, so they should also be included in the diet especially during periods of rapid growth. Given adequate amounts of essential amino acids, cells can make any other amino acids they need.

Amino acids from the diet and from protein turn-over form an **amino acid pool,** which is an assortment of amino acids in cells. Cells use amino acids to make pro-teins, nucleic acids, and other substances (Figure 3.14). Liver cells manufacture blood proteins such as albumins and blood clotting factors. Connective tissue cells make collagen and elastin, and muscle cells produce large quan-tities of actin and myosin. Amino acids also are used to make neurotransmitters in neurons, hormones in glands, and hemoglobin in erythrocytes.

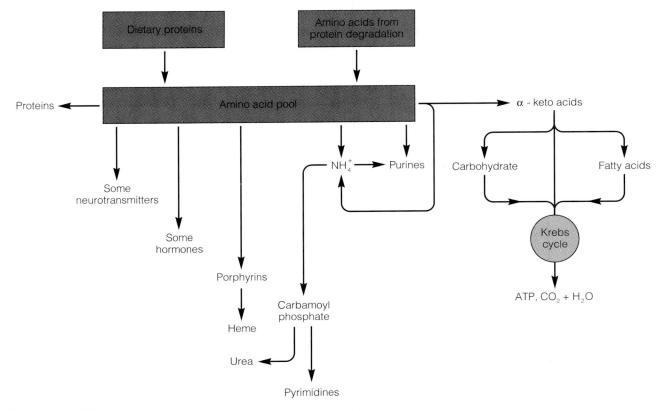

Figure 3.14

Metabolism of amino acids from the amino acid pool.

Protein synthesis is an important component—maybe the most important component—of protein metabolism. Because of its significance in cell function it was discussed in chapter 2, but should be reviewed at this time.

Before they can be used for energy, amino acids undergo transamination or deamination (Figure 3.15). **Transamination** (tranz-am-in-a'shun) is the exchange of functional groups between an amino acid and a keto acid (an acid that has a ketone group). Transamination does not change the total number of amino acid molecules in a cell, and it never produces an essential amino acid. It can increase the supply of one amino acid at the expense of a more abundant one thus allowing a cell to make a protein it otherwise may not make. In one kind of **deamination** (de-am-in-a'shun), an amino group and 2 hydrogen atoms are removed from an amino acid yielding an alpha-keto acid, ammonia, and NADH + H⁺. Deamination, which decreases a cell's amino acids, occurs when amino acids are used for energy or for the synthesis of glucose or fatty acids.

Nearly all tissues produce ammonia from deamination of amino acids, and the ammonia must be removed before it builds to toxic levels. Ammonia is transported as ammonium ions and is carried in the amino acids, aspartate and glutamine. When it reaches the liver it is incorporated into urea, a much less toxic waste, by way of the **urea cycle** (Figure 3.16):

1. An ammonium ion combines with carbon dioxide using energy and phosphate from ATP to form carbamoyl phosphate.
2. Carbamoyl phosphate combines with ornithine to form citrulline. (The amino acids ornithine and citrulline participate in the urea cycle but are not found in proteins.)
3. An amino group from aspartate is added to citrulline using energy from ATP to form arginosuccinate.
4. Arginosuccinate breaks down to arginine and fumarate. (Fumarate can enter the Krebs cycle.)
5. Arginine combines with water to form ornithine and urea. Ornithine repeatedly combines with carbamoyl phosphate and is recycled. The relatively nontoxic urea travels in the blood and is excreted by the kidneys.

The **carbon skeletons** of amino acids, many of which are Krebs cycle intermediates such as pyruvate; oxaloacetate; and alpha-ketoglutarate; can be metabolized

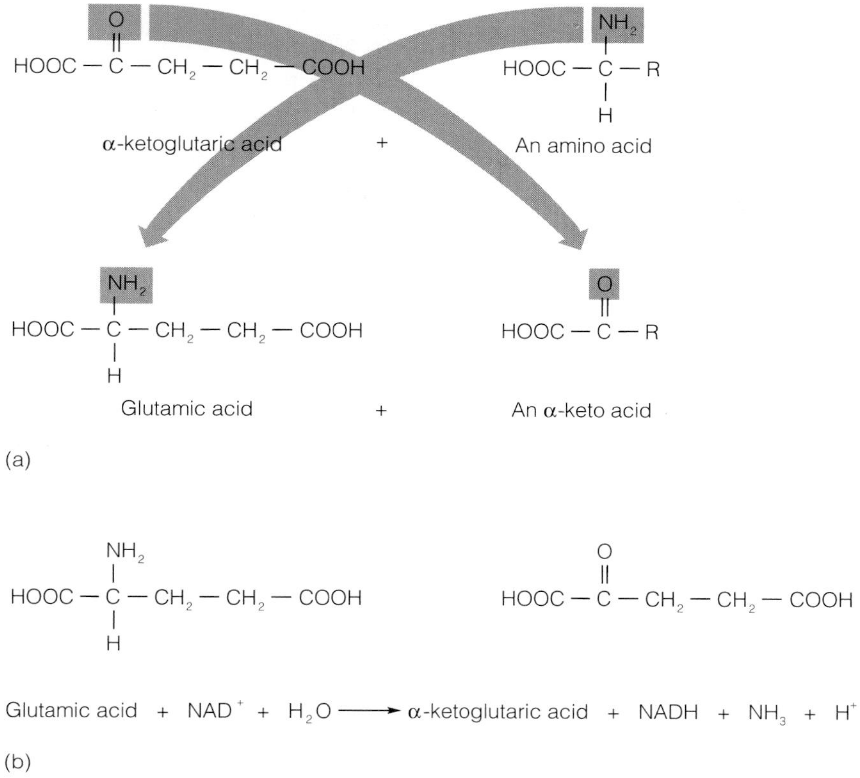

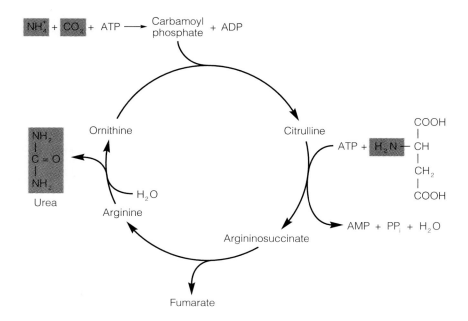

Figure 3.15

(*a*) Transamination, and (*b*) deamination.

Figure 3.16

The urea cycle.

Questions

Objective 10

(a) What is nitrogen balance?

(b) What conditions affect nitrogen balance?

(c) Define essential amino acid and amino acid pool.

(d) Distinguish between deamination and transamination.

Objective 11

(a) How are amino acids metabolized for energy and what happens to the nitrogen they contain?

(b) What is gluconeogenesis and what is its significance?

for energy after deamination. This occurs in starvation as carbohydrate and lipid stores are depleted, in protein turnover as small amounts of protein are degraded and some amino acids deaminated, and in positive nitrogen balance when protein intake is excessive. Compared to carbohydrates and lipids, proteins provide the least efficient source of metabolic energy.

Gluconeogenesis (glu-ko-ne-o-jen'ĕ-sis) is the synthesis of glucose from noncarbohydrate sources. It occurs between meals when glycogen is depleted and helps keep the blood glucose level in a narrow range. Gluconeogenesis provides glucose to brain cells and red blood cells that cannot use other nutrients. It accelerates during starvation when up to 60 percent of the tissue proteins can be degraded before death ensues. Gluconeogenesis occurs mainly in the liver and kidney cells, which contain the appropriate enzymes. Carbon skeletons are metabolized in the Krebs cycle and by cytosol enzymes to form pyruvate and energy from GTP is used to form phosphoenolpyruvate (PEP). Energy in PEP drives glycolysis in reverse to make glucose.

See Questions—Objectives 10 and 11

Nucleotide Metabolism

Though no nucleic acid components are essential nutrients, they are absorbed as nucleotides and nucleosides (a nucleoside is a base and a sugar), and most are degraded to purines, pyrimidines, ribose, deoxyribose, and phosphates. Of these products, phosphates, sugars, and the purine adenine can be used to make nucleic acids. Most other purines and pyrimidines are catabolized for energy and resynthesized when cells make nucleic acids and nucleotides. Most cells can both synthesize and degrade purines and pyrimidines.

Nucleotide metabolism is carefully regulated to allow nucleic acid synthesis in preparation for cell division and protein synthesis as needed. This regulation also allows synthesis of nucleotides such as ATP, GTP, and other energy-storing molecules; FAD, NAD, and other electron transport molecules; and control molecules such as cAMP and cGMP.

See Questions—Objective 12

Regulation of Metabolism at the Cellular Level

Metabolism is regulated within cells mainly by controlling the activity of enzymes, which are localized in particular organelles or other specific sites. For example, adenylate cyclase is found in plasma membranes; enzymes that transfer energy to ATP are bound to the cristae of mitochondria; those that synthesize proteins are bound to ribosomes; and those that synthesize or degrade fats are bound to the smooth endoplasmic reticulum. Enzyme concentrations are regulated according to cellular needs by controlling protein synthesis. Enzyme activity is regulated by the action of ligands on allosteric enzymes, feedback inhibition, and actions of isozymes.

Allosteric (al-o-ster′ik) **enzymes** are subject to regulation through an **allosteric site,** in addition to the substrate binding site. (Allosteric means "other shape.") Ligands that bind to an allosteric site change the shape of the molecule, including its active site. Such reversible changes turn enzymes on or off depending on whether a ligand binds to the allosteric site.

Feedback inhibition, or end product inhibition, may or may not involve an allosteric enzyme. As noted in chapter 1, it regulates the synthesis of various substances in living cells. The inhibiting product attaches to an allosteric site or otherwise inhibits an enzyme when it is plentiful and detaches when it is in short supply. Because feedback inhibition acts quickly and directly, the cell wastes no energy making an unnecessary product or even a special inhibitor; it simply uses the product of a reaction as its inhibitor.

Certain biochemical reactions are catalyzed by **isozymes,** which are multiple forms of the same enzyme. Though these forms catalyze the same reaction, they can have different effects. For example, the lactic dehydrogenase (LDH) isozyme in the heart muscle typically oxidizes lactate to pyruvate, whereas the isozyme in skeletal muscle typically reduces pyruvate to lactate, particularly during exercise. Because of chemical differences, isozymes are subject to allosteric regulation, as when a ligand inhibits one isozyme and has no effect on another. Such selective inhibition serves to regulate cellular metabolism.

Metabolism of major nutrients is summarized in Figure 3.17.

See Questions—Objective 13

Metabolic Effects of Damaged and Deficient Enzymes

Metabolic disturbances can result from damaged or deficient enzymes. Agents that damage enzymes are environmental toxins such as heavy metals, cyanide ions, carbon monoxide, insecticides, and some medicines. Enzyme deficiencies usually are due to abnormal genes that make a defective enzyme or no enzyme.

Heavy metals denature proteins, including enzymes, by distorting the molecular shape. Such distortion is especially detrimental to enzymes because it changes the shape of the active site, so the enzyme cannot form a complex with its substrate.

Cyanide ions and carbon monoxide bind with high affinity to molecules such as cytochromes and hemoglobin. Both specifically block oxidative phosphorylation at the last site in the electron transport chain (the cytochrome a-a$_3$ complex), thereby decreasing the amount of energy captured in ATP. Carbon monoxide also binds to

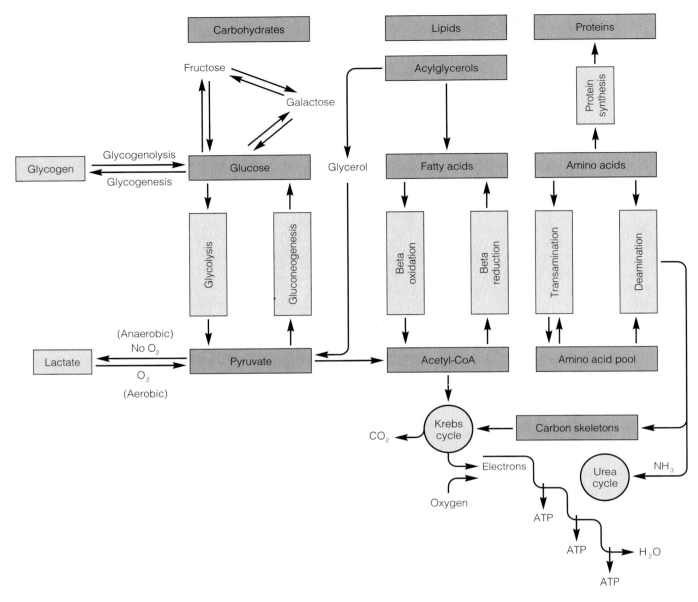

Figure 3.17

Summary of metabolism of major nutrients.

hemoglobin much more tightly than oxygen. Hemoglobin molecules carrying carbon monoxide cannot also carry oxygen, so oxygen transport to cells is severely compromised.

Barbiturates, some tranquilizers, and the insecticide rotenone block oxidative phosphorylation at the first site (NAD to FMN) and thereby decrease the amount of energy captured in ATP. In addition, certain insecticides specifically inhibit cholinesterase, an enzyme that inactivates acetylcholine. When the enzyme is inhibited, acetylcholine remains active and causes violent spasms. In respiratory muscles such spasms can be fatal.

Enzyme deficiencies can interfere with the metabolism of all types of nutrients—carbohydrates, lipids, and proteins. These effects include inability to metabolize stored glycogen, alterations in the blood lipoprotein concentrations, and accumulation of toxic levels of some amino acids.

Glycogen-storage diseases are caused by deficiencies in enzymes that break down glycogen. Affected cells metabolize glucose and store it as glycogen but cannot break it down. Excessive glycogen deposits interfere with liver and kidney functions. The inability to release glucose from glycogen increases fat metabolism, which can lead

Questions

Objective 13

(a) How does the location of enzymes and metabolites regulate metabolism?

(b) What regulatory processes control the activity of cellular enzymes?

to ketosis, and causes low blood glucose, which leads to muscle weakness.

Hyperlipidemias are excesses of lipoproteins or other blood lipids. One such disorder, an absence of lipoprotein lipase, prevents chylomicron digestion. It becomes apparent as early as the first week after birth and can be treated only with a fat-restricted diet.

Hypolipidemias (deficiencies of blood lipids) usually are caused by a defective enzyme that prevents synthesis of a protein normally found in lipoproteins. Patients deficient in proteins needed to make chylomicrons, VLDLs, or LDLs often have only minor symptoms, but those deficient in proteins to make HDLs usually have excessive cholesterol deposits in their tissues.

Several disorders result from deficiencies in urea cycle enzymes. Hyperammonemia, a high-blood ammonia concentration, is caused by a lack of the enzyme that makes carbamoyl phosphate or a lack of the enzyme that makes citrulline. Citrullinuria (accumulation of citrulline in the urine) is caused by a lack of the enzyme that acts on citrulline. Arginosuccinic aciduria (accumulation of arginosuccinic acid in the urine) is caused by a lack of the enzyme that acts on arginosuccinic acid. Patients with these disorders must limit their protein intake to minimize the amount of waste nitrogen to be removed by the urea cycle.

Several disorders result from a deficiency of an enzyme required to metabolize a specific amino acid. Normally, the essential amino acid phenylalanine is used to make proteins or is converted to tyrosine and used to make the pigment melanin and certain hormones and neurotransmitters. In patients with phenylketonuria, the enzyme to convert phenylalanine to tyrosine is deficient, so phenylalanine is converted to other metabolites such as phenylketones. The accumulation of phenylketones in the blood causes irreversible mental retardation. Restricting the diet from early infancy to the small quantity of phenylalanine needed for protein synthesis and supplying tyrosine for synthesis of other products can prevent mental retardation.

In humans (and most other mammals), lactase activity is high at birth, declines during growth, and remains low in adulthood. Exceptionally low lactase activity causes a condition called **lactose intolerance**—the inability to digest lactose. In the United States, 20 percent of white adults and 70 percent of black adults suffer from lactose intolerance. When milk or milk products are ingested, people who suffer from lactose intolerance develop osmotic diarrhea and intestinal gas. Osmotic diarrhea results from the high intestinal concentration of lactose that draws water into the lumen. Gas is produced by certain intestinal bacteria that metabolize lactose.

Gout is a severe, painful arthritis often caused by a genetic defect that allows uric acid deposits to develop in the joints. The treatment of gout illustrates how knowledge of biochemical pathways can be used to thwart disease processes. Limiting the quantity of purines in the diet by limiting meats, especially organ meats, reduces the quantity of amino acids available to synthesize uric acid. The drugs probenecid and sulfinpyrazole increase urate excretion and prevent reabsorption from the kidneys into the blood, but they can foster the formation of urate kidney stones. Phenylbutazone increases urate excretion and reduces inflammation. Allopurinol specifically inhibits xanthine oxidase, so xanthine, a soluble precursor of urate, is excreted before urate can form and precipitate in kidneys or joints.

See Questions—Objective 14

Questions

Objective 14

(a) How do the metabolic disorders described in this chapter disrupt homeostasis?

(b) What kinds of substances damage enzymes?

(c) What causes enzyme deficiencies?

(d) How can disorders associated with damaged or deficient enzymes be treated?

(e) Can such disorders be prevented? If so, how?

Study and Review

Did you get the essentials?

The following summary contains the basic concepts from chapter 3 except for the key terms that have been omitted. Read the chapter. Then try to fill in the blanks from memory. If you cannot fill in all the blanks, review the chapter paying particular attention to boldface terms.

Metabolism includes _____ reactions that break down molecules and _____ reactions that synthesize more complex molecules. Many metabolic pathways involve a series of reactions and products along the way that are called _____ . Chemical reactions in the body are controlled by catalysts called _____ and some of them require _____ , such as hydrogen or electron carriers in reactions involving _____ (a gain of electrons) and _____ (a loss of electrons).

After entering cells glucose molecules are _____ . In the liver and muscle cells, glucose is used to make glycogen by a process called _____ . Glycogen is broken down by _____ . In cells that need energy, glucose enters _____ , a metabolic pathway that produces _____ _____ under aerobic conditions and _____ _____ under anaerobic conditions. Other products include 4 molecules of _____ level _____ , 2 of which are used for _____ , and reduced _____ . In the absence of oxygen, reduced _____ repeatedly transfers hydrogen to _____ _____ , thereby keeping _____ operating. When oxygen again becomes available hydrogen is transferred back to _____ , which is decarboxylated to become _____ . The production of lactic acid in _____ and its use to make glucose in _____ cells is called the _____ cycle.

Krebs cycle reactions take place in the _____ of _____ . With each turn of this cycle a molecule of _____ _____ is metabolized and the products include release of 2 molecules of _____ _____ , reduction of 3 molecules of _____ and 2 of _____ , and making of 1 molecule of _____ level _____ . Reduced hydrogen carriers carry hydrogen to the _____ _____ system on the _____ _____ membrane. Each pair of hydrogens on reduced NAD generates _____ molecules of _____ as it undergoes energy-capturing _____ _____ , but those on reduced FAD generate only _____ molecules of _____ .

As an alternative means of metabolizing glucose the _____ _____ pathway produces _____ for nucleotide synthesis and _____ for fatty acid synthesis but does not produce _____ directly.

Lipids in _____ are digested to _____ and _____ _____ by the enzyme _____ _____ . Fatty acids are metabolized for energy by _____

Continued on next page

Study and Review

_____ , which breaks them into molecules of _____ _____ (destined for the _____ cycle) and transfers hydrogen to both _____ and _____ (destined for _____ transport and _____ _____). Molecules incompletely metabolized by this process are called _____ bodies. Fatty acids are synthesized from acetyl-CoA by _____ _____ , which occurs in the _____ on _____ _____ _____ . A fatty acid the body cannot make but that it needs is the _____ fatty acid _____ acid. The body can make _____ from

Molecules that transport lipids in the blood are called _____ . In addition to chylomicrons, they include _____ that prevent cholesterol deposition, _____ made in the liver, _____ made in the blood and largely responsible for cholesterol deposits in blood vessels.

The body is constantly breaking down and resynthesizing proteins in a process called _____ _____ . The body is said to be in _____ balance when _____ intake equals _____ _____ . It is in _____ _____ balance during _____ when it loses more _____ than it takes in. Conversely, it is in _____ _____ balance during _____ when it takes in more _____ than it loses. All _____ _____ needed to make proteins can be synthesized from the _____ _____ _____ . Exchange of amino and keto groups is _____ , and the removal of amino acids is _____ . The body gets rid of ammonia by way of the _____ cycle. The remains of amino acids called _____ _____ can be used to make _____ in _____ .

The body can break down and make _____ such as DNA and RNA as needed.

Metabolism is regulated by such processes as _____ inhibition, _____ , and enzymes with _____ sites.

Enzymes can be damaged by _____ and made deficient by _____ . Metabolic disorders include _____ _____ (a deficiency of lactase) and _____ (uric acid accumulation).

Chemistry of Bone, Muscle, and Nerve Tissue

4

OBJECTIVES

Physiology of Bone

1. Describe how bone is maintained and how exercise, mechanical stress, vitamins, and hormones affect bones.

Contraction at the Molecular Level

2. Describe the gross and microscopic structure and specific properties of skeletal muscle and relate them to body function.
3. Explain the sliding filament theory of contraction and how contraction is regulated at the molecular level.

The Motor Unit

4. Describe the function of motor units and how drugs alter it.

Muscle Metabolism

5. Explain how metabolic processes contribute to muscle function.

Excitability of Neurons

6. Explain how neurons display excitability and how excitability contributes to neural function.

Synaptic Transmission

7. Explain how signals are relayed across synapses and how such transmissions contribute to neural function.

Neurotransmitters

8. Summarize the properties of well-know neurotransmitters and how they control function at the cellular level.
9. Compare neuroactive peptides and neuropharmacological agents to well-known neurotransmitters.
10. Explain the functions to autonomic neurotransmitters and their receptors.
11. Summarize the effects of neuropeptides and drugs on autonomic functions.

Chemistry of Vision

12. Relate the structure and function of rods to vision in dim light and that of cones to color vision.

Endorphins and Enkephalins

13. Characterize the endorphins and enkephalins.

Study and Review

Physiology of Bone

Bones are constantly being remodeled—a little bone tissue added along one surface and a little reabsorbed along another. By this means certain areas of bones are thickened and reinforced in response to exercise and mechanical stress as discussed below. In childhood and adolescence more bone is formed than is reabsorbed, and these young bones contain large amounts of collagen. In adulthood up to about age 40, the processes are in equilibrium, but after age 40 more bone is reabsorbed than is formed. By old age, the marrow cavity has enlarged and minerals and collagen have been lost from the matrix, making the bones brittle and subject to fracture.

Mechanical stress contributes to bone remodeling by creating pressure on crystalline hydroxyapatite, the mineral portion of blood. Such pressure causes convex (outwardly curving) surfaces to become negatively charged and attract positive ions, including H^+ in extracellular fluids. Removing the H^+ makes the extracellular fluid alkaline and increases the activity of the enzyme **alkaline phosphatase** (fos'fa-tāsz) in osteoblasts and osteocytes. This enzyme makes calcium phosphate and fosters bone formation. Lack of pressure causes concave (inwardly curving) surfaces to become positively charged. The extracellular fluid becomes acidic and increases the activity of the enzyme **acid phosphatase** in osteoclasts. This enzyme digests calcium phosphate and fosters bone degradation. Together these processes thicken bone that is under pressure (where outward curving occurs) and reduce the thickness of bone not under pressure (where inward curving occurs).

Vitamins and hormones also play important roles in bone growth and maintenance. Three vitamins (A, C, and D) and many hormones are known to be involved in this complex regulatory activity. Vitamins A and C are required in synthetic processes. Vitamin A facilitates synthesis of chondroitin sulfate, a substance intermingled with collagen fibers that gives plasticity to the organic bone matrix. Vitamin C (ascorbic acid) facilitates the addition of hydroxyl groups to the amino acid proline and the formation of cross-linkages between collagen molecules; cross-linkages add tensile strength to the bone matrix.

Vitamin D, after activation by enzymes in the kidney or liver, travels in the blood to the intestinal mucosa where it facilitates Ca^{2+} absorption by inducing synthesis of a calcium carrier protein and increasing the active transport of Ca^{2+} into mucosal cells. Though vitamin D is necessary to maintain adequate blood Ca^{2+}, an excess can cause bone destruction by stimulating osteoclasts to remove Ca^{2+} from bone. Vitamin D synthesis is stimulated by parathormone, or parathyroid hormone (PTH), and inhibited by high-blood Ca^{2+} concentration.

Parathormone increases blood Ca^{2+}. Low-blood Ca^{2+} stimulates its release and high-blood Ca^{2+} inhibits its release by negative feedback. Parathormone, in addition to activating vitamin D, causes the kidneys to return Ca^{2+} to the blood, maintaining sufficient blood Ca^{2+} to meet the needs of all cells. Though parathormone can remove Ca^{2+} from bone to maintain adequate blood Ca^{2+}, it normally facilitates bone deposition by making Ca^{2+} available to osteoblasts.

The hormone calcitonin, which opposes parathormone, reduces the blood Ca^{2+} concentration by decreasing osteoclast activity, by preventing formation of new osteoclasts, and by increasing osteoblast activity, at least for a few days. Calcitonin is more effective in children because their osteoclasts are far more active than those in adults. In children, osteoclasts release up to 5 grams of Ca^{2+} into extracellular fluids daily, but in adults osteoclasts only release 0.8 grams. Calcitonin acts for 1 to 2 hours as a Ca^{2+} regulator until its effects are overridden by parathormone.

Growth hormone, which stimulates protein synthesis in general, has special effects on bone. It stimulates division of chondroblasts and osteoblasts, causes osteoblasts to secrete collagen, and promotes Ca^{2+} absorption from the intestine. Thyroid hormones and insulin also appear to work with growth hormone to promote collagen synthesis.

The steroid hormones cortisol and aldosterone from the cortex of the adrenal glands, testosterone from the testes, and estradiol from the ovaries also affect bone metabolism. Cortisol can cause collagen degradation with concurrent reabsorption of Ca^{2+} into the blood. Aldosterone stimulates the reabsorption of Na^+ and other minerals from the kidneys and makes these minerals available for deposition in bone. (Though bone contains mainly hydroxyapatite, it also contains small quantities of other minerals.) Testosterone promotes proteins synthesis during bone development, calcium retention, and the deposition of Ca^{2+} in bones. Estradiol promotes protein synthesis and increased bone density by Ca^{2+} accumulation. It can cause osteoclasts to convert to osteoblasts and thereby foster bone deposition.

See Questions—Objective 1

Contraction at the Molecular Level

Each whole skeletal muscle contains many bundles of fibers. Each muscle fiber is a long, cylindrical, multinucleated cell 10 to 100 μm in diameter that can extend the full length of a muscle (up to 30 cm). Within each fiber are many small **myofibrils** (mi-o-fi'brilz), and each myofibril contains many still smaller **myofilaments** (mi''o-fil'ă-mentz).

Questions

Objective 1

(a) How does mechanical stress affect bone?

(b) How do vitamins help to maintain bone?

(c) How do hormones help to maintain bone?

The functional unit of a myofibril is a **sarcomere** (sar'ko-mēr) (Figure 4.1a), which extends from one horizontal line called a Z-line to the next, a distance of about 2 μm in a resting sarcomere. Thick filaments are located entirely within the A-band and thin filaments are anchored to and imbedded in Z-lines. As shown in Figure 4.1b, the plasma membrane of a muscle fiber is called the **sarcolemma** (sar-ko-lem'ah) and its cytoplasm is called **sarcoplasm** (sar'ko-plazm). The **sarcoplasmic reticulum** (sar'co-plaz-mik rĕ-tik'u-lum) corresponds to the endoplasmic reticulum in other cells. **Transverse (T) tubules,** which are folds of sarcolemma filled with extracellular fluid, create channels through which the extracellular environment extends deep inside muscle fibers. Near the T tubules, the sarcoplasmic reticulum ends in saclike structures called the **terminal cisternae.** A T tubule and 2 adjacent terminal cisternae form a **triad.**

Skeletal muscle is stimulated to contract (produce tension) when signals from nerves spread throughout the sarcoplasmic reticulum. Such signals cause the terminal cisternae to release calcium ions, which diffuse through the sarcoplasm and bind to a regulatory protein. Such calcium binding initiates contraction. Subsequent active transport of calcium ions back to the sarcoplasmic reticulum initiates relaxation.

Though contraction of skeletal muscles can move the whole body, the contraction process itself occurs at the molecular level. According to the **sliding filament theory** of contraction, proposed in the 1950s by H. E. Huxley and based on electron micrograph studies, muscle contraction occurs when thin filaments slide over thick ones, thereby increasing the overlap of filaments and shortening the sarcomeres without changing the length of the filaments themselves.

Myofilaments contain four important proteins, **myosin** (mi'o-sin), **actin** (ak'tin), **troponin** (tro-po'nin), and **tropomyosin** (tro''po-mi'o-sin) as shown in Figure 4.1c. Myosin is found in thick filaments; actin forms the basic structure of thin filaments; and troponin and tropomyosin are associated with it. Myosin and actin are **contractile proteins,** that is, they create tension in the contraction process. Troponin and tropomyosin are **regulatory proteins,** that is, they help to control the contraction process.

The thick filaments each consist of several hundred myosin molecules held together mainly by hydrogen bonds and hydrophobic forces. Each myosin molecule has two globular heads, or **crossbridges,** oriented toward the ends of A-bands. Crossbridges have 2 functional sites. One site acts as a ligand (a part of a molecule that can bind to another molecule). It attaches to binding sites on the protein actin. The other site binds ATP. Finally, crossbridges have ATPase activity.

Like the thick filaments, the thin filaments also are composed of protein molecules, actin, tropomyosin, and troponin. Actin in a muscle cell is **F actin,** a polymer of globular units of **G actin.** Two molecules of F actin are wound about each other in a helix that forms the core of a thin filament. Actin has surface binding sites that myosin binds to during the contraction process. Tropomyosin, once thought to be a precursor of myosin, is a fibrous, double-stranded helical protein. Overlapping tropomyosin molecules form a continuous, flexible structure. Troponin, an aggregation of smaller molecules sometimes called **troponin complex,** binds Ca^{2+}. When the troponin complex lacks Ca^{2+}, it inhibits ATPase activity in the heads of myosin molecules. When it has bound Ca^{2+}, it can no longer inhibit that enzyme.

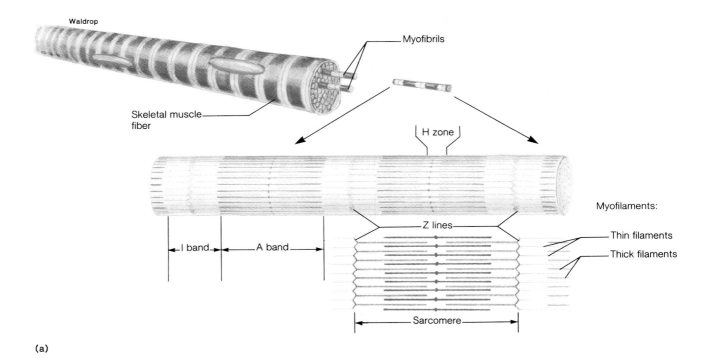

(a)

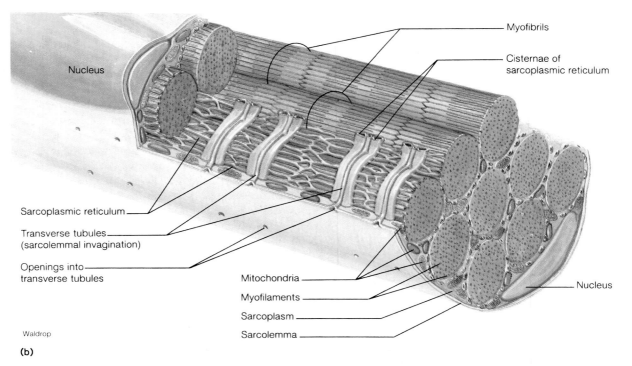

(b)

Figure 4.1

(*a*) A diagram of skeletal muscle showing striations, sarcomeres, and (*b*) a cutaway view of a muscle cell, showing the arrangement of cytoplasmic structures. (*c*) The structure of thick and thin filaments, including the locations and orientations of sites where ATP and actin bind to myosin heads.

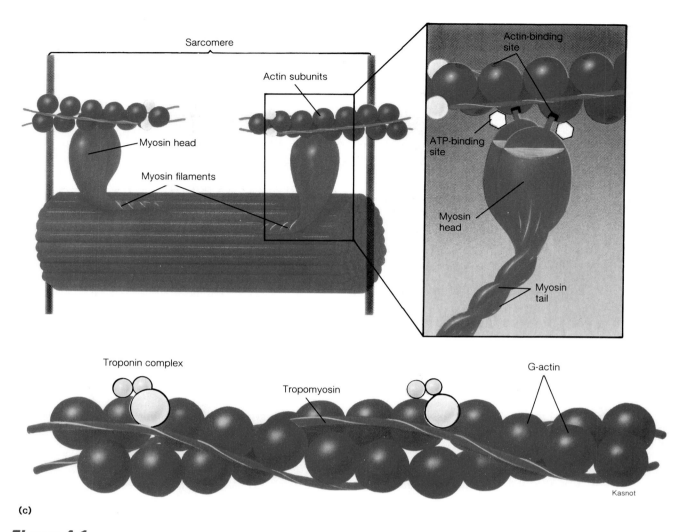

(c)

Figure 4.1

Continued

In a thin filament, the actin molecules are arranged as 2 coiled chains with a groove between them; fibrous tropomyosin molecules lie in a groove created by the coiling of a double strand of actin molecules. When a muscle fiber is relaxed tropomyosin apparently covers binding sites on actin and prevents them from binding with myosin. When stimulation releases Ca^{2+} to bind to troponin, tropomyosin slides into the groove between the actin chains exposing sites that can bind with myosin.

Thick and thin filaments are arranged as hexagons within each myofibril—6 thin filaments surround each thick filament and form a small hexagon, and 6 thick filaments surround each small hexagon and form a large hexagon. Every thin filament lies adjacent to crossbridges on each of 3 myosin filaments, so that actin and myosin can easily interact during contraction.

According to the sliding filament theory, molecular movements account for a **contraction cycle** with attach, pull, and release phases. Crossbridges of myosin

attach to actin and pull the thin filaments toward the H-zone, as the crossbridges swivel in an oarlike movement. The crossbridges release their hold on actin and return to their original position. This **attach-pull-release cycle** occurs asynchronously—first, one fiber and then another crossbridge attaches to actin. It is repeated over and over again during the contraction of a whole muscle. Many small, discrete movements at the molecular level produce a smooth and continuous sliding of the thick and thin filaments past each other. Though some of the details of the contraction process are conjectural, the steps in the contraction cycle are based on evidence from biochemical analysis, electron microscopy, X-ray diffraction, and other techniques.

As the contraction cycle (Figure 4.2) begins, ATP binds to the myosin head in the presence of magnesium ions. The binding of ATP causes the myosin head to dissociate from actin and drift to a 90° angle with the actin filament. ATPase in the crossbridge hydrolyzes ATP, but

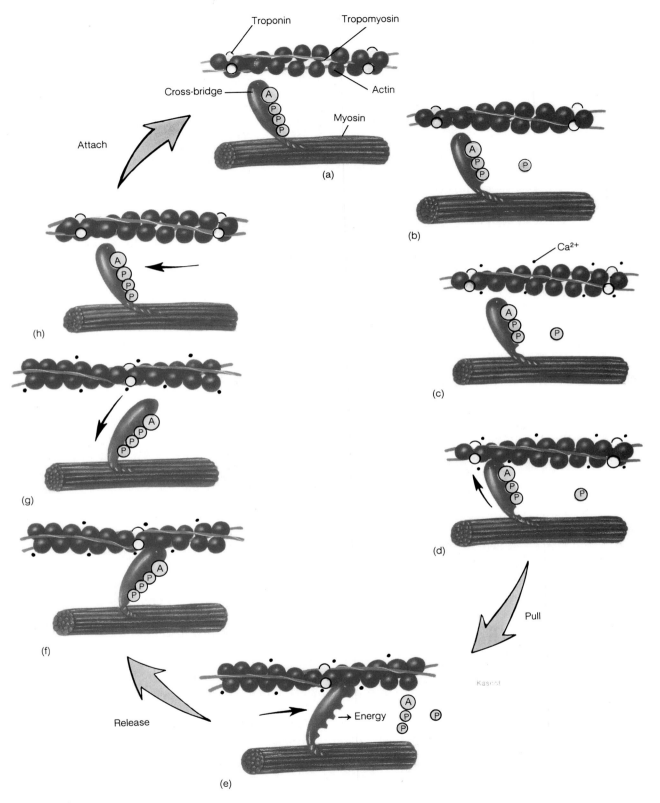

Figure 4.2

The steps in the contraction cycle.

the ADP and P_i (inorganic phosphate) remain bound to the crossbridge and do not yet release energy. Ca^{2+} binds to troponin, which causes tropomyosin to expose the myosin binding sites on actin. The activated binding site on the myosin head can now bind actin. As long as Ca^{2+}-troponin deflects tropomyosin, myosin will continue to bind to actin. Steps a through d constitute the attach phase of the cycle.

Binding of the myosin head to actin releases ADP, P_i, and energy. The energy causes a change in the shape of the myosin head, so that it swivels to a 45° angle with actin and forces actin and myosin to slide over each other. The ends of the actin filaments move toward the center of the A-band. This step constitutes the power stroke, or pull, phase of the cycle.

When ADP and P_i have been released and swiveling has occurred, a second molecule of ATP binds to the myosin head. Binding of ATP inactivates the actin binding site and frees the myosin head. The myosin head returns to its original position and is ready to undergo another crossbridge cycle. Steps f through h constitute the release phase of the cycle. No additional energy from ATP is required for the release cycle.

Repetitions of the contraction cycle shorten sarcomeres as fixed-length filaments slide over one another. H-zones shorten and may disappear, but the A-band width remains constant. The most significant event is that filaments slide because of cyclic changes in the shape and position of myosin heads.

The direct connection between neural signals reaching the sarcolemma and the sliding of myofilaments in contraction is called **excitation-contraction coupling.** This coupling is accomplished through controlled release of calcium ions, which exert a powerful regulatory effect. A cytoplasmic Ca^{2+} concentration as low as 1 micromole per liter stimulates contraction, though the Ca^{2+} concentration in the sarcoplasmic reticulum is 2,000 times that amount. The idea that Ca^{2+} activates contraction has been confirmed by experiments using aequorin, a bioluminescent protein from jelly fish that emits light when it binds with Ca^{2+}. When muscle fibers are injected with aequorin and stimulated under proper conditions, they glow during the time between stimulation and the production of tension.

Like the membrane of a neuron (described below), the sarcolemma is an excitable membrane. Stimulation of the membrane causes a signal called an **action potential** to be propagated along the membrane and down into T tubules. Once an action potential reaches the sarcolemma it travels along T tubules to the region of terminal cisternae of the sarcoplasmic reticulum where Ca^{2+} is stored. The action potential in the T tubule membrane increases the permeability of the terminal cisternae to Ca^{2+}, and Ca^{2+} diffuses into the sarcoplasm near the sarcomeres. This increased permeability is probably due to a chemical messenger molecule called inositol phosphate

that initiates mechanical opening of calcium pores in the sarcoplasmic reticulum. Once released, Ca^{2+} diffuses only a short distance to reach calcium binding sites on troponin. When sufficient Ca^{2+} binds to troponin, tropomyosin moves and myosin binding sites on actin are exposed and contraction can occur.

When stimulation of the muscle cell membrane stops, the terminal cisternae cease to release Ca^{2+}. An ATP-driven calcium pump in the sarcoplasmic reticulum continuously transports Ca^{2+} from the sarcoplasm back inside the sarcoplasmic reticulum. After Ca^{2+} release ceases, the calcium pump quickly lowers the Ca^{2+} concentration in the sarcoplasm. As the sarcoplasmic Ca^{2+} concentration decreases, Ca^{2+} dissociates from troponin. Tropomyosin again blocks myosin binding sites on actin and contraction ceases. The events in excitation-contraction coupling can be summarized as follows:

1. The action potential travels along the T tubules to the sarcoplasmic reticulum.
2. Stimulation of the sarcoplasmic reticulum causes Ca^{2+} to leak into the sarcoplasm.
3. Ca^{2+} diffuses to and binds with troponin, thereby allowing contraction to occur.

Contraction, which is slow to start, is even slower to stop. It continues for 50 to 100 msec after the last action potential, and until the calcium pump lowers the sarcoplasmic Ca^{2+} level enough so that tropomyosin can block myosin binding sites on actin.

It may be surprising that muscle relaxation requires some energy—at least indirectly. The breaking of actin-myosin binding in contracted muscle fibers is a passive process, but it can occur only when energy from ATP is available to operate the calcium pump. Though this pump operates continuously, a contracted muscle fiber cannot relax until sufficient Ca^{2+} has been removed from troponin. Also, ATP must bind to myosin to separate myosin and actin and to allow relaxation. ATP so bound is subsequently hydrolyzed and its energy used in later contraction cycles. Excitation, contraction, relaxation, and the role of calcium ions in these processes are all summarized in Figure 4.3.

See Questions—Objectives 2 and 3

The Motor Unit

A **motor unit,** the functional unit of contraction, consists of a single motor neuron, including all the branches of its axon, and the muscle fibers those axon branches innervate (Figure 4.4). Each muscle fiber has a terminal branch of an axon lying in a groove in the sarcolemma. The sarcolemma beneath nerve endings is called the **motor end plate;** the membrane within a motor end plate is folded to form

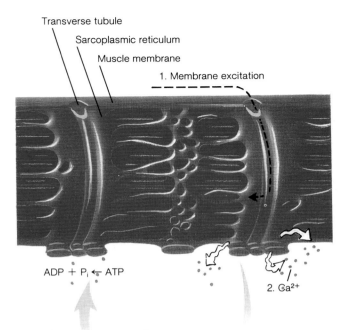

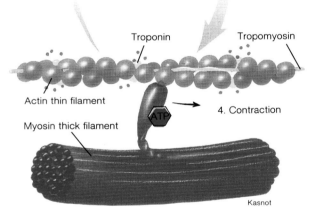

Figure 4.3

The role of calcium ions in the concentration and relaxation processes: (*1*) Excitation of the membrane. (*2*) Calcium ions released from sarcoplasmic reticulum. (*3*) Calcium ions diffuse through sarcoplasm and some bind to troponin and cause tropomyosin to move away from the myosin binding sites of actin. (*4*) Contraction occurs. (*5*) Calcium ions are actively transported back to the sarcoplasmic reticulum and tropomyosin again covers the binding sites on actin; covering the binding sites inhibits contraction.

Questions

Objective 2

(a) How is skeletal muscle organized from the protein filament level to the whole muscle level?

(b) Which proteins are contractile proteins and which are regulatory proteins?

Objective 3

(a) How does the sliding filament theory explain muscle contraction?

(b) How is contraction regulated at the molecular level?

(c) What is excitation-contraction coupling and what is its significance?

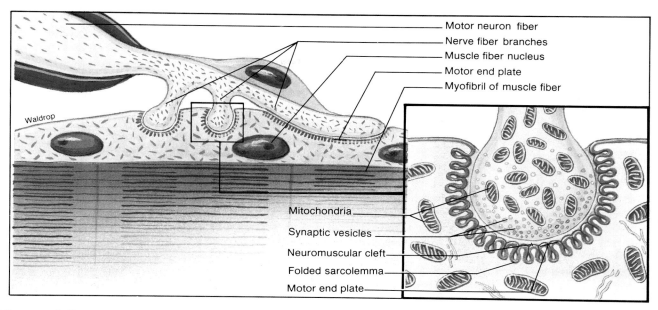

Figure 4.4

How an axon innervates a muscle fiber: In a motor unit of an axon, each branch of a motor nerve goes to a muscle fiber where the axon terminal and the sarcolemma form a myoneural junction. The axon terminal lies in a groove in the sarcolemma; the portion of the sarcolemma beneath the axon is the motor end plate.

subneural clefts. The terminal ends of the axon and the motor end plate together constitute the **neuromuscular junction,** or **myoneural** (mi-o-nu'ral) **junction.**

A signal traveling along a motor neuron (an action potential) causes all its axon terminals to release a chemical substance called acetylcholine (ACh). ACh diffuses to receptor sites on the motor end plates of all muscle fibers of that motor unit, thus causing the motor end plate to generate a membrane potential called the **end plate potential,** which initiates an action potential in the sarcolemma. Signals at the end plate are always excitatory and a single action potential in a motor neuron generally causes skeletal muscle contraction.

The membranes of the motor end plates also contain the enzyme **cholinesterase** (ko-lin-es'ter-ās), which breaks down acetylcholine. This causes the action potential to subside, and allows the membrane at the end plate to return to its resting potential until it receives another signal from the motor neuron. Processes involved in muscle contraction are summarized in Table 4.1 and their control is summarized in Figure 4.5.

A variety of drugs and diseases interfere with events at the myoneural junction. Curare (d-turbocurarine), a poison put on arrowheads by some South American Indians, can be used to prevent muscle contraction during surgery, but it has been replaced by other drugs in most surgical procedures. Nerve gases, pesticides, and toxins also affect neuromuscular junctions. Some nerve gases and certain pesticides inactivate cholinesterase and cause spasmodic muscular contractions by prolonging acetylcholine activity. Botulism toxin, from the bacterium *Clostridium botulinum,* is commonly found in inadequately processed canned goods and paralyzes muscles by inhibiting acetylcholine release from axon terminals.

See Questions—Objective 4

Muscle Metabolism

Muscle action requires a continuous supply of ATP and the transformation of its chemical energy to mechanical energy for the swiveling of myosin heads. Energy from ATP also powers the calcium pump to return calcium to the sarcoplasmic reticulum and the Na^+-K^+ pump to maintain the sarcolemma ready to receive an action potential.

Oxygen and nutrients are carried in the blood to all cells, including muscle cells, where they can be used in cellular metabolism to make ATP. Metabolism in muscle cells has some special attributes.

When an organism is at rest or exercising moderately, oxygen and glucose are plentiful. Large quantities of ATP are synthesized from ADP and P_i in the mitochondria in **aerobic metabolism.** Some energy is transferred from ATP to another energy storage molecule called

Table 4.1
Processes Involved in Muscle Contraction from Stimulation through Contraction to Relaxation

1. An action potential reaching axon terminals of a motor neuron initiates the release of acetylcholine from synaptic vesicles.

2. Acetylcholine diffusing to and binding to receptor sites on the motor end plate creates a local end plate potential.

3. The end plate potential initiates an action potential that is propagated in all directions along the sarcolemma and into the T tubules.

4. Depolarization of the T tubules causes calcium ions to leak out of the terminal cisternae of the sarcoplasmic reticulum that surrounds the myofibrils.

5. Calcium ions bind to receptor sites on the troponin complex.

6. The troponin complex causes tropomyosin to recede into a groove between chains of actin molecules, exposing binding sites on actin. Myosin heads with hydrolyzed ATP (ADP + P_i + energy) bind to actin. This is the attach phase of the contraction cycle.

7. Using energy from ATP, the myosin heads swivel and pull the actin filaments closer to the middle of the sacromere. This is the pull phase of the contraction cycle.

8. The ATP binding sites on myosin are exposed and new molecules of ATP bind to myosin; the actin binding sites are inactivated and the myosin heads return to their original positions. This is the release phase of the contraction cycle.

9. The contraction cycle is repeated continuously as long as ATP is available and sufficient calcium is bound to troponin to allow myosin binding sites on actin to remain exposed.

10. The action potential lasts about 2 msec, but the contraction cycle continues for 50 to 100 msec as long as adequate calcium is bound to troponin.

11. When the action potential ceases, calcium release from the sarcoplasmic reticulum also ceases. The calcium pump removes calcium from the sarcoplasm and depletes calcium bound to troponin.

12. Tropomyosin again blocks binding sites on actin; myosin can no longer bind to actin so contraction ceases and the muscle fibers relax.

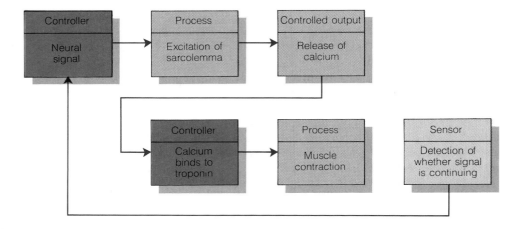

Figure 4.5

Summary of control processes in skeletal muscle contraction and relaxation. Cessation of signal turns process off.

phosphocreatine (fos-fo-kre′at-in), or **creatine phosphate,** which consists of a nitrogenous substance creatine combined with a phosphate group:

$$ATP + creatine \rightleftharpoons phosphocreatine + ADP$$

Phosphocreatine and creatine kinase, the enzyme that catalyzes the above reaction, are plentiful in both the skeletal and heart muscles. When heart tissue is damaged in a heart attack, the enzyme leaks into the blood where its concentration can be measured to estimate the degree of tissue damage.

In addition to regulating contraction as described above, Ca^{2+} also stimulates oxidation in muscle mitochondria. It acts by increasing the rate at which pyruvate (a metabolic product of glucose) enters mitochondria and by stimulating the activity of some of the oxidative enzymes in mitochondria.

During strenuous exercise, oxygen is used as rapidly as it is delivered to the muscles and ATP levels increase to their maximum. Even so, too little energy is available to maintain maximally strenuous activity for more than a minute or so. Energy stored in phosphocrea-

Questions

Objective 4

(a) What are the functional properties of a motor unit?

(b) How can events at the myoneural junction be altered by drugs and disease?

$$Glucose + 6H_2O + 2ATP + 6O_2 \longrightarrow 6CO_2 + 12H_2O + 40ATP$$

Net: 38 ATP

(a)

$$Glucose + 2ATP \longrightarrow 2\ Lactic\ acid + 2CO_2 + 4ATP$$

Net: 2 ATP

(b)

Figure 4.6

Summary reactions for (*a*) aerobic metabolism and (*b*) anaerobic metabolism.

tine is quickly transferred back to ATP, but this adds only about 3 seconds to the duration of contraction. Then the muscle must switch to anaerobic metabolism.

In **anaerobic metabolism,** (Figure 4.6) when muscles lack sufficient oxygen to metabolize glucose aerobically, they metabolize it anaerobically and obtain a net of 2 ATP per glucose molecule—only 1/19 of that obtained in aerobic metabolism. Because so little energy is obtained from each glucose molecule, large quantities of glucose are required.

If muscle fibers were dependent entirely on oxygen delivered moment by moment, the ATP supply would be depleted and contraction would cease in less than 1 second. Three mechanisms, myokinase activity, oxygen storage in myoglobin, and the oxygen debt mechanism, prolong contraction when oxygen from the blood is insufficient to support aerobic metabolism.

The enzyme **myokinase** (mi-o-kin′ās) moves a phosphate from one ADP to another: 2ADP $\rightharpoonup$ ATP + AMP. It makes more energy available as ATP but causes AMP to accumulate. Cells must subsequently add 2 phosphates to AMP to generate new ATP.

Myoglobin (mi-o-glo′bin), a pigment in muscle similar to hemoglobin in red blood cells, binds 1 molecule of oxygen (hemoglobin binds 4). Myoglobin indirectly increases the rate at which oxygen diffuses into muscle fibers from the blood by binding oxygen and lowering the concentration of free oxygen in intracellular fluid. Muscle fibers vary in myoglobin content, but even in fibers with large quantities of myoglobin, it plays only a small role in prolonging contraction.

When these mechanisms for supporting aerobic metabolism fail to meet the demands of muscle fibers for ATP, metabolism becomes anaerobic. Pyruvate from glycolysis is converted to lactic acid. Much of the lactic acid diffuses into the blood but enough remains in muscle cells to lower the pH of muscle tissue. The lower pH releases additional oxygen from hemoglobin in blood but also inhibits enzymes in cells and makes anaerobic metabolism a self-limiting process. At best, anaerobic metabolism provides energy for only a few seconds of maximal exertion.

Anaerobic metabolism incurs an **oxygen debt,** an amount of oxygen required to metabolize lactic acid and resaturate the myoglobin with oxygen. We "pay" the oxygen debt after a period of exercise when we breathe rapidly and deeply and use up to 6 times the normal amount of oxygen. Lactic acid in heart muscle cells is converted to pyruvate and metabolized there, but lactic acid in skeletal muscle is transported to the liver and converted back to glucose. Trained athletes incur a smaller oxygen debt than untrained individuals for a given amount of exertion because training increases the amount of blood the heart can pump per minute and usually causes muscle hypertrophy. Improved heart performance allows the heart to pump more blood to active muscles if it is needed to provide these muscles with oxygen and nutrients and waste removal. Hypertrophied cells contain somewhat more myoglobin and can store a bit more oxygen.

In addition to compensating for inadequate oxygen delivery, muscle tissue also compensates to some degree for inadequate nutrient delivery. During rest, when glucose is plentiful, muscle tissues use some glucose to store

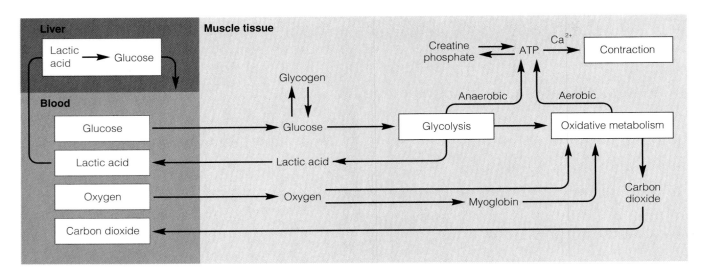

Figure 4.7

Main steps in muscle metabolism.

energy in phosphocreatine and more glucose to synthesize glycogen. During exercise, the glycogen is broken down to glucose. Along with initiating contraction, Ca^{2+} from the sarcoplasmic reticulum binds to **calmodulin** (kal-mod′u-lin), an intracellular calcium-binding protein. This binding activates enzymes, some of which lead to increasing the ATPase activity of myosin and some of which catalyze the breakdown of glycogen. The glycogen stored in all muscle cells combined provides enough glucose to supply energy for a person to run about 10 miles and glycogen stored in the liver supplies energy for another 3 miles.

Fatty acids stored in the body's fat deposits also supply energy for muscle contraction. When a person is exercising at 50 to 60 percent of maximum effort about half the energy used by muscles comes from fatty acids and about half from glycogen as long as glycogen lasts. When a person is exercising at 85 percent of maximum effort almost all the energy comes from glucose derived from glycogen. Only glucose can be metabolized during strenuous exercise because too little oxygen is available to use fatty acids, which are metabolized aerobically. Epinephrine, a hormone released from the adrenal gland during stress, causes muscles to start using fatty acids instead of glucose as their energy source. Epinephrine, like calmodulin-Ca^{2+}, makes glucose available at the same time oxygen depletion has reduced fatty acid metabolism. However, when glycogen becomes depleted muscles depend entirely on energy from fatty acids and can occur only as fast as oxygen can be supplied. When strenuous exercise ceases, oxygen availability more nearly matches oxygen demand, and epinephrine is no longer secreted. Muscles then obtain relatively more energy from fatty acids. Muscle metabolism is summarized in Figure 4.7.

See Questions—Objective 5

Excitability of Neurons

In neurons, **excitability,** the ability to respond to a stimulus, causes a change in a membrane that initiates or helps to initiate a signal. To understand excitability and signals we need to review some fundamental principles of electricity.

Electrical charges are arbitrarily designated as either positive or negative. Among charged particles, like charges repel and unlike charges attract, creating electrical force between charged particles. The attracting force between particles of unlike charge increases as the quantity of charges increases or as the square of the distance between particles decreases. The repelling force between particles of like charge increases as the quantity of charge increases or as the square of the distance between the particles decreases.

Electrical force is related to energy and work. **Energy** is the capacity to do work, and **work** is force times the distance over which the force acts. (Work = force × distance.) Energy is needed to hold oppositely charged molecules apart, and they have the potential to do work when they come together. This is comparable to magnets doing work when opposite poles attract each other.

The force between any particles of unlike charge is called a **potential difference,** or simply a **potential.** Potential is measured between two points and is expressed in units called **volts.** Small potentials across cell membranes are measured in thousandths of volts, or **millivolts** (mV). When a potential exists, it causes charged particles to flow if a conductor is present. This flow is called

Questions

Objective 5

(a) How does muscle metabolism differ during rest and activity?

(b) What is oxygen debt and how is it paid?

current. According to Ohm's Law, current intensity equals **voltage** (potential in volts) divided by the **resistance** of the medium through which the current flows:

$$current = voltage/resistance.$$

Diffusion of charged particles creates **diffusion potentials** and is important in neuron function. Let us first see how diffusion produces potentials in a nonliving system with solutions of 0.5 molar NaCl and 0.1 molar NaCl separated by a membrane freely permeable to both Na^+ and Cl^- (Figure 4.8a). Each compartment is electrically neutral (has the same number of positive and negative charges), but because of the concentration gradient between the two compartments ions will diffuse from compartment 1 to compartment 2. Both sodium and chloride ions have hydration shells—water molecules attached to them. Chloride ions have fewer water molecules, weigh less, and diffuse more rapidly than sodium ions.

Both sodium and chloride ions diffuse down their concentration gradients, but the more rapid diffusion of chloride ions leads to a surplus of negative charges in compartment 2 and a surplus of positive ions in compartment 1 (Figure 4.8b). This creates a temporary potential be-

tween the two compartments that lasts only until Na^+ diffusion comes to equal Cl^- diffusion. A potential difference created by differential diffusion of charged particles between two solutions is a **diffusion potential.**

As negative ions increase in compartment 2, an electrical gradient is created such that some Cl^- ions are repelled away from compartment 2 and return to compartment 1. Ion movements due to the electrical gradient counteract those that occur because of the concentration gradient until an equilibrium is reached and there is no net flow of ions. In a system where the membrane is permeable to both Na^+ and Cl^- (Figure 4.8c), equilibrium is reached when the ion concentrations in the two compartments are equal. At equilibrium, the potential is zero.

In contrast, living systems have selectively permeable membranes—membranes that allow some substance to pass through and block others. In such a system an **equilibrium potential** exists when the system has no net flow of a given ion. In a hypothetical membrane permeable to K^+ but impermeable to Na^+ and Cl^- (Figure 4.9), K^+ diffuses down its concentration gradient until it has established an equal but opposite electrical gradient. The resulting potential is the equilibrium potential—the potential at which no net diffusion occurs.

Every cell has a membrane potential, but membrane potentials vary from -5 to -100 mV with the inside of the cell negative to the outside. In nerve and muscle cells this potential varies significantly with stimulation. The potential across a membrane not recently stimulated is the **resting membrane potential**—about -70 mV in a typical resting neuron. This potential and how it can be altered are directly related to neural signals. Different concentrations of ions inside and outside a membrane give rise to the resting membrane potential. Na^+, K^+, Ca^{2+}, Mg^{2+}, Cl^-, HCO_3^-, PO_4^{3+}, and SO_4^{2-}, and charged protein molecules are present on both sides of the membrane in different concentrations. Differences in the Na^+, K^+, and Cl^- concentrations account for much of the potential between intracellular and extracellular fluids.

Neuron membranes are selectively permeable— freely permeable to K^+ and Cl^-, relatively impermeable to Na^+, and totally impermeable to large, negatively charged intracellular protein molecules. Such selective permeability contributes to the resting membrane potential. According to Table 4.2, K^+ would produce a potential of -90 mV, Na^+ a potential of $+60$ mV, and Cl^- a potential of -70 mV. Cl^- diffuses down its concentration gradient into the cell until an equal electrical gradient develops in the opposite direction. Because Cl^- moves passively and has an equilibrium potential equal to the resting membrane potential, it contributes little to generating that potential.

K^+ and Na^+ contribute significantly to the resting membrane potential as diffusion occurs along concentration and electrical gradients. K^+ moves out of the cell along

Compartment 1
0.5 M NaCl

Compartment 2
0.1 M NaCl

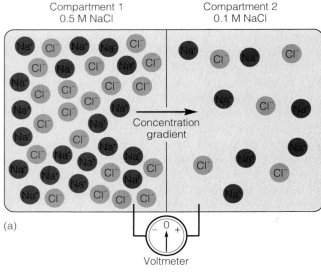

(a)

Voltmeter

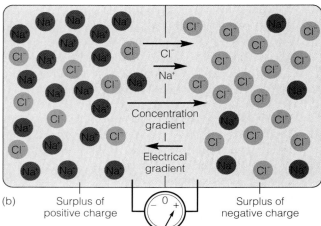

(b)

Surplus of
positive charge

Surplus of
negative charge

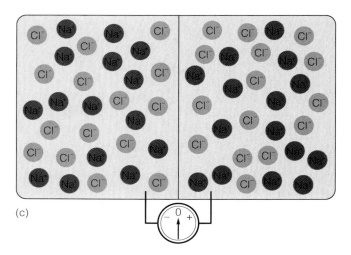

(c)

the concentration gradient and into the cell along the electrical gradient. Na^+ moves into the cell along both gradients. However, membrane impermeability slightly impedes K^+ movement and greatly impedes Na^+ movement. Thus, the net movement of K^+ and Na^+ is determined by the combined effects of concentration and electrical gradients and by selective membrane permeability.

Because the membrane is more permeable to K^+ than to Na^+, its resting potential is nearer the equilibrium potential for K^+ than for Na^+ (-70 mV is nearer to -90 mV than to $+60$ mV). The resting potential is due mainly, but not entirely, to the high membrane permeability to K^+ and the net diffusion of K^+ out of the cell. At -70 mV, net diffusion of K^+ is out of the neuron, Cl^- has no net diffusion, and Na^+ diffusion is impeded by low membrane permeability.

If no energy were expended to maintain the resting membrane potential, diffusion would create a membrane potential of about -60 mV. Many cells maintain a resting potential of -70 mV by using energy to operate Na^+-K^+ pumps in the membrane. This pump is **electrogenic,** that is, it contributes to an electrical gradient across the membrane by moving different numbers of the 2 ions. Typically, for each molecule of ATP hydrolyzed, 3 Na^+ are moved out of the neuron and only 2 K^+ are moved into the neuron. As positive charges accumulate outside the neuron, it becomes more negative inside, and the electrical gradient down which Na^+ and K^+ diffuse into the neuron increases.

In summary, several forces interact to maintain a resting membrane potential. K^+ concentrations inside and outside a resting neuron remain constant because of the following:

1. Fairly large quantities of K^+ diffuse down its concentration gradient out of the neuron.
2. Small quantities of K^+ diffuse down the electrical gradient into the neuron.
3. The membrane pump actively transports sufficient K^+ into the neuron to maintain constant but different concentrations of K^+ on each side of the membrane.

Figure 4.8

Diffusion potential and subsequent equilibrium: (*a*) The concentration gradient causes ions to diffuse from compartment 1 to compartment 2. (*b*) Because Cl^- diffuses more rapidly than Na^+, an electrical gradient opposing the concentration gradient develops. (*c*) The system eventually comes to equilibrium when concentrations of each of the ions Na^+ and Cl^- are the same in both compartments and no net movement occurs.

Compartment 1
0.1 M KCl

Compartment 2
0.1 M NaCl

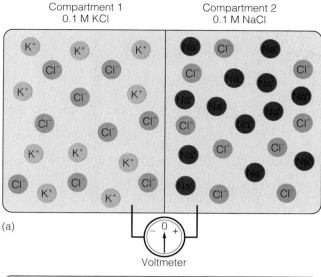

(a)

Voltmeter

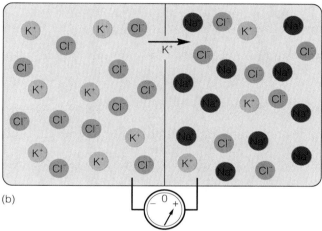

(b)

Figure 4.9

An equilibrium potential across a membrane permeable only to K^+: (*a*) Compartment 1 contains KCl and compartment 2 contains NaCl, both in 0.1 molar concentrations.
(*b*) Equilibrium is reached when K^+ has moved down its concentration gradient to the point at which the diffusion potential remaining equals the electrical potential created. The potential difference at this point is the equilibrium potential.

Table 4.2

Ion Concentrations In and Around a Typical Neuronal Membrane in Millimoles/Liter of Solution at Equilibrium

Ion	Inside Cell	Outside Cell	Potential
Na^+	15	150	+60
K^+	150	5	−90
Cl^-	9	125	−70

Na^+ concentrations inside and outside a resting neuron remain constant because of the following:

1. Small quantities of Na^+ diffuse into the neuron along its concentration and electrical gradients.
2. The membrane pump actively transports sufficient Na^+ out of the neuron to maintain constant but different concentrations of Na^+ on each side of the membrane.

See Questions—Objective 6

Questions

Objective 6

(a) How do electrical principles relate to neuron function?

(b) What is a resting membrane potential and how is it produced?

(c) How does the membrane pump affect neural function?

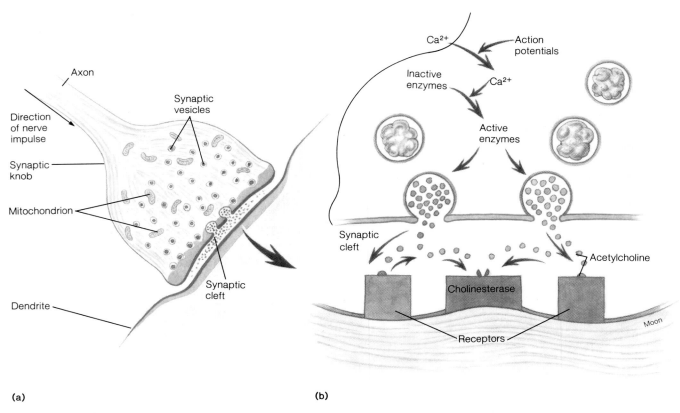

(a)   (b)

Figure 4.10

A synapse: (*a*) a general diagram, (*b*) a detailed view of neurotransmitter release.

Synaptic Transmission

Most signals reaching axon terminals are transmitted to another neuron across **chemical synapses** by a chemical neurotransmitter. A few signals cross **electrical synapses** by ion flow across a cytoplasmic bridge, as in gap junctions in cardiac and smooth muscle.

In chemical synapses, the **presynaptic** (pre-sin-ap'tik) **neuron** releases neurotransmitter and the **postsynaptic** (post-sin-ap'tik) **neuron** receives it. In a typical synapse (Figure 4.10a), a **presynaptic knob** of an axon is separated from the **postsynaptic region** of a dendrite, cell body, or axon by a small space called the **synaptic cleft.**

Mechanism of Synaptic Transmission

Axon terminals continuously synthesize neurotransmitter and store it in **vesicles,** which aggregate near the surface of the presynaptic knobs. When a nerve impulse arrives at a presynaptic knob, it increases the membrane permeability to Ca^{2+} via voltage-sensitive calcium channels, which remain open throughout depolarization. As Ca^{2+}

enters the presynaptic knob it stimulates neurotransmitter release, probably by causing vesicles to fuse with the cell membrane.

Once released, neurotransmitter molecules diffuse across the narrow synaptic cleft and bind with receptor proteins of the postsynaptic membrane. Such binding directly opens gated pores in some membranes, but it may act in other ways. Opening gated channels lets ions flow across the membrane, altering the voltage. The kinds of ions moving and their direction of movement depend on which neurotransmitter acts as the ligand. Stimulatory neurotransmitters cause voltage changes that result in synaptic potentials. The chemical neurotransmitter signal is changed into a graded synaptic potential. This is called **transduction.** Inhibitory neurotransmitters cause opposite voltage changes and interfere with signaling. Finally, the neurotransmitter dissociates from the receptor and is inactivated by a postsynaptic membrane enzyme or is returned to the presynaptic knob for reuse. The steps in the transmission of a signal across a synapse are summarized in Figure 4.10b.

See Questions—Objective 7

Questions

Objective 7

What events take place at a chemical synapse?

Neurotransmitters

Chemical substances called neurotransmitters are released from presynaptic neurons and interact with specific receptor sites on postsynaptic neurons. At least 30 known or putative neurotransmitters have been found. A **putative neurotransmitter** satisfies some but not all criteria for being a neurotransmitter. Many neurotransmitters have been found in brain tissue, and more are likely to be discovered.

Early in embryological development, neurons have the potential to synthesize several chemical transmitters, but as development proceeds most neurons come to produce only 1 neurotransmitter. For many years it was believed that all mature neurons produce a single neurotransmitter, but it now appears that some mature neurons contain 2, or occasionally, 3 chemical transmitters. Much remains to be learned about multiple transmitter neurons, and the **Dale principle of neuronal specificity**—that mature neurons produce a single specific neurotransmitter—still applies to the vast majority of neurons.

In most neurons, neurotransmitters bind to specific receptors and alter membrane potentials, but more complicated arrangements have been found especially in the brain. For example, some neurotransmitters act on more than one kind of receptor. Also, it now appears that many neurotransmitters act directly or indirectly on voltage-sensitive ion channels and that effects on ion channels occur over a much longer time span than those of transient synaptic potentials.

The 6 neurotransmitters that have some reasonably well-established functions display a wide diversity in chemical structure (Figure 4.11). Each has particular properties worthy of special consideration (Table 4.3).

The neurotransmitter **acetylcholine** (as″e-til-ko′lin) is released at neuromuscular junctions, that is, junctions between motor neurons and skeletal muscle cells, at certain brain synapses, and at many synapses in the autonomic nervous system. Like other neurotransmitters, acetylcholine's action depends on its effect on postsynaptic membranes. For example, at receptors on neuromuscular junctions it increases permeability to Na^+ and K^+ and its excitatory effect leads to muscle contraction. At receptors in a specialized conduction system in the heart, it increases membrane permeability to K^+ but not to Na^+, so it causes hyperpolarization and decreases the rate of firing of portions of the heart's signal conducting system.

Acetylcholine is inactivated by the postsynaptic membrane enzyme **cholinesterase** (ko-lin-es′ter-ās), which catalyzes hydrolysis to acetate and choline. Some acetate and choline molecules are actively transported back to the presynaptic neuron and resynthesized into acetylcholine; the rest enter the bloodstream and are carried to other cells.

Another neurotransmitter, **norepinephrine** (nor″ep-ĕ-nef′rin), or noradrenalin, is secreted by many neurons of the sympathetic nervous system and by some brain neurons. Depending on the receptor to which it binds, norepinephrine can be excitatory or inhibitory. This transmitter is important in regulating the activity of visceral (internal) organs and in controlling certain brain functions. Norepinephrine can be inactivated at the postsynaptic neuron by the enzyme monoamine oxidase, reabsorbed back into the presynaptic knob, or inactivated in the liver by the enzyme catecholamine O-methyl transferase (COMT).

The neurotransmitter **dopamine** (do′pă-men) is chemically related to norepinephrine and to epinephrine, which is the predominant secretion of the adrenal medulla. All these substances are **catecholamines** (kat-eh-kōl′ă-menz), which are synthesized from the amino acid tyrosine. Dopamine is a neurotransmitter of certain brain neurons involved in motor control. The related substance epinephrine appears to act as a neurotransmitter in some neurons of the medulla oblongata, where many vital functions are controlled.

Dopamine deficiency is associated with Parkinson's disease, a disorder characterized by muscle rigidity,

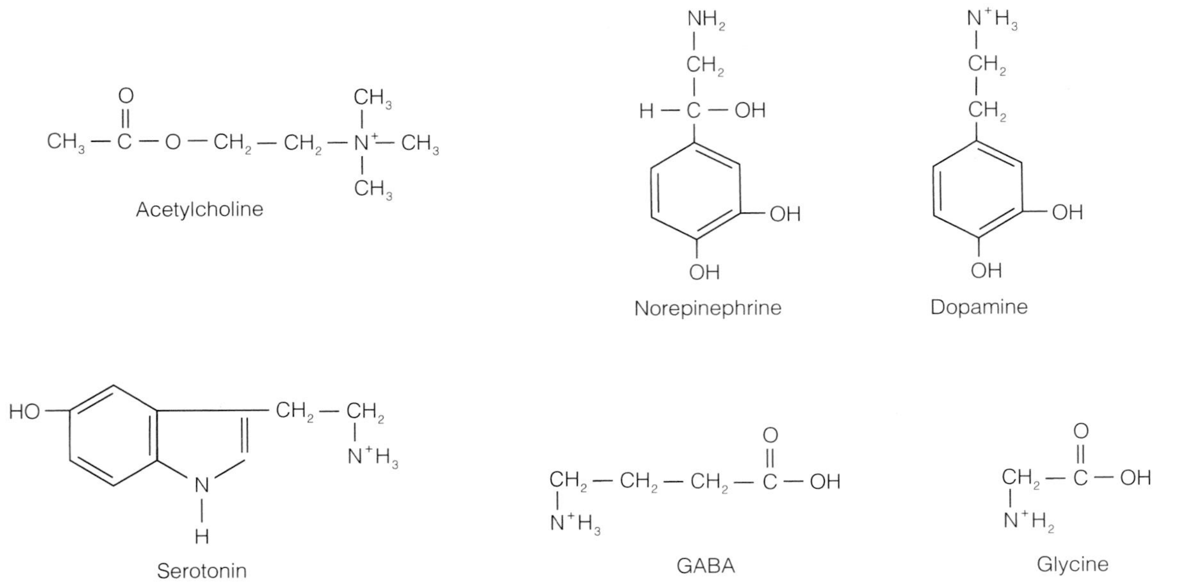

Figure 4.11

Neurotransmitters.

Table 4.3 Properties of Neurotransmitters		
Neurotransmitter	**Where Found**	**Actions**
Acetylcholine	Neuromuscular junctions, autonomic nervous system, and brain	Excites muscles, decreases heart rate, and relays various signals in the autonomic nervous system and the brain
Norepinephrine	Sympathetic nervous system and brain	Regulates activity of visceral organs and some brain functions
Dopamine	Brain	Involved in control of certain motor functions
Serotonin	Brain and spinal cord	May be involved in mental functions, circadian rhythms, and sleep and wakefulness
Gamma-aminobutyric acid	Brain and spinal cord	Inhibits various neurons
Glycine	Spinal cord	Inhibits neurons

tremors, and an uncoordinated, forward-leaning gait. Excesses of dopamine may be involved in mental disorders such as schizophrenia.

Serotonin (ser-o-to'nin), a neurotransmitter synthesized from the amino acid tryptophan, is released by certain neurons in the brain and spinal cord. It occurs in higher concentrations in blood platelets and in certain cells of the digestive tract, where it apparently has functions other than neurotransmission. The functions of serotonin as a transmitter are poorly understood, but it may play a role in the regulation of wakefulness and sleep and other circadian rhythms.

Gamma-aminobutyric acid (GABA), found only in the central nervous system, has an inhibitory effect at its receptors. GABA makes postsynaptic membrane potentials more negative by increasing the membrane's permeability to K^+ and Cl^-. Thus, the membrane becomes hyperpolarized.

The simple amino acid **glycine** is a transmitter in certain inhibitory synapses in the spinal cord. Though glycine is probably present in all cells, it acts as a neurotransmitter only when it is released from a neuron.

In recent years, about 25 small diffusible molecules called **neuroactive peptides** have been shown to relay signals between neurons. These molecules are stored and released by presynaptic neurons onto adjacent neurons, which can respond if they have the appropriate receptor. Some serve as hormones outside the brain, but many now appear to play a role in neurotransmission, too. Information about selected neuroactive peptides is summarized in Table 4.4.

Table 4.4
Neuroactive Peptides

Substance	Where Secreted*	Where Discussed in This Text
Cholecystokinin-pancreozymin (CCK)	Digestive tract, maybe cerebral cortex	
Endorphins and enkephalins	Various parts of central nervous system, digestive tract	Chapter 4
Hypothalamic hormones (CRH, GRH, LRH, PRH, somatostatin, TRH)	Hypothalamus, maybe other parts of the brain; somatostatin from digestive tract	Chapter 5
Oxytocin	Posterior pituitary gland, maybe parts of brain	Chapter 5
Substance P	Digestive tract, some afferent neurons, many other brain neurons	Chapter 4
Vasoactive intestinal peptide (VIP)	Maybe digestive tract and hypothalamus	Chapter 4
Vasopressin	Posterior pituitary gland, maybe other parts of the brain	Chapter 5

*Sites of secretion most clearly established are listed first, though these sites may involve some function other than release of neurotransmitters.

In addition to endogenous neurotransmitters normally found in the body, certain exogenous substances from outside the body affect neuron excitability by mimicking or blocking the action of a neurotransmitter. Substances that mimic a neurotransmitter fit its receptor so well that a neuron acts as if the neurotransmitter itself were present. Nicotine in small quantities has that effect on some acetylcholine receptors. Substances that block the action of a neurotransmitter bind with receptors but fit poorly and fail to function. Curare paralyzes skeletal muscles by blocking acetylcholine receptors in this way.

Certain substances increase neuron excitability. Those that affect membrane permeability to Ca^{2+} and maybe other ions include substances found in beverages—theobromine in cocoa, theophylline in tea, and caffeine in coffee. Strychnine and tetanus toxin (a poison made by certain bacteria) increase neuron excitability by blocking the action of inhibitory transmitters such as glycine and can cause convulsions.

Other substances decrease excitability. Dilantin possibly stabilizes the threshold of neurons against hy-perexcitability by promoting Na^+ efflux. It is used to prevent seizures—sudden discharges from groups of neurons in the brain.

Some anesthetics increase the stimulation required to reach the threshold for an action potential. Lipid-soluble anesthetics such as ether dissolve in neuronal membranes and increase K^+ permeability, causing hyperpolarization and making membranes less responsive to stimulation. The tranquilizer and muscle relaxant diazepam (Valium) enhances the inhibitory effects of GABA. The hallucinogen lysergic acid diethylamide (LSD) is chemically similar to serotonin and binds to its receptors. However, its effects—hallucinations and bizarre behaviors—are quite different from those of serotonin.

Efferent autonomic neurons synthesize and secrete neurotransmitters that must be inactivated to prevent continuous stimulation and to allow repolarization of stimulated neurons. In these respects, autonomic neurons are no different from other neurons. Differences in sympathetic and parasympathetic functions are determined by which neurotransmitter is released and how that transmitter interacts with the receptor to which it binds (Figure 4.12).

The autonomic nervous system produces 2 neurotransmitters. **Acetylcholine** is released by all presynaptic neurons and by all parasympathetic postganglionic neurons. It also is released by a few sympathetic postganglionic neurons, notably those that innervate sweat glands and those that dilate blood vessels in skeletal muscle. Neurons that release acetylcholine are said to be **cholinergic** (ko-li-ner'jik). **Norepinephrine** (noradrenalin) is released by most sympathetic postganglionic neurons; such neurons are said to be **adrenergic** (ad-ren-er'jik).

When acetylcholine is released by cholinergic neurons, its effects are determined by the nature of the receptor with which it interacts (Figure 4.13). Acetylcholine receptors are of two kinds: **nicotinic** (nik-o-tin'ik) **receptors** and **muscarinic** (mus-kar-in'ik) **receptors.** (These receptor names derive from early investigations in which nicotine and muscarine, a toxin from toadstools, were used to study cholinergic neurons.) The actions of acetylcholine are said to be nicotinic when they mirror the effects of nicotine or muscarinic when they duplicate the effects of muscarine.

Nicotinic receptors are found on both sympathetic and parasympathetic postganglionic neurons—the neurons that are stimulated by signals from preganglionic neurons. Such receptors are said to be type 1 nicotinic receptors. (Type 2 nicotinic receptors are found in the cell membranes of skeletal muscles.) How acetylcholine interacts with nicotinic receptors depends mainly on the acetylcholine concentration. Small amounts stimulate such receptors and large amounts inhibit them.

Muscarinic receptors are found at several sites—on organs innervated by postganglionic parasympathetic

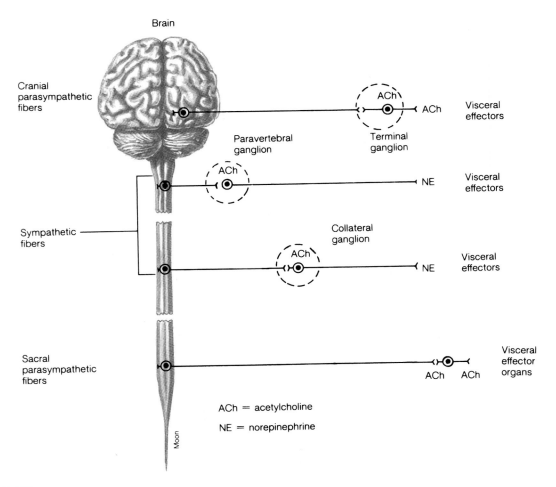

Brain

Cranial parasympathetic fibers

Paravertebral ganglion

Terminal ganglion

Visceral effectors

Sympathetic fibers

Collateral ganglion

Visceral effectors

Visceral effectors

Sacral parasympathetic fibers

Visceral effector organs

ACh = acetylcholine

NE = norepinephrine

Moon

Figure 4.12

The fibers of the autonomic nervous system and their usual neurotransmitters.

neurons (heart and most smooth muscle); or on organs innervated by sympathetic cholinergic neurons (sweat glands and some smooth muscle in blood vessels that supply skeletal muscles). Acetylcholine acting on muscarinic receptors reduces the heart rate, increases intestinal motility (movements that mix and propel food), causes sweating, and dilates blood vessels in skeletal muscles.

Comparable to cholinergic effects, adrenergic effects also are determined by the nature of the receptor with which the neurotransmitter, in this case, norepinephrine, interacts. However, the action of norepinephrine from neurons is augmented by norepinephrine and epinephrine secreted as hormones from the adrenal medulla. The adrenal medulla (the core of the adrenal gland) develops embryologically from the same tissue as autonomic neurons. When stimulated by adrenergic neurons, the adrenal medulla releases about 4 times as much epinephrine as norepinephrine. Compared with the rapid and short-lived action of norepinephrine at synapses, the action of epinephrine and norepinephrine released into the blood is slow and long-lasting.

Receptors for epinephrine and norepinephrine are found in most internal organs, especially in smooth muscle, and they respond to these substances regardless of whether they come from neurons or the adrenal medulla. Four types of receptors—$alpha_1$, $alpha_2$, $beta_1$, and $beta_2$ have been identified. Researchers have found certain chemical substances that bind to particular receptors, and they use these to study receptor properties. Epinephrine generally excites both alpha and beta receptors; norepinephrine excites mainly alpha receptors. The specific effects of either substance on an organ depends on the type, quantity, and sensitivity of the organ's receptors (Figure 4.13).

Many organs have both alpha and beta receptors, and the effect of stimulating one type can be directly opposite to those of stimulating the other type. For example, in arterial smooth muscle stimulating $alpha_1$ receptors generally causes muscle contraction and blood vessel constriction, whereas stimulating $beta_2$ receptors causes muscle relaxation and blood vessel dilation. Similarly, stimulating $beta_1$ receptors accelerates the breakdown of

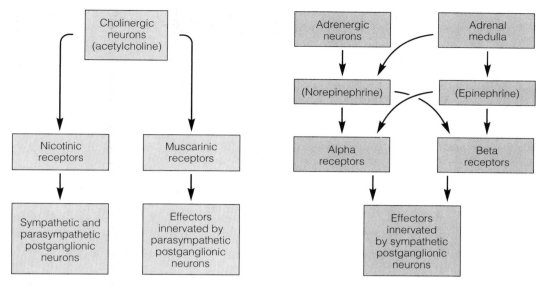

Figure 4.13

Summary of neurotransmitters and receptors involved in the function of the autonomic nervous system.

fat deposited in adipose tissues and stimulating alpha₂ receptors prevents it. Actions are not always opposite. Stimulating either alpha or beta receptors in intestinal smooth muscle reduces contractility, thereby decreasing motility.

Differences in the effects of stimulating various receptors appear to depend on what kind of response occurs in the cell membrane of a stimulated cell. Stimulating beta receptors, either type 1 or type 2, activates the enzyme adenylate cyclase. This, in turn, releases a second messenger molecule called cyclic AMP. (Second messengers receive a message delivered to a cell membrane and relay it to appropriate cell components [chapter 5].) In contrast, stimulating alpha₁ receptors increases the permeability of the cell membrane to calcium ions and stimulating alpha₂ receptors inhibits adenylate cyclase and prevents the release of cyclic AMP. These observations explain how stimulation of alpha or beta receptors can have opposite effects.

According to the classic view, autonomic function involves one transmitter-one receptor synapses, but more complex relationships exist. Finding that some autonomic neurons produce effects not attributable to either cholinergic or adrenergic effects led to the discovery of multiple transmitters in a single neuron and neuropeptides that modulate autonomic functions. Some neuropeptides act on either a special presynaptic receptor or a postsynaptic receptor. Sometimes two substances act on their respective receptors in a complex synapse, or one acts as a modulator of the other.

Neuropeptides have been found in all efferent components of the autonomic system—sensory, preganglionic, postganglionic sympathetic, and postganglionic parasympathetic—and in the adrenal medulla. For example, many sensory neurons contain the neuropeptide **substance P,** an 11-amino acid molecule, first discovered in horse tissues in 1931. Substance P is a powerful **vasodilator** (va″so-di-la′tor), a substance that relaxes smooth muscle and increases blood vessel diameters. Substance P also plays a role in the body's response to tissue injury.

Autonomic neurons, including those in many local ganglia and in the vagus nerve, release enkephalins. Which kinds of enkephalins are released and where they come from varies among species. Much more research is needed to determine their effects.

Some postganglionic sympathetic neurons release **neuropeptide Y,** which has been found to have two actions—constriction of some blood vessels and inhibition of signals initiated by norepinephrine at organ receptors.

Some postganglionic parasympathetic neurons release **vasoactive intestinal peptide** (VIP). VIP probably modulates the effects of acetylcholine on blood vessel walls and smooth muscle in various internal organs. In salivary glands, it increases secretion and dilates blood vessels probably by increasing the ability of acetylcholine to bind to muscarinic receptors.

Neuropeptides also affect the adrenal medulla. Substance P prolongs epinephrine and norepinephrine output during times of stress. Enkephalins are stored in adrenal medulla cells in a concentration of about 6 nanomoles (10^{-9} or billionths of a mole) per gram of wet tissue. They are stored as large protein molecules called **polyproteins,** from which enkephalins and other opiatelike peptides can be derived. Adrenal enkephalins may modulate blood pressure by inhibiting release of epinephrine and norepinephrine. They may contribute to stress-induced analgesia seen in victims of serious injuries who feel no pain for some time after the injury.

Table 4.5
Drugs That Affect Adrenergic Effector Organs

Site of Action	Drug	Action	Effect on Sympathetic Activity	Medical Application
Axon terminals of postganglionic neurons	Ephedrine	Augment release of norepinephrine	Augment	Nasal decongestant, antiasthmatic
	Guanethidine (Ismelin)	Deplete stores and prevent release of norepinephrine	Depress	Lower blood pressure
Alpha receptors	Methoxamine (Vasoxyl)	Excite α receptors, especially in blood vessels	Augment	Increase blood pressure in shock by vasoconstriction without affecting heart
	Phenoxybenzamine (Dibenzyline)	Block α receptors	Depress	Lower blood pressure
Beta receptors	Isoproterenol (Isuprel)	Excite β receptors	Augment	Bronchodilation, increase heart rate in patients with heart block
	Propanolol (Inderal)	Block β receptors	Depress	Lower blood pressure by decreasing cardiac output and suppressing renin release
	Practolol	Block β_1 receptors	Depress heart receptors	Not in general use
	Butoxamine	Block β_2 receptors	Depress receptors except in heart	Not in general use

Drugs are used in humans to augment or block autonomic nervous system function at two points: (1) at cholinergic effector organs, which have receptors for acetylcholine and (2) at adrenergic effector organs, which have receptors for norepinephrine. The action of drugs at adrenergic effector organs varies, depending on whether they alter norepinephrine synthesis or release, and on which receptors they affect.

At cholinergic effector organs, muscarinic receptors can respond to either parasympathetic or sympathetic stimulation. Both pilocarpine and methacholine excite these receptors. Drugs called **parasympathomimetics** (par-ah-sim″path-o-mim-et′ikz) increase parasympathetic activity. Pilocarpine is used to reduce pressure within the eyeball in diseases such as glaucoma. This drug dilates the canal that drains aqueous humor from the eyeball.

Drugs called **parasympatholytics** (par-ah-sim″path-o-lit′ikz) decrease parasympathetic activity. Atropine, a widely used drug, blocks the action of acetylcholine at muscarinic receptors by binding specifically to the receptors. It prevents the normal transmitter from reaching the receptor, so the effector organ is not stimulated. Atropine has no effect on nicotinic receptors because it fails to bind to them. Thus, atropine specifically suppresses parasympathetic activity. It is used during eye examinations to dilate pupils and to relax the ciliary muscles. Dilating the pupil allows examination of the retina; relaxing the ciliary muscles allows assessment of vision in

unstrained eyes. Atropine stops muscle spasms in the digestive tract and in the ducts of certain organs; however, it is no longer used for this purpose. In the dose required, it has undesirable side effects such as an increased heart rate, decreased salivation, and ciliary muscle relaxation.

Many drugs are available to augment or suppress adrenergic effects (Table 4.5). Augmenting drugs are called **sympathomimetics** (sim-path″o-mim-et′ikz); suppressing drugs are called blocking agents or **sympatholytics** (sim′path-o-lit′ikz). Some drugs have highly specific effects—they excite or block a particular type of receptor and are exceedingly important in modern medicine.

Four sympathomimetics currently in use are phenylephrine, clonidine, phentolamine, and isoproterenol. Phenylephrine, which stimulates $alpha_1$ receptors, is a commonly used bronchodilator. It is found in Chlor-Trimeton, Dimetapp, and some other medicines used to alleviate symptoms of upper respiratory infections and certain allergies. Clonidine, which stimulates $alpha_2$ receptors, can be used to treat hypertension (high blood pressure). It appears to exert its effects on adrenergic receptors in the central nervous system and to inhibit centers in the medulla that would accelerate the heart rate and constrict blood vessels. Phentolamine stimulates both kinds of alpha receptors, but it is used mainly to diagnose pheochromocytoma, a disorder in which adrenal medulla tissue proliferates and releases epinephrine and norepinephrine, thereby causing extremely high blood pressure.

Phentolamine quickly reduces high blood pressure caused by pheochromocytoma. Isoproterenol (Isuprel), which excites beta receptors, relaxes bronchial spasm in asthma, emphysema, and bronchitis. It also relaxes digestive tract smooth muscle and it increases the quantity of blood pumped by the heart.

The drug propanolol (Inderal) is a **beta blocker,** that is, its sympatholytic action is due specifically to its ability to bind to beta receptors and prevent them from being stimulated. Propanolol is effective in treating hypertension and some cases of angina pectoris (chest pain

due to reduced blood flow in the heart muscle) and cardiac arrhythmias (irregular heart beats). Its antihypertensive action may be due to its ability to block sympathetic signals in vasomotor centers in the brain or in the heart muscle itself. It also may act by inhibiting the release of a substance called renin from the kidneys, which ordinarily initiates a sequence of reactions that elevate blood pressure.

See Questions—Objectives 8 through 11

Questions

Objective 8

(a) Define neurotransmitter.

(b) List the properties of each of the 6 main neurotransmitters.

1.

2.

3.

4.

5.

6.

Objective 9

(a) What are neuroactive peptides?

(b) How do exogenous substances affect neuronal membrane permeability?

Objective 10

(a) What neurotransmitters are released at various sites in the autonomic nervous system?

(b) How do cholinergic and adrenergic receptors differ?

Continued on next page

Questions Continued

Objective 10 Continued

(c) What would be the effect of blocking cholinergic receptors?

(d) What would be the effect of blocking adrenergic receptors?

Objective 11

(a) What neuropeptides have autonomic effects?

(b) By what mechanisms do they act?

(c) What specific effects are known for substance P and for endorphins and enkephalins?

(d) How do drugs affect the autonomic nervous system?

Chemistry of Vision

The human retina contains two kinds of receptor cells—about 125 million rods and about 5.5 million cones (Figure 4.14). A rod has a long outer segment, an inner segment, and a synaptic body. The outer segment contains a light-sensitive pigment, **rhodopsin** (ro-dop′sin), which is stored in disks of intracellular membranes. The inner segment contains organelles, especially mitochondria. The synaptic body transfers signals from the rod to bipolar cells.

Cones are similar to rods, except that they are shorter and thicker and contain 1 of 3 color-sensitive pigments. Rods and cones stimulated by light hyperpolarize instead of depolarize as do most other receptors. Hyperpolarization, which results from decreased permeability of receptor cell membranes to Na^+, lasts as long as stimulation persists.

Rods, being much more light-sensitive than cones, allow vision in dim light, but they do not detect colors. In dim light we see black, shades of gray, and white according to the light intensities detected by the rods.

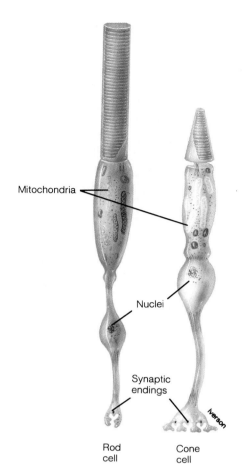

Mitochondria

Nuclei

Synaptic
endings

Rod
cell

Cone
cell

Figure 4.14

The internal structure of a rod and a cone.

Rhodopsin consists of the enzyme **opsin** (op'sin) and **retinal** (ret'in-al), a carotenoid pigment derived from vitamin A. The tips of rods are imbedded in the pigmented epithelium, which stores vitamin A. Rhodopsin is continuously synthesized in the inner segment, migrates to the base of the outer segment, and is incorporated into membrane discs. New discs produced hourly replace those used up at the tip of the outer segment.

Eliciting signals from light receptors is a photochemical process (Figure 4.15a). Inactive rhodopsin contains the isomer *cis*-retinal. A single **photon** (the smallest unit of light) can activate a molecule of rhodopsin, converting retinal to its *trans* isomer and activating the enzyme **transducin** (tranz-du'sin). Inactive transducin has guanosine diphosphate (GDP) bound to it, and during activation GDP is converted to guanosine triphosphate (GTP). Transducin-GTP behaves as an intracellular transmitter and moves from the disks to the cell membrane, where it activates the enzyme phosphodiesterase (PDE). In an unstimulated cell membrane, cyclic guanosine monophosphate (cGMP) maintains a resting membrane potential by keeping sodium pores open. PDE

hydrolyzes cGMP, allowing sodium pores to close and the cell membrane to depolarize. Hyperpolarization spreads along the membrane to synaptic terminals and initiates a signal in a bipolar cell.

Amplification, which occurs in two reactions, allows a single photon of light to elicit a neural signal. First, each rhodopsin activates many molecules of transducin by converting GDP to GTP. Second, the powerful enzyme PDE hydrolyzes over 4,000 molecules of cGMP per second. Deactivation of transducin and PDE turns off receptors and makes membranes ready for restimulation (Figure 4.15b). A chemical timer component of transducin initiates deactivation by converting GTP to GDP. This reaction also makes energy available for future reactions. Another component of transducin inhibits PDE, thereby allowing sodium pores to open and restoring the resting membrane potential. Finally, a kinase deactivates rhodopsin by adding phosphates to amino acids in opsin. Then rhodopsin forms a complex with **arrestin** (ah-res'tin) that prevents it from binding to transducin until it is again stimulated by light.

Humans perceive light of wavelengths between 400 and 700 nanometers (nm) as a spectrum of color. Properties of pigments in colored objects determine how objects absorb or reflect light, and thus, dictate the color they appear to have. For color to be perceived, certain wavelengths must reach light-sensitive pigments in cones. Cones are named by the pigment they contain. Blue cones have the pigment **cyanolabe** (si-an'o-lāb), green cones **chlorolabe** (klo'ro-lāb), and red cones **erythrolabe** (e-rith'ro-lāb). Each pigment absorbs light of particular wavelengths. All the pigments contain retinal, and cones apparently function in much the same way as rods do. Because cones require a higher intensity of light than rods to become excited, they are not active in dim light, so colored objects appear as shades of gray in dim light.

See Questions—Objective 12

Endorphins and Enkephalins

Opiates have relieved pain—and created addicts—for centuries. During the search for nonaddictive substitutes the discovery of opiate receptor sites on brain cells raised an interesting question: How can such sites exist if opiates are not natural body substances? The first clue came from beta-lipotropin, a protein made and degraded in the anterior pituitary gland to **endorphins** and **enkephalins.** Two enkephalins, 3 endorphins, and **dynorphin** (dīn-or'fin), a polypeptide that includes an enkephalin, have been identified and shown to bind to opiate receptors. These substances sometimes are referred to as endogenous (from inside the body) opiates. Further confirmation of the actions of endogenous opiates came from work with **naloxone** (nal-oks'ōn), or Narcan, which has the same effects

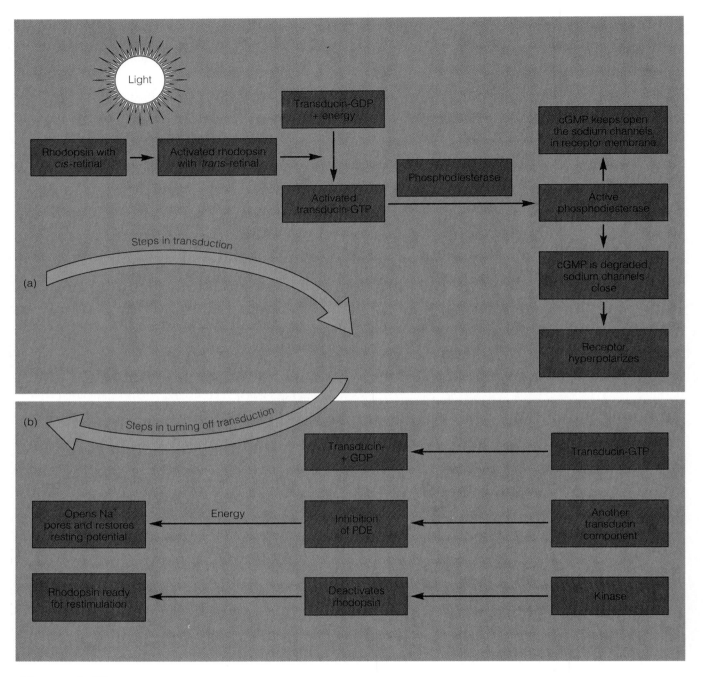

Figure 4.15

(*a*) Events in transduction of light energy to a neural signal. (*b*) Events in turning off transduction.

on endogenous opiates. Naloxone reverses effects of drug overdoses by replacing opiates bound to receptors.

Endogenous opiates play important roles in analgesia (pain relief). Electrical stimulation of certain regions of brain stem gray matter produces analgesia, possibly because it releases endogenous opiates. Studies in animals provide additional evidence: (1) Morphine and brain stimulation, both at nonanalgesic level, together produce analgesia; (2) Tolerance for opiates (the need for larger doses to achieve the same effect) develops not only from repetitive doses of morphine but also from repetitive brain stimulation; (3) Naloxone partially antagonizes the effects of stimulation.

In studies of central nervous system peptides, cholecystokinin (CCK), already known as a gastrointestinal hormone, was found in the brain. CCK acts as an opiate antagonist, but it also provides negative feedback to restore normal pain sensitivity after the release of endogenous opiates.

Questions
Objective 12

(a) How does dim light stimulate receptors and what reactions lead to rods relaying signals?

(b) How are the receptors prepared for restimulation?

(c) How do cones function in color vision?

Questions
Objective 13

(a) What events led to the discovery of endorphins and enkephalins?

(b) How are opiates thought to relieve pain?

(c) Of what clinical value is naloxone?

(d) What is presynaptic inhibition?

How endorphins and enkephalins produce their effects is unclear, but postsynaptic inhibition (hyperpolarization of postsynaptic neurons) does not seem to be involved. One proposed mechanism for enkephalin action is **presynaptic inhibition:** when an enkephalin is released at a synapse, it binds to presynaptic terminals of adjacent neurons and inhibits their release of an excitatory neurotransmitter.

Efforts to use endorphins and enkephalins in pain therapy have yet to achieve much success. One problem may be their short-activity span that requires them to be released continuously to maintain analgesia. Another may be that they are rapidly counteracted by CCK. Future research may lead to their use in alleviating pain without risk of addiction.

See Questions—Objective 13

Study and Review

Did you get the essentials?

The following summary contains the basic concepts from chapter 4 except for the key terms that have been omitted. Read the chapter. Then try to fill in the blanks from memory. If you cannot fill in all the blanks, review the chapter paying particular attention to boldface terms.

The functions of bone include _____ for all other body parts, providing for muscle attachment and thereby _____ , _____ for internal organs, _____ of minerals, and production of _____ cells.

The enzyme _____ fosters bone formation, whereas the enzyme _____ fosters bone degradation.

Factors that contribute to healthy bones include _____ , _____ , _____ , and _____ .

A muscle cell is called a _____ and it contains many small _____ , which consist of smaller _____ . The functional unit in a muscle is a _____ . The membrane of muscle cell is the _____ , the cytoplasm is _____ , the endoplasmic reticulum is the _____ _____ . Of the 4 proteins in myofilaments, _____ forms thick filaments and has heads that form _____ . In thin filaments, _____ is the contractile protein, _____ binds calcium, and _____ changes shape, exposing _____ binding sites on _____ .

In the _____ cycle, _____ binds to _____ in the _____ phase, the _____ swivel in the _____ phase, and a second molecule of _____ binds to _____ in the _____ phase. The explanation of the contractile process is called the _____ _____ theory.

The functional unit of contraction is the _____ _____ . A _____ junction includes the ends of motor axons and the _____ _____ _____ of the sarcolemma. Neural signals in axons initiate the release of _____ stored in axon tips. This substance diffuses across the _____ _____ and binds to _____ on the sarcolemma. It stimulates the sarcolemma until it is broken down by the enzyme _____ . The drug _____ competes with _____ for receptor sites.

Muscles store a little energy in _____ and they can make a molecule of ATP from 2 molecules of _____ . Muscles also contain _____ , which stores a little oxygen. When enough oxygen is available muscles can extract energy from _____ _____ , but must use _____ when oxygen is limited.

The force between particles of unlike charge is called a _____ _____ and in the body is expressed in _____ . Such a difference in solutions containing diffusible particles is a _____ _____ . When no net diffusion occurs an _____ _____ has been reached. Neurons ready to respond to a stimulus have a _____ _____ _____ of about −70 mV.

The relay of signals in a chemical _____ occurs as a _____ released from a _____ _____ diffuses across a _____ _____ to a _____ _____ . Changing a chemical signal to a potential is called _____ . Such a potential is an _____ _____ _____ when in _____ it leads to an action potential. An opposite potential is an _____ · _____ _____ , which in _____ makes a neuron more difficult to stimulate.

Among well-established neurotransmitters _____ is released at neuromuscular junctions and inactivated by the enzyme _____ . _____ is released in the sympathetic nervous system and is inactivated by the neuronal enzyme _____ _____ . Other neurotransmitters include _____ , _____ , _____ , and _____ .

Rods contain the visual pigment _____ , whereas cones contain 1 of 3 similar _____ , each sensitive to particular _____ of light. When a rod is activated *cis*-_____ becomes _____ _____ and the enzyme _____ is activated. GDP becomes _____ and other reactions cause the closing of _____ _____ and the development of _____ .

_____ and _____ are opiatelike molecules that play a role in _____ .

Hormones and Chemical Aspects of Reproduction

5

Hormones and the Endocrine System

To qualify as a **hormone** (hor′mŏn), a chemical substance must be synthesized and secreted by a cell, travel in the blood, and bind to specific receptors of **target cells** to exert its regulatory effects. Cells that secrete regulatory substances into the blood constitute the **endocrine system.** Not all regulatory substances are hormones. Some are neurotransmitters or metabolic products such as carbon dioxide. Norepinephrine released from a synapse is a neurotransmitter; when secreted by cells of the adrenal medulla it is a hormone. Carbon dioxide, a waste product of cellular metabolism, travels in the blood and is detected by cells in certain brain centers that regulate breathing. It is not considered a hormone because all cells release it.

The endocrine system (Figure 5.1) includes the pituitary (hypophysis), thyroid, parathyroid, adrenals, the islets of Langerhans of the pancreas; secretory cells of the hypothalamus, gonads, thymus, pineal gland, heart, kidneys, and placenta, and the **diffuse endocrine system** (DES)—an assortment of endocrine cells found in epithelial tissues, such as the digestive tract lining.

See Questions—Objective 1

Properties of Hormones

Some hormones are synthesized as inactive molecules called **prohormones** and are later converted to their active form. For example, insulin is synthesized as **preinsulin** on ribosomes, transported through the endoplasmic reticulum, and converted to **proinsulin** by the removal of a small chain of amino acids. Proinsulin is stored in the Golgi apparatus. When proinsulin is secreted, a segment called **C-peptide** is cleaved from it to make active insulin.

Most small hormone molecules are transported in the blood by nonspecific globulins, albumins, or specific proteins such as **thyroxine-binding globulin,** which carries thyroid hormones, and **transcortin,** which carries hydrocortisone from the adrenal cortex. Progesterone from the ovary or placenta can be carried on transcortin or on albumin. These proteins help to distribute a hormone throughout the body and prevent the hormone from being released from the blood too rapidly.

When the blood concentration of a hormone increases above the amount needed, the increased concentration tends to suppress further secretion of the hormone. Then the blood concentration of the hormone begins to decrease as some of it binds to receptors. As the blood concentration decreases, suppression also decreases, and

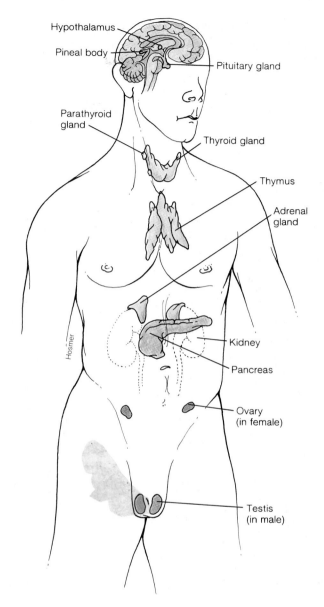

Figure 5.1

The glands of the endocrine system are located in various parts of the body. In addition to discrete glands, many organs contain diffuse endocrine structures.

hormone secretion begins again. This self-regulation of the blood concentration of a hormone is **negative feedback.**

Hormonal control systems are similar to neural control systems in some respects, but they also have some special properties as shown in Table 5.1.

See Questions—Objective 2

Questions

Objective 1

(a) What is a hormone?

(b) What is the general function of the endocrine system and what are its components?

Table 5.1
A Comparison of Neural and Hormonal Control Systems

Characteristic	Neural Control Systems	Hormonal Control Systems
Nature of message	Action potential and neurotransmitter	Hormone
Mode of transmission	Conduction on neuron and transmission at synapses	Through blood
Effects	Change membrane potential of affected cells; carry afferent and efferent signals; involved in perception, consciousness, learning, memory, emotion, and control of movements	Change rates of cellular metabolic activities of target cells
Time for effects to occur	Usually less than one second	Usually several minutes
Duration of effects	Milliseconds to seconds (except memory storage)	Minutes to hours, sometimes even days

Mechanisms of Hormone Action

Though endocrine glands can release hormones spontaneously, certain stimuli—blood concentrations of some substances, neural signals, or hypothalamic releasing or inhibiting hormones—can increase or decrease the rate of secretion. Once in the blood a hormone can reach almost any cell in the body. However, each hormone affects only its target cells, the cells that have receptors for it. Nearly all cells have receptors for thyroid and growth hormones, but only intestinal epithelium, bone, and certain kidney cells have receptors for parathyroid hormone.

Hormones exert their effects at the cellular level by changing the rate of certain cellular processes. For example, the pancreatic hormone insulin increases the rate at which glucose enters cells (except those of the liver and the brain). How hormones produce their physiological effects is not fully understood, but it is possible to describe in a general way some mechanisms by which hormones act.

Steroids, being relatively small molecules synthesized from cholesterol, are lipid soluble and easily pass through cell membranes. They diffuse into all cells and apparently diffuse out of nontarget cells without having any effect. In target cells, steroids bind to protein receptors in the nucleus. The mechanism of action of thyroid hormone and vitamin D now appears to be the same as that of steroid hormones. All bind to receptors and activate or inactivate a segment of DNA usually 50 to 100 genes long. Activated DNA directs mRNA transcription and the mRNA moves to the ribosome where it directs protein synthesis (chapter 2). We will call this the **protein synthesis mechanism** of hormone action (Figure 5.2). Hormones that act by this mechanism show their effects relatively slowly because of the time required for protein synthesis.

This understanding of steroid hormone action allows a new interpretation of some endocrine abnormalities. Instead of occurring because of excesses or deficiencies of the hormones themselves, some abnormalities may be due to variations in the activity of the receptor. Testicular feminization, in which male secondary sex characteristics fail to develop, may be caused by deficient or defective receptors for the male sex hormones. Similarly, cells of hormone-dependent tumors, such as certain breast tumors, have estrogen receptors, and estrogen stimulates the tumors to grow. If a means of blocking or destroying the receptors could be found, the tumor would lack stimulation and cease to grow. In the future, contraceptives that act by blocking receptors for reproductive hormones may be available.

Questions

Objective 2

(a) What is a prohormone?

(b) How are hormones transported?

(c) Briefly explain the role of negative feedback in controlling hormone secretion.

(d) What are similarities and differences in neural and hormonal control systems?

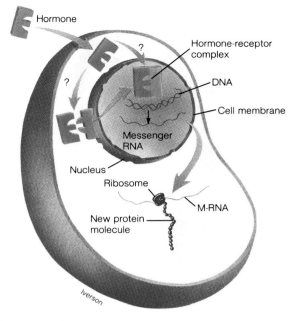

Figure 5.2

Steps in the action of steroid and thyroid hormones: The hormone diffuses through the cell membrane and across the cytoplasm to the nuclear membrane. It binds to a receptor on the membrane or in the nucleus and activates a gene. Gene activation causes transcription of messenger RNA and synthesis of a protein.

Many protein and polypeptide hormones, such as those from the pituitary and parathyroid glands and glucagon and insulin from the pancreas, act by the cyclic adenosine monophosphate mechanism, or the **cAMP mechanism** (Figure 5.3). In the cAMP mechanism, a hormone (the first messenger) binds to receptor on the outer surface of the cell membrane. Each receptor has an associated G protein. Some G proteins are excitatory and others are inhibitory. An excitatory G protein activates **adenylate cyclase** (ad″en-il′āt si′klās) in the cell membrane. An inhibitory G protein inhibits adenylate cyclase. Active adenylate cyclase converts ATP to cAMP, and cAMP acts as a **second messenger.**

One role of the second messenger cAMP can be to activate an enzyme called a protein kinase. **Protein kinase** (ki′nās), using phosphate and energy from ATP, phosphorylates another enzyme and thereby activates or inactivates it. When an enzyme is activated it catalyzes a specific reaction—often a rate-limiting reaction in a metabolic pathway. If the rate-limiting reaction occurs, all the reactions in the pathway occur and these reactions constitute the physiological effects of the hormone. Many hormone (and neurotransmitter) effects occur via enzyme phosphorylation.

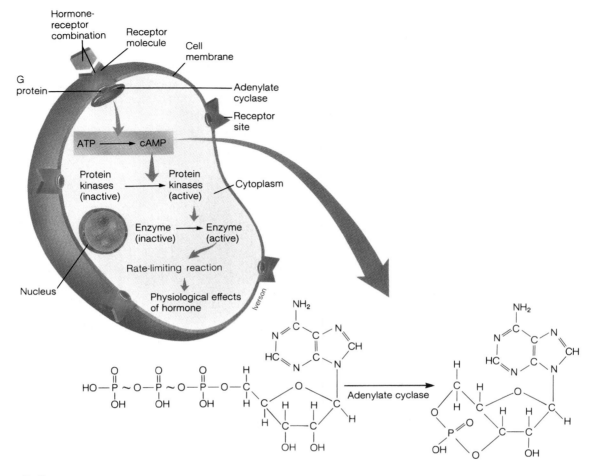

Figure 5.3

Steps in the action of some polypeptide hormones: The hormone binds to a receptor on the cell membrane and thereby activates the enzyme adenylate cyclase. Prostaglandins sometimes amplify the activity of adenylate cyclase. The second messenger cAMP is formed; cAMP acts directly or indirectly by activating a kinase to produce the physiological effects of the hormone.

In some membranes, cyclic nucleotide **cGMP** (cyclic guanosine monophosphate) is produced from GTP (guanosine triphosphate) by **guanylate** (gwan′il-āt) **cyclase.** The cGMP mechanism operates in the same way as the cAMP mechanism. Atrial natriuretic hormone from the heart acts by the cGMP mechanism to promote sodium excretion.

An enzyme called **hormone-sensitive lipase** is activated by phosphorylation in response to any of several hormones including epinephrine, glucagon, thyroid hormones, and the pituitary hormone ACTH. It is inhibited by insulin. Activation of hormone-sensitive lipase occurs as follows:

1. The hormone binding to a receptor activates adenylate cyclase.

2. The messenger cAMP is released and activates a protein kinase.
3. The protein kinase phosphorylates (activates) the hormone-sensitive lipase.
4. The lipase initiates the breakdown of fats in adipose tissues.

Some hormone actions accelerate processes and some inhibit them. Such actions constitute a complex set of on-off switches that regulate cell function. Cells respond to different hormones according to the kinds of receptors they have. Because specific events follow the binding of a hormone to its receptor, each hormone produces particular physiological effects.

In some instances neurotransmitters have actions similar to hormones. For example, both the hormone glucagon and the neurotransmitter norepinephrine can trigger a series of chemical reactions that degrade glycogen. The fact that both hormone and neurotransmitter effects can occur by similar mechanisms shows the close relationship between the body's major control systems.

Similar to neural signals and hormones, calcium ions play a significant regulatory role in a variety of physiological processes, such as blood clotting, bone formation, movements within cells, release of neurotransmitters, and muscle contraction and relaxation. Calcium is now recognized as an important second messenger in cells. It binds to **calcium-binding proteins,** which are cytosol proteins similar to hormone receptors.

The calcium-binding protein **calmodulin** (kal-mod'u-lin) binds or releases calcium ions depending on conditions within the cell. In unstimulated mammalian cells, the concentration of free Ca^{2+} in the cytosol is exceedingly small—10^{-7} moles per liter. A neural or hormonal stimulus to a cell can cause an influx of Ca^{2+} into the cytosol from outside the cell or from within the endoplasmic reticulum, thus raising the concentration to 10^{-6} moles per liter or even higher. The increase in the Ca^{2+} concentration causes the formation of a Ca^{2+}-calmodulin complex. The Ca^{2+}-calmodulin complex can activate enzymes such as adenylate cyclase or phosphorylase kinase. Other calcium-binding proteins include troponin in skeletal and cardiac muscle and a vitamin D-dependent calcium-binding protein that binds Ca^{2+} from food in the small intestine and facilitates its absorption.

Recent studies suggest that mechanisms involving calcium and those involving cAMP often work together to create a complex cellular regulating system. Ca^{2+} regulates enzymes that synthesize and degrade cAMP, and cAMP regulates the movement of Ca^{2+} into and out of cells. Both cAMP and Ca^{2+} control intracellular events by controlling protein kinases.

Receptors are as necessary as hormones for hormone actions, so the number of receptors affects hormone activity. Cells can react to high hormone concentrations by reducing their number of active receptors. Receptors can become inactive by sinking below the surface or by changing their molecular shape.

Some hormones affect not only the number of their own receptors but also those for other hormones. For example, excess thyroid hormones increase epinephrine receptors on heart cells in proportion to the degree of excess thyroid hormones. The effect is to increase the heart rates, presumably because greater numbers of epinephrine receptors are stimulated.

The ability of a small amount of a hormone to exert a significant effect is explained by **enzyme amplification,** a cascade effect that occurs in the cAMP mechanism as follows: A single hormone molecule binding to a single receptor site activates several molecules of adenylate cyclase. Each adenylate cyclase molecule produces many molecules of cAMP, and each molecule of cAMP activates many molecules of the appropriate protein kinase. Each protein kinase molecule, in turn, activates many molecules of a particular enzyme, such as hormone-sensitive lipase. Each lipase molecule breaks down many lipid molecules.

Steroid hormones display similar amplification as one molecule of messenger RNA directs the synthesis of many molecules of a particular protein (usually an enzyme), and each enzyme molecule produces many product molecules.

To serve as regulators, hormones, much like neurotransmitters, must be inactivated once they have exerted their effects. Most hormones are inactivated or degraded by liver enzymes and then excreted. Thyroid hormones are excreted in bile, whereas steroid hormones are excreted in urine. Certain end products of steroid hormones appear in urine in amounts proportional to the amount secreted. Thus, end product concentrations in urine often can be used clinically to estimate the amount of the hormone being secreted. Such tests are used to identify athletes who are using anabolic steroids. Epinephrine and norepinephrine (either as a hormone or a neurotransmitter) are degraded by the enzymes monoamine oxidase and catechol-o-methyltransferase. End products of the degradation are excreted in the urine and can be measured. Peptide and protein hormones are inactivated by proteolytic enzymes. The protein insulin cannot be given by mouth because enzymes digest the insulin before it can be absorbed into the blood.

See Questions—Objectives 3 and 4

Prostaglandins

Discovered in the 1930s and named for a presumed association with the prostate gland, **prostaglandins** (pros-tah-glan'dinz) are now known to be important chemical regulators of cellular function. They are derived from arachidonic acid (a 20-carbon unsaturated fatty acid), contain a 5-carbon ring, and 2 side chains (Figure 5.4). Prostaglandins are synthesized and secreted as needed. They act by binding to extracellular receptors either near the cells that made them or after traveling in the blood. They are extremely potent—some cause smooth muscle fibers to contract at a concentration of only 1 nanogram per milliliter.

The physiological effects of prostaglandins are extremely varied and can be antagonistic. Some augment the action of other regulatory molecules. Much remains to be learned about prostaglandin actions but when a hormone alters the cAMP concentration, prostaglandins augment that effect. When insulin increases cAMP and causes cells to take up glucose, prostaglandins further increase

Questions

Objective 3

(a) Compare and contrast the mechanisms of action of steroid and thyroid hormones with that of polypeptide hormones.

(b) How do calcium and calmodulin function as regulators?

Objective 4

(a) How do properties of receptors affect hormone actions?

(b) How does enzyme amplification affect the body's response to certain hormones?

(c) How are hormones inactivated?

the cAMP concentration. In fat cells where insulin decreases the concentration of cAMP and prevents the release of fatty acids, prostaglandins further decrease cAMP. Some prostaglandins cause fever during infections, and aspirin reduces the fever by inhibiting prostaglandins.

Two groups of prostaglandins (PGs), designated by their chemical structure as PGEs and PGFs, usually have antagonistic actions. For example, PGEs dilate blood vessels and bronchial passages, whereas PGFs constrict them. Thus, the balance between these two types of pros-

taglandins helps to maintain homeostasis in certain physiological processes. However, both PGEs and PGFs cause uterine smooth muscle to contract, and both have been used to induce abortions. In fact, during the first 6 months of pregnancy, the uterus is more sensitive to prostaglandins than to oxytocin (a hormone that stimulates uterine contraction during labor). PGEs also increase blood vessel permeability, and inhibit aggregation of blood platelets.

See Questions—Objective 5

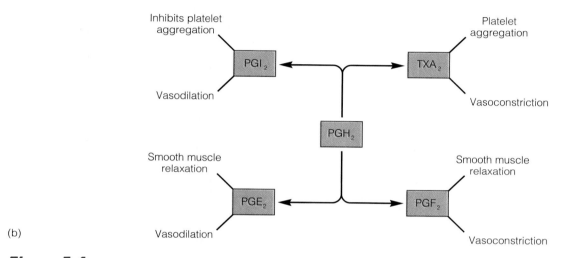

(a)   Prostaglandin E_1 (PGE_1)

Prostaglandin E_2 (PGE_2)

(b)

Figure 5.4

(*a*) In naming prostaglandins (PG), a letter such as E indicates the kind of ring it contains, and the subscript indicates the kind of side chains it has. (*b*) Some actions of prostaglandins.

Questions

Objective 5

(a) What are the main properties of prostaglandins?

(b) Give some examples of the functions of prostaglandins.

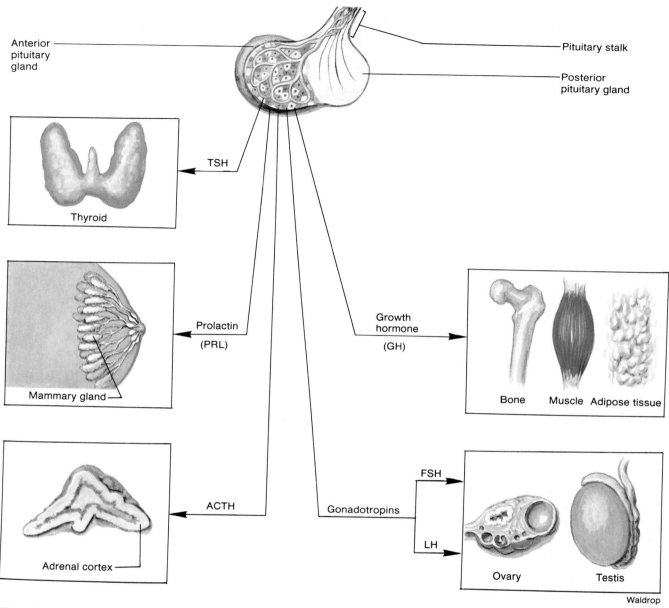

Figure 5.5

The hormones of the adenohypophysis and their effects.

Pituitary Gland

A small structure lying beneath the hypothalamus, the **pituitary gland** or hypophysis, consists of the **anterior pituitary** (adenohypophysis) and the **posterior pituitary** (neurohypophysis). Both are regulated by the hypothalamus, which provides anatomical and physiological connections between the nervous and endocrine systems. The adenohypophysis is connected with the hypothalamus via the **hypophyseal portal system,** which consists of venules that carry blood between capillaries of the hypothalamus

and those of the adenohypophysis. Releasing and inhibiting hormones from the hypothalamus travel through these vessels and regulate adenohypophyseal functions. The neurohypophysis and hypothalamus are connected by the **hypophyseal tract,** a nerve tract of axons whose cell bodies are in the hypothalamus. These cells synthesize hormones that travel down axons and are stored in the neurohypophysis, which is a storage area and not a gland.

Several hormones of significance in humans are synthesized and secreted from the adenohypophysis (Figure 5.5). They are growth hormone, prolactin, follicle-stimulating hormone (FSH), luteinizing hormone (LH), thyroid stimulating hormone (TSH), and adrenocorticotropic hormone (ACTH). Beta-lipotropin, whose action is less clearly understood, also is synthesized and released from the adenohypophysis. Also discussed here is melanocyte—stimulating hormone (MSH) from the poorly developed intermediate lobe of the pituitary gland. Among these hormones, all of which are polypeptides, only growth hormone has a general effect on nearly all cells. Several are **tropic** (trop'-ik) **hormones**—hormones that act on other glands. They include the gonadotropins, FSH and LH, and the chemically related TSH. Tropic ACTH is synthesized along with beta-lipotropin and MSH as a single polyprotein.

In general, a tropic hormone from gland A stimulates gland B to release its secretion. The hormone from gland B acts on its target cells and it also acts by negative feedback to inhibit hormone secretion from gland A. This mechanism serves to maintain the concentration of the hormone from gland B fluctuation within a narrow range (Figure 5.6).

Somatotropin (so-mat-o-trōp'in), also called **growth hormone** (GH), acts directly on certain cells to stimulate growth and indirectly on others to release proteins, such as insulin-like growth factor I. Growth hormone also causes cartilage and bone cells to divide and osteoblasts to secrete collagen. It promotes Ca^{2+} absorption from the intestine. Thyroid hormones and insulin work with growth hormone to promote collagen synthesis.

Growth hormone and growth factors are essential to stimulate growth to adult size and to maintain adult size thereafter. Together they increase the rate at which cells take up amino acids and use them to synthesize proteins. They also stimulate free fatty acid release from fat cells and glycogen breakdown in the liver, which releases glucose into the blood. These functions provide adequate nutrients to cells between meals and during stress.

Growth hormone has been synthesized in the laboratory using recombinant DNA. The human gene for synthesis of the hormone is combined with bacterial DNA, and the bacterial cells then produce human growth hormone, which can be isolated from other metabolic products and purified for therapeutic use in humans.

The protein hormone **prolactin** (pro-lak'tin), or lactogenic hormone, stimulates milk secretion in mammary glands previously prepared for milk production by other hormones—estradiol, progesterone, corticosteroids, and insulin. Suckling of the infant is the primary stimulus for prolactin secretion and thus leads to lactation. Suckling stimulates mechanoreceptors in the nipples and signals from them go to the hypothalamus. These signals probably cause PRH secretion and inhibit PIH secretion. The anterior pituitary secretes prolactin, which stimulates the mammary glands to synthesize milk for the next feeding. As long as suckling occurs periodically, milk se-

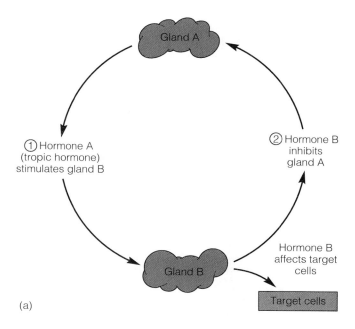

(a)

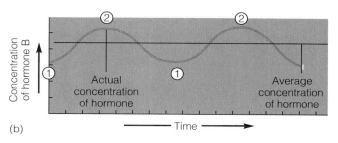

(b)

Figure 5.6

(*a*) Negative feedback from a tropic gland to the gland that produces the tropic hormone. (*b*) Fluctuation of the blood concentration of a hormone around its set point.

cretion will also continue unless it is interrupted by pregnancy. Suckling inhibits FSH and LH release from the hypothalamus and suppresses ovulation in some women, but it is not a reliable means of birth control.

The glycoprotein **follicle-stimulating hormone** (FSH) stimulates the maturation of an ovum (egg) each month during a female's reproductive years. It also stimulates maturation of sperm in males. When FSH is released in females it initiates maturation of 10- to 20-groups of cells called follicles. One of these follicles dominates the others, matures, and produces estrogen hormones. The estrogens act by negative feedback to suppress further release of FSH.

Another glycoprotein hormone involved in reproduction is **luteinizing** (lu'te-in-īz''ing) **hormone** (LH). In females, it stimulates ovulation (release of an ovum) and causes follicular cells to produce progesterone, which among other things, stimulates mammary gland development. In males, LH stimulates development of interstitial cells (Leydig cells) of the testes and causes them to produce testosterone.

The glycoprotein called **thyroid-stimulating hormone** (TSH), or thyrotropin, stimulates the thyroid gland to synthesize and secrete its hormones, triiodothyronine (T_3) and thyroxine (T_4). When these hormones reach sufficient concentration in the blood, they act by negative feedback to suppress further release of TSH.

Adrenocorticotropic (ad-re″no-kor″tik-o-trōp′ik) **hormone** (ACTH), or corticotropin, is a polypeptide that acts on the cortex of the adrenal gland to regulate synthesis and secretion of several of its hormones, especially the glucocorticoids. Glucocorticoids, in turn, stimulate the release of fatty acids and glucose into the blood and help the body to resist stress and inflammation. One of the glucocorticoids, cortisol, is especially important in regulating ACTH secretion. When the adrenal cortex has been sufficiently stimulated by ACTH, cortisol acts by negative feedback to suppress further release of ACTH.

Although not proven to have a specific effect in humans, **beta-lipotropin** (lip-o-trop′in) has been isolated from the anterior pituitary glands of various animals. Endorphins and enkephalins can be made from beta-lipotropin in the anterior pituitary and in the brain directly (chapter 4).

The intermediate lobe of the pituitary gland, a rudimentary structure in humans, secretes small quantities of **melanocyte** (mel-an′o-sīt) **-stimulating hormone** (MSH). MSH is identical to the first 13 amino acids of ACTH, and MSH effects in humans are probably masked by the greater effect of ACTH. Patients with Cushing's disease, a disorder caused by excess ACTH, sometimes develop excessive skin pigment. This hyperpigmentation occurs because of an excess of MSH produced along with the excess ACTH.

Several adenohypophyseal hormones are regulated by small peptide hormones from neurosecretory cells of the hypothalamus. They are released into the hypothalamic capillaries and travel through the hypophyseal portal vein to the adenohypophysis. Each regulatory hormone stimulates or inhibits release of an anterior pituitary hormone (Table 5.2). **Somatostatin** (so-mat-o-stat′in), or growth-hormone-inhibiting hormone (GHIH), also stimulates the making of glucose from noncarbohydrate sources when blood sugar falls below the normal concentration.

Hormonal regulation operates at several levels. Hypothalamic hormones regulate hormones of the anterior pituitary. Tropic hormones from the anterior pituitary regulate other glands such as the thyroid, adrenal cortex, and gonads. Hormones from these glands act by negative feedback to inhibit the release of both tropic and hypothalamic hormones (Figure 5.7). Complex relationships

Table 5.2

Hypothalamic Hormones that Regulate Secretion of Hormones from the Adenohypophysis

Hormone	Abbreviation	Function
Thyrotropin-releasing hormone	TRH	Stimulates release of TSH
Corticotropin-releasing hormone	CRH	Stimulates release of ACTH
Gonadotropin-releasing hormone	GnRH	Stimulates release of FSH and LH, depending on concentration
Growth-hormone-releasing hormone	GHRH	Stimulates release of GH
Growth-hormone-inhibiting hormone (somatostatin)	GHIH	Inhibits release of GH
Prolactin-releasing hormone	PRH	Stimulates release of prolactin
Prolactin-inhibiting hormone	PIH	Inhibits release of prolactin

between hormones of the gonads, adenohypophysis, and hypothalamus help to regulate reproductive functions. Neural signals elicited by sexual arousal, stress, anxiety, or trauma, variations in the light-dark cycle, and the sucking of a breastfed infant, also regulate hypothalamic hormones.

Certain hypothalamic hormones such as TRH appear to be widely distributed and to possess multiple functions. TRH was characterized as a tripeptide in 1969 and a few years later was shown to stimulate the release of TSH from the anterior pituitary by activating adenylate cyclase. Since then it has been shown to release certain other pituitary hormones such as prolactin, and in psychiatric and metabolic disorders, growth hormone as well. In addition to the hypothalamus, it has been found in other parts of the brain—hippocampus, amygdala, pons, medulla, and cerebral cortex. TRH is present in high concentration in brain tissue in Huntington's chorea, schizophrenia, and sometimes in Alzheimer's disease and depression. When administered experimentally to animals, TRH elicits many effects like those resulting from stimulation of cholinergic receptors (chapter 4).

Two chemically similar peptide hormones, oxytocin and antidiuretic hormone (Figure 5.8) are called **neurosecretions** because they are synthesized in hypothalamic neurons and stored in the neurohypophysis. When action potentials cause their release from axons, they enter the blood and act as hormones.

The hormone **oxytocin** (oks-e-to′sin) stimulates contraction of smooth muscle in the uterus and the con-

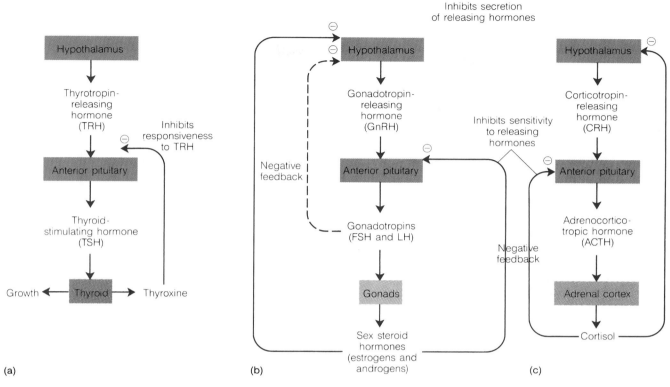

Figure 5.7

Examples of negative feedback in the regulation of function of the anterior pituitary gland and glands affected by tropic hormones: (*a*) thyroid, (*b*) gonads, and (*c*) adrenal cortex.

tractile cells around mammary gland ducts. Distention of the cervix and vagina or stimulation of the nipple cause oxytocin release, therefore the hormone is released in large quantities during sexual intercourse, labor, and lactation. Oxytocin can be used to induce labor in near-term pregnancy, but it is not known whether oxytocin plays a significant role in the natural onset of labor.

Oxytocin release during lactation is reasonably well understood. An infant's sucking of the nipple sends neural signals that are relayed to the neurohypophysis and cause oxytocin release. Within 30 to 60 seconds oxytocin causes smooth muscle contraction in mammary gland ducts and milk is ejected. The release of milk by the sucking of an infant is an example of neural-hormonal **positive feedback** (Figure 5.9). Positive feedback, which can be responsible for emptying body cavities but is otherwise unusual in a living organism, accelerates a process. For control by positive feedback to be effective, there must be a mechanism outside the feedback loop to stop the process, otherwise the process would accelerate indefinitely. In this example, the infant's sucking accelerates milk flow by positive feedback and milk flow stops when the sucking ceases or the breast is empty.

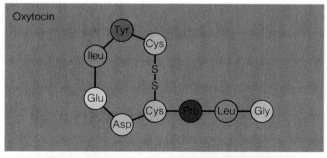

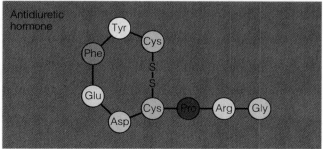

Figure 5.8

The amino acid composition of polypeptide hormones oxytocin and antidiuretic hormone is quite similar.

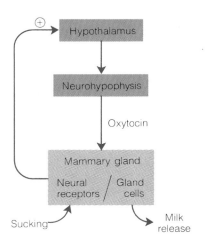

Figure 5.9

A neural-hormone positive feedback loop. The sucking of an infant stimulates receptors in the nipple of the mammary gland. Neural signals travel to the hypothalamus, where they stimulate the release of oxytocin from the neurohypophysis. Oxytocin travels in the blood to the mammary gland and stimulates the release of milk.

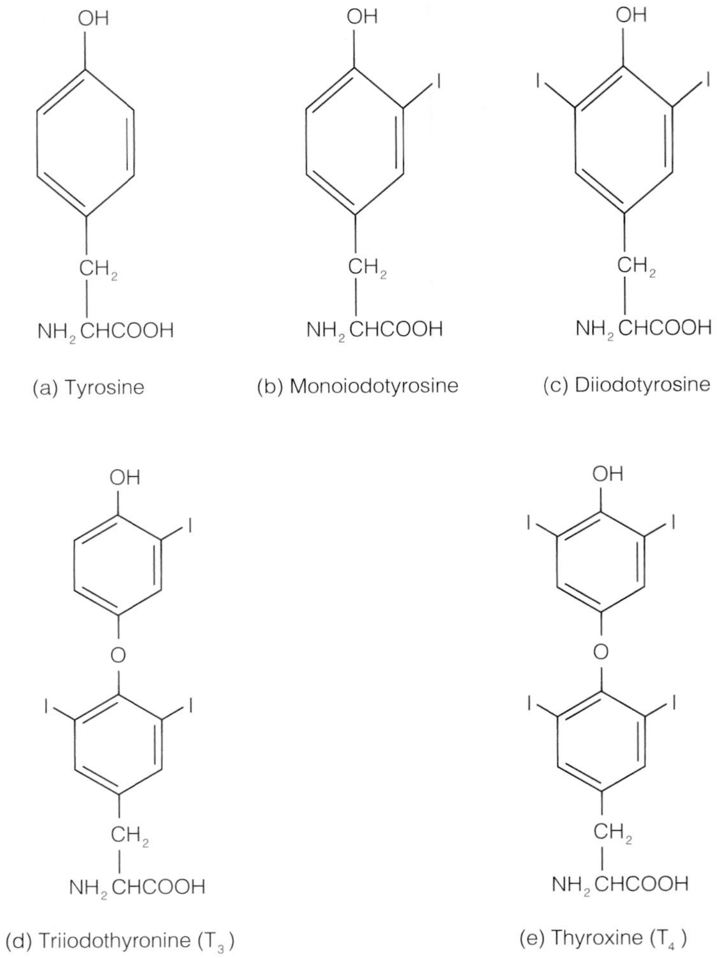

Antidiuretic (an-ti-di-u-ret'ik) **hormone** (ADH) prevents excess water loss in urine. This hormone is also called **vasopressin** (va-so-pres'in) because in an abnormally high concentration it can constrict blood vessels. ADH is secreted when hypothalamic osmoreceptors detect that the blood and cerebrospinal fluid become concentrated (exert increased osmotic pressure). ADH causes the kidneys to return more water to the blood. This decreases osmotic pressure, increases the blood volume, and decreases the urine volume (chapter 9).

See Questions—Objectives 6 and 7

Thyroid Gland

The **thyroid gland** is located in the neck below the larynx (voice box) and is filled with **colloid,** which is a thyroglobulin with thyroid hormone molecules attached. The thyroid gland can store enough hormone in the colloid to supply the body for about 2 months. Some of the cells around the follicles, the parafollicular C cells, produce the hormone calcitonin.

Thyroid hormones are amino acid derivatives synthesized by iodination of the amino acid tyrosine (Figure 5.10). The thyroid gland removes iodide ions from the blood and converts them to molecular iodine. It then replaces certain hydrogen atoms of tyrosine molecules with iodine. Precursor molecules called monoiodotyrosine (MIT) and diiodotyrosine (DIT) are the first to be syn-

(a) Tyrosine (b) Monoiodotyrosine (c) Diiodotyrosine

(d) Triiodothyronine (T_3) (e) Thyroxine (T_4)

Figure 5.10

Thyroid hormones and their precursors: (*a*) tyrosine, (*b*) monoiodotyrosine, (*c*) diiodotyrosine, (*d*) triiodothyronine (T_3), and (*e*) thyroxine (tetraiodothyronine or T_4).

thesized. Precursor molecules are coupled to make the active thyroid hormones **triiodothyronine** (tri-i-o''-do-thi'ron-ēn), or T_3, and **thyroxine** (T_4). The hormones are stored and attached to thyroglobulin within follicles.

When the gland is stimulated to secrete hormones, T_3 and T_4 are separated from the thyroglobulin and released into the blood where they are carried by a thyroxine-binding globulin or an albumin. Far more T_4 than T_3 is released, and T_4 binds more tightly to the blood proteins. However, T_3 has much greater hormone activity and most of the T_4 is converted to T_3 by the endoplasmic reticulum of liver, kidney, and other cells. Nearly all cells are target cells for thyroid hormones. In cells that convert T_4 to T_3, some T_3 goes to the nucleus and some is released into the blood plasma and transported to other cells.

Thyroid hormones increase oxygen consumption and metabolism in most tissues and also stimulate protein

Questions

Objective 6

(a) What are the functions of the hormones secreted by the anterior pituitary gland?

(b) How are anterior pituitary hormones regulated?

(c) What are some effects of excesses and deficiencies of anterior pituitary hormones?

Objective 7

(a) What are the functions of the hormones released from the posterior pituitary gland?

(b) Where do posterior pituitary hormones come from?

(c) How are posterior pituitary hormones regulated?

synthesis. They accelerate oxygen consumption and metabolism by increasing the concentration and activity of mitochondrial enzymes. The higher metabolic rate produces more ATP for use by the Na^+-K^+ pump and other cellular activities. The presence of T_3 in the nucleus of a cell and the binding of T_3 to nuclear receptors initiates protein synthesis in a manner similar to that of steroid hormones.

The primary regulator of thyroid hormones is TSH (thyroid-stimulating hormone) from the anterior pituitary. When blood hormone concentrations decrease,

TSH is released and travels to the thyroid gland where it stimulates the synthesis and release of the hormones. As blood concentrations of the hormones increase, they suppress TSH by negative feedback. TRH can stimulate TSH release, and it can be inhibited by thyroid hormones, but it is most effective in adapting the body to the environment. Living in a cold environment stimulates TRH release, and living in a warm climate suppresses it. During pregnancy a placental TSH-like hormone increases thyroid hormone release and accelerates the metabolic rate sufficiently to maintain growth and fetal development.

Questions

Objective 8

(a) How are the hormones of the thyroid gland synthesized?

(b) How are they stored and released?

Objective 9

(a) What are the effects of thyroid hormones?

(b) How are thyroid hormones regulated?

Calcitonin (kal-sit-o′nin) is secreted from the C cells of the thyroid gland when the blood calcium concentration is high.

See Questions—Objectives 8 and 9

Parathyroid Glands

The **parathyroid** (par-ah-thi′roid) **glands** consist of four groups of cells imbedded in the thyroid gland. These cells synthesize and secrete **parathormone** (par-ah-thor′mōn), or parathyroid hormone (PTH), a large polypeptide that helps to regulate the metabolism of calcium and certain other minerals.

With respect to calcium metabolism, PTH and calcitonin have antagonistic effects. Their actions along with those of vitamin D maintain calcium homeostasis. When blood calcium decreases, the parathyroids release PTH. It increases calcium absorption from the intestine by activating vitamin D, which stimulates synthesis of a calcium carrier protein in the small intestine. PTH also decreases calcium excretion by the kidneys and can remove

calcium from bones if necessary to maintain a normal blood calcium concentration. When blood calcium increases, calcium acts by negative feedback and suppresses PTH secretion. At the same time, it stimulates calcitonin secretion by thyroid C cells. Calcitonin lowers blood calcium by inhibiting reabsorption from bones, causing kidneys to excrete it, and suppressing its absorption.

See Questions—Objective 10

Adrenal Glands

Located superior to each kidney, the **adrenal** (ad-re′nal) **glands** each have a small inner medulla and a thick outer cortex. The **adrenal medulla,** being derived from the same tissue as sympathetic neurons, secretes hormones in response to sympathetic stimulation. The **adrenal cortex** secretes more than 25 different corticosteroid hormones.

Medullary hormones, **epinephrine** (adrenalin) and **norepinephrine** (noradrenalin), are synthesized from the amino acid tyrosine via dopamine. Vitamin C is required for activity of the enzyme that converts dopamine to nor-

Questions

Objective 10

(a) What is the function of parathormone and how is it regulated?

(b) What is the source of calcitonin and how is its function related to that of parathormone?

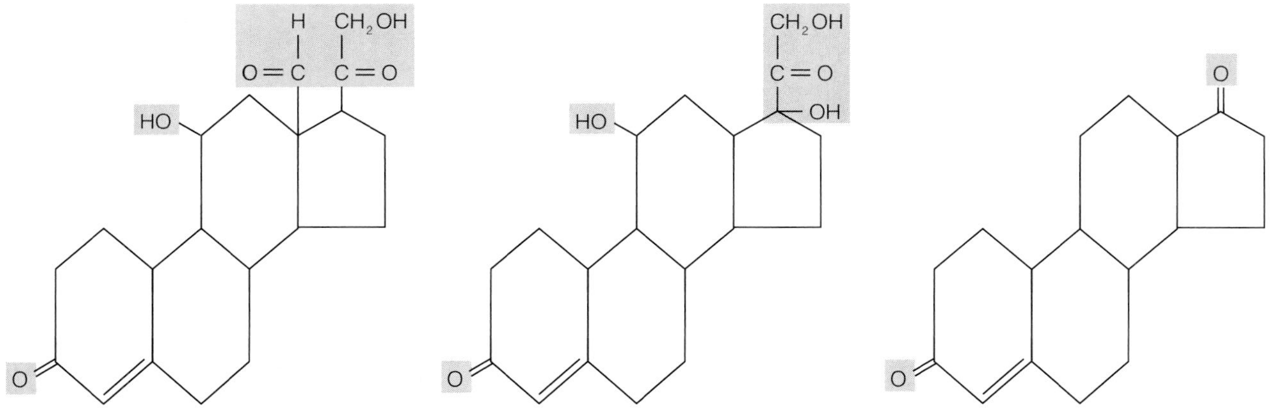

(a) Aldosterone   (b) Cortisol   (c) An androgen

Figure 5.11

Note the small differences in the structure of the steroid hormones secreted by the adrenal cortex: (*a*) aldosterone, (*b*) cortisol, and (*c*) an androgen.

epinephrine. About four-fifths of the norepinephrine is converted to epinephrine by the addition of a methyl group. Compared with the brief action of norepinephrine released at synapses, it and epinephrine released as hormones can act for up to 3 minutes—the time required for enzymes in liver and other tissues to inactivate them.

Adrenocortical hormones, which are corticosteroids synthesized from cholesterol, fall in three groups—**mineralocorticoids, glucocorticoids,** and **androgens** (Figure 5.11). Small chemical differences in these hormones account for their vastly different functions.

The major mineralocorticoid **aldosterone** (al-dos′ter-ōn) regulates the sodium and potassium content of

body fluids (Figure 5.12). Aldosterone is secreted when sodium decreases or potassium increases in the blood. It acts on the Na^+-K^+ pump in kidney tubule cells and causes sodium to be returned to the blood and potassium to be excreted. As the sodium concentration in the blood increases, water follows it by osmosis, and the blood volume also increases. Overall, aldosterone increases blood sodium and blood volume and decreases blood potassium.

The **glucocorticoids** (glu-ko-kor′te-koidz) help to regulate not only carbohydrate metabolism as the name suggests, but also protein and fat metabolism. Though glucocorticoids regulate specific metabolic processes, their effects are far reaching because altering one metabolic

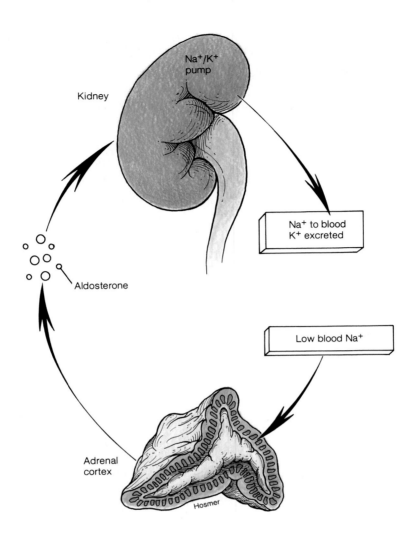

Figure 5.12

Aldosterone regulates sodium, potassium, and water in body
fluids.

process usually alters others, too. Glucocorticoids of the
human adrenal cortex include the potent **cortisol** (kor'tis-
ol), or hydrocortisone, and the less potent **corticosterone**
(kor-tik-os'ter-ón). These hormones are secreted after
several hours without food when liver stores of glucose have
been depleted. They stimulate synthesis of glucose from
noncarbohydrate sources, cause fat cells to break down
fats and release fatty acids into the blood, and cause many
cells to break down proteins and release amino acids into
the blood. Increasing blood fatty acids causes many cells
to use more fatty acids and less glucose to meet their
energy needs.

Cortisol and corticosterone counteract some effects of inflammation and help the body to respond to
stress. In general, inflammation increases capillary permeability and causes lysosomes of injured cells to release
enzymes. How cortisol counteracts inflammation is not
well understood, but it seems to reduce permeability of

capillaries and make lysosomal membranes more stable.
Cortisol helps the body to respond to stress by mobilizing
fatty acids and other nutrients that provide energy to meet
the demands of a stressful situation. It also stimulates the
adrenal medulla to release epinephrine and norepinephrine. Cortisol is exceedingly important in maintaining homeostasis through its continuous regulation of metabolism
and through its ability to reduce inflammation and stress.

Adrenal sex hormones, or gonadocorticoids, are
mostly **androgens** (an-dro'jenz), chemically similar to testosterone and convertible to it in some tissues. These hormones stimulate pubic hair development in females, but
they normally have no effects in males. Adrenal tumors
can produce large amounts of androgens that can cause
masculinization of a female or precocious sexual development in a young male.

Aldosterone secretion is regulated by the blood
potassium concentration, a blood peptide **angiotensin** (an-

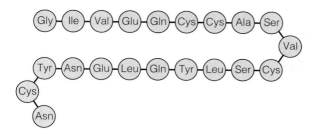

(a) One Polypeptide chain of the small protein Insulin

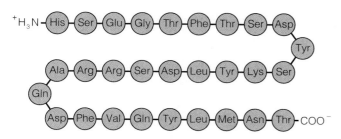

(b) The polypeptide glucagon

Figure 5.13

(*a*) One polypeptide chain of the small protein insulin, and (*b*) the polypeptide glucagon.

je-o-ten′sin), ACTH, and to a lesser degree, by blood sodium concentration. Cortisol and corticosterone secretions are controlled mainly by ACTH. ACTH also stimulates aldosterone release, but only cortisol exerts significant feedback on ACTH. A high cortisol level suppresses release of both ACTH and CRH. The adrenal cortex receives less stimulation as cortisol secretion decreases. CRH and ACTH are not suppressed and they stimulate secretion of cortisol and other adrenocortical hormones.

See Questions—Objective 11

Endocrine Functions of the Pancreas

The **pancreas** produces both digestive and endocrine secretions. Its endocrine cells, which form clusters called **islets of Langerhans** (lan′ger-hanz), are about three-fourths insulin-producing **B cells,** less than one-fourth glucagon-producing **A cells,** and a few cells that produce the metabolic regulators somatostatin and pancreatic polypeptide. **Insulin** is a small protein containing 51 amino acids, and **glucagon** is a polypeptide containing 29 amino acids (Figure 5.13). Insulin and glucagon constitute one of the body's important mechanisms for regulating the

Questions
Objective 11

(a) What hormones does the adrenal medulla secrete?

(b) What do adrenal medulla hormones do and how are they regulated?

(c) What hormones does the adrenal cortex secrete?

(d) What do adrenocortical hormones do and how are they regulated?

blood glucose concentration (Figure 5.14). When the blood glucose rises, insulin enters the blood and binds to cell surface receptors where it facilitates the entry of glucose into cells. When the blood glucose drops, insulin secretion stops and glucagon facilitates the breakdown of glycogen, thereby increasing the blood glucose level. Both hormones are regulated by negative feedback according to the blood glucose level.

See Questions—Objective 12

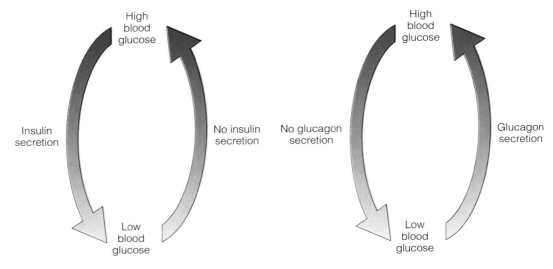

Figure 5.14

Insulin and glucagon regulate blood glucose concentration.

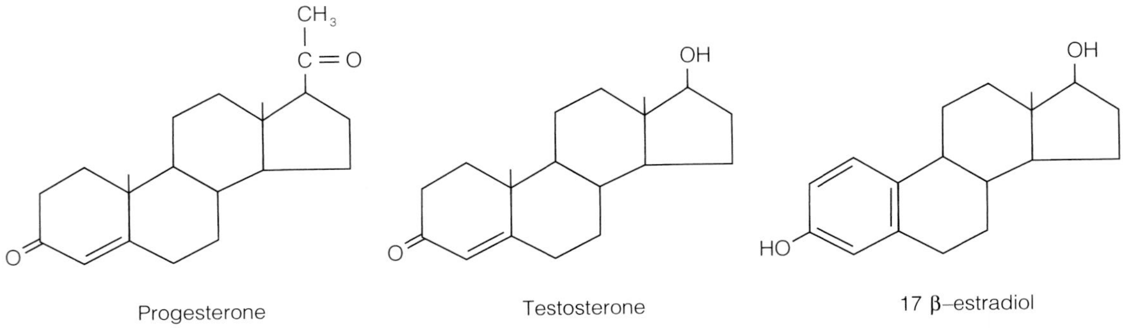

Progesterone   Testosterone   17 β–estradiol

Figure 5.15

Steroid hormones.

Chemical Aspects of Reproduction

The primary reproductive organs are **gonads** (go'nadz)—female ovaries and male testes. Gonads produce sex hormones and **gametes** (gam'ētz)—ova in females and sperm in males. Each reproductive system also has ducts that transport gametes and glands that release secretions into the ducts. **Sex hormones**—testosterone, estradiol, and progesterone (Figure 5.15), which have only small chemical differences, stimulate maturation of reproductive organs and development of secondary sexual characteristics—male musculature and hair pattern and female breasts and body contours—during puberty. Puberty, the period during which fertility is attained, probably results from hypothalamic maturation and gonadotropin-releasing hormone (GnRH) secretion.

Sex hormones are synthesized from cholesterol or acetyl-CoA and are transported in the blood on specific globulins or in loose association with plasma albumins. These steroid hormones diffuse through membranes and bind to nuclear receptors in target cells. Hormone-receptor complexes act on DNA to induce transcription and protein synthesis. Hormone molecules that do not enter target cells are removed from the blood mostly by the liver and to a small extent directly by the kidneys. The liver converts steroid hormones to inactive substances, excreting some with bile and releasing some into the blood for excretion by the kidneys.

See Questions—Objective 13

Questions
Objective 12

(a) What are chemical properties of the major pancreatic hormones?

(b) What do pancreatic hormones do?

(c) How are they regulated?

Male Reproductive Physiology

Testosterone and **dihydrotestosterone** are responsible for the development and maintenance of masculine characteristics. During embryonic development, chorionic gonadotropin from the placenta stimulates the testes to secrete testosterone from about the seventh week of development until a few weeks after birth. At the onset of puberty gonadotropin-releasing hormone (GnRH) causes the anterior pituitary gland to release luteinizing hormone

Questions
Objective 13

(a) How do male and female hormones differ in chemical structure and function?

(b) Where are sex hormones produced and from what are they synthesized?

(LH), which stimulates interstitial cells to make testosterone and dihydrotestosterone. Male hormone secretion rises sharply during puberty, peaks in the early 20s, declines gradually until the early 40s and more rapidly thereafter. Secretion stays at about one-fifth the peak rate even at age 80. High dihydrotestosterone acts by negative feedback to suppress LH and GnRH release.

During puberty testosterone stimulates the growth and development of male sex organs and the development of male secondary sexual characteristics—male body hair pattern, muscular development, skeletal growth, and basal metabolic rate. Dihydrotestosterone controls hair growth, which occurs not only in the pubic area, but also on the face, chest, limbs, and axillae (armpits). Increased protein synthesis accounts for a higher metabolic rate, muscular development, thicker skin, and probably an increase in red blood cell numbers. Increased melanin synthesis deepens the skin hue. Bone matrix synthesis and mineral deposition account for bone growth. Enlargement of the larynx, especially the thyroid cartilage, or "Adam's apple," and corresponding lengthening of the vocal cords deepens the voice.

Once the male reproductive system is fully developed, male hormones are still needed to maintain mature organs and produce sperm. Testosterone increases muscle strength by increasing muscle cell size and glycogen storage. It stimulates the pituitary gland to release

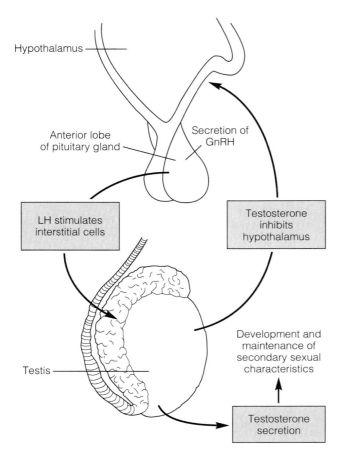

Figure 5.16

The regulation of testosterone secretion.

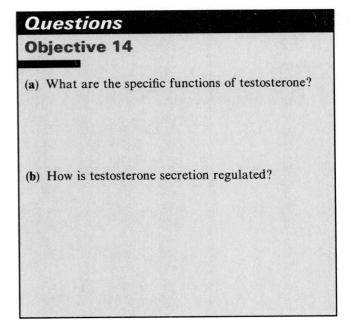

(a) What are the specific functions of testosterone?

(b) How is testosterone secretion regulated?

involved. When sperm are being produced, Sertoli cells release inhibin, which acts by negative feedback to inhibit both GnRH and FSH release.

See Questions—Objective 14

growth hormone and prepares the heart to grow in response to it. It also fosters sperm maturation by stimulating Sertoli cells. Dihydrotestosterone stimulates the prostate gland and sebaceous glands. The skin of acne patients contains 2 to 20 times the normal amount of this hormone.

Spermatogenesis (sper″mat-o-jen′es-is), the process by which functional sperm are produced, takes place in the testes from puberty until testosterone secretion becomes too low to stimulate it. Though testosterone acts directly on spermatocytes undergoing meiosis, FSH binds to Sertoli cell receptors, causing those cells to mediate sperm maturation. FSH may be needed to initiate sperm maturation but not to maintain it once it starts.

Carefully regulated release of GnRH and LH are needed for normal testosterone secretion and sperm production (Figure 5.16). When the blood testosterone level rises, testosterone acts by negative feedback to inhibit GnRH release. As GnRH decreases, LH decreases and interstitial cells are not stimulated. When the testosterone level is too low to inhibit it, GnRH secretion again releases LH. FSH regulation is not fully understood, but GnRH and the hormone **inhibin** (in-hib′in) appear to be

Female Reproductive Physiology

During embryonic development of a human female several million **oogonia** (o-o-go′ne-ah), cells capable of producing ova, migrate from the yolk sac to gonads that become ovaries. Many oogonia degenerate and surviving cells undergo **oogenesis** (o-o-jen′is-is), becoming **primary oocytes** (o-o-sītz). At birth about 2 million remain and by puberty only 400,000 remain. All the ova a woman can ever produce come from these primary oocytes. At puberty, a few oocytes develop each month to become **secondary oocytes,** and one reaches full maturity and is released into a uterine tube, where it can be fertilized if the female has recently had intercourse. If a sperm penetrates a secondary oocyte, it divides and produces a mature **ovum** (o′vum). Its nucleus and the sperm nucleus fuse. Cytoplasm provides nutrients until it implants in the uterus. Unfertilized oocytes degenerate.

After puberty an ovarian cycle of follicle development and degeneration (Figure 5.17) occurs monthly. FSH stimulates follicular cells to divide, mature, and secrete hormones and fluid to form **primary follicles.** Among several such follicles, one competes most favorably for available FSH and becomes the dominant follicle. Partially developed follicles degenerate. As the primary fol-

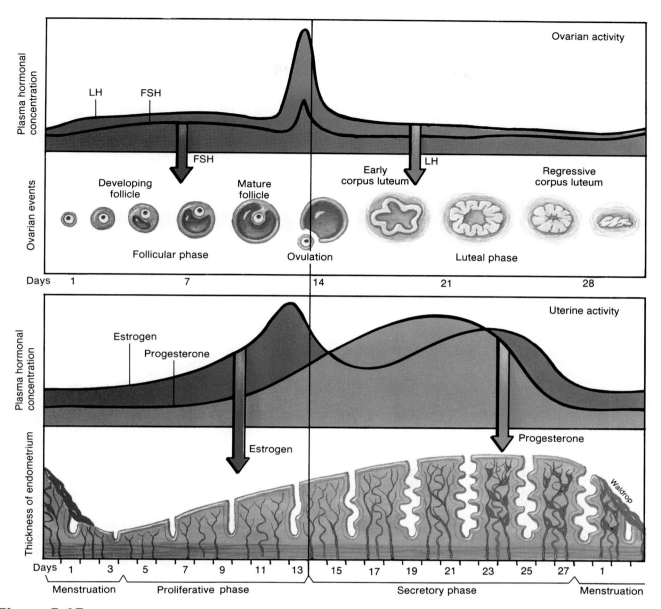

Figure 5.17

Main events in the menstrual cycle related to the ovarian
cycle and hormone concentrations.

licle develops, a fluid-filled cavity surrounds the oocyte and
adjacent follicle cells divide and protrude into the cavity.
At **ovulation,** the mature follicle ruptures, releasing the
oocyte. If a sperm penetrates the oocyte, it completes maturation, and its nucleus fuses with the sperm nucleus.

After ovulation, LH stimulates the cavity to fill
with a semiclotted fluid and remaining follicular cells to
enlarge and fill with a yellow pigment, **lutein** (lu'te-in).
Such a follicle is a **corpus luteum** (lu'te-um)—literally,
yellow body. If the ovum is not fertilized, the corpus luteum
degenerates after about ten days, leaving a scarlike **corpus
albicans** (al'bik-anz)—literally, white body. Degeneration

of the corpus luteum marks the end of one ovarian cycle,
but another soon begins. If the ovum is fertilized, the
corpus luteum enlarges and persists as an endocrine gland
for the first 6 to 8 weeks of pregnancy.

In addition to releasing ova, maturing follicles secrete **estrogens,** mainly **estradiol,** which help to regulate
the menstrual cycle, a monthly sequence of events that
includes sloughing part of the uterine lining. After ovulation, corpus luteum cells produce estrogens and **progesterone.** Progesterone maintains pregnancy and helps to
regulate the menstrual cycle.

Only small amounts of estradiol are released before puberty. After puberty, estradiol secretion increases to about 20 times the prepubertal level. Estradiol stimulates reproductive organs to grow and become functional and develops the secondary sexual characteristics, such as the female pattern of fat distribution. The vaginal epithelium becomes more resistant to infections and trauma. The uterus doubles or triples in size in the first few years after puberty and its endometrium proliferates. Ciliated epithelial cells of uterine tubes increase in number and in the activity of cilia, which assists ova in reaching the uterus. Estradiol also stimulates duct development and fat deposition in mammary glands.

In addition to its effects on the reproductive system, estradiol affects the skeleton, skin, metabolism, and electrolyte balance. It stimulates bone growth during the pubertal growth spurt and then stops growth by causing epiphyses to fuse with the shafts of the long bones. Women usually are not as tall as men because estradiol has a faster, more potent effect on epiphyseal fusion than testosterone. Estradiol causes skin to become soft and smooth but thicker and more vascular than in childhood. Growth of pubic hair and the development of axillary sweat glands probably are controlled, not by estradiol, but by adrenal androgens, which also account for acne in females. Estradiol increases the metabolic rate—but by only about one-third as much as testosterone, and it causes a smaller increase in total body protein than testosterone. As a result of these differences, women generally have slower metabolism, smaller bodies, less muscle, and more body fat than men. Fat is deposited in the breasts, thighs, and buttocks, producing feminine body contours.

Progesterone prepares the uterus for implantation of a fertilized ovum and the secretory portions of the mammary glands for milk production. It causes uterine tubes to release secretions that help to nourish the ovum and also suppresses uterine contractions during pregnancy. Progesterone slightly increases protein catabolism, making amino acids from body tissues available for protein synthesis in a growing fetus if dietary nutrients do not satisfy the need.

Several hormones coordinate uterine functions with ovarian functions in the monthly female **menstrual** (men'stru-al) **cycle** (refer back to Figure 5.17). In this cycle, **menstruation** (men-stru-a'shun) is the shedding of the uterine lining and the bleeding that accompanies it. Blood loss ranges from 25 to 65 milliliters, and clotting is prevented as fibrinolysin degrades fibrin as it forms. At **menarche** (men-ar'ke), the onset of menstruation, ovarian and uterine functions also begin. Hypothalamic maturation leads to episodic GnRH secretion, which causes the anterior pituitary to secrete pulses of FSH and LH.

A menstrual cycle lasts from the beginning of one menstrual flow to the beginning of the next, typically about 28 days. Variations from 24 to 35 days are common, with

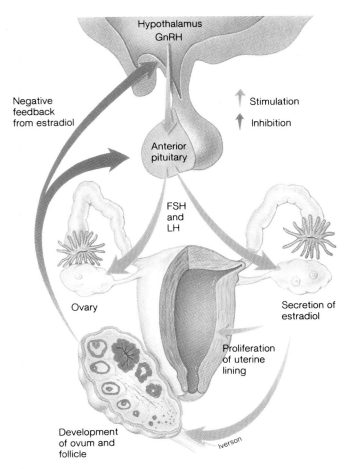

Figure 5.18

Summary of the preovulatory portion of the menstrual cycle.

most variation occurring between menstruation and ovulation. Differences in GnRH, FSH, LH, estrogen, and progesterone secretion account for some variations, but stress also accounts for variations in cycle length. Stress may cause signals from other parts of the brain to be relayed to the hypothalamus, where they alter GnRH release. Regulation of the preovulatory portion of the menstrual cycle is summarized in Figure 5.18 and that of the postovulatory portion is summarized in Figure 5.19. Hormone actions are summarized in Table 5.3.

Menstrual cycles normally continue at more or less regular intervals from puberty to menopause, except during pregnancy. **Menopause** (men'o-pawz), the cessation of menstruation, occurs around age 50 with the gradual cessation of ovarian function. Follicles fail to develop, estradiol and progesterone secretion decreases, and menstrual periods become irregular and eventually cease. Estradiol production decreases markedly throughout most of the fifth decade of a woman's life in spite of the fact that more FSH, LH, and GnRH are secreted than before menopause. Decreasing estradiol provides less and less

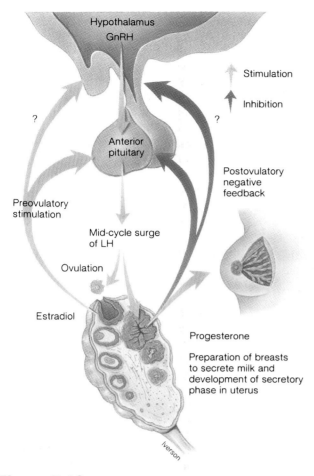

Figure 5.19

Summary of ovulation and the postovulatory portion of the menstrual cycle.

Table 5.3

Actions of Hormones Involved in the Female Reproductive Cycle

Hormone	Actions
Follicle-stimulating hormone (FSH)	Stimulation of follicle growth Stimulation of estradiol release
Estradiol (low level)	Increase in sensitivity of anterior pituitary to GnRH Stimulation of anterior pituitary to emphasize LH synthesis over FSH synthesis Inhibition of LH release Proliferation of cells of uterine lining
Estradiol (high level)	Inhibition of FSH release Stimulation of LH release
Luteinizing hormone (LH)	Induction of ovulation Stimulation of follicle growth Formation and maintenance of corpus luteum (source of more estradiol and progesterone) Stimulation of progesterone release
Progesterone (high level)	Further development of uterus in preparation for implantation Increase in size and sensitivity of breasts (with estradiol) Inhibition of LH release Inhibition of further follicle development
Progesterone (low level)	Degeneration of corpus luteum Increase in release of FSH (thus cycle begins again)

negative feedback inhibition of the other hormones. "Hot flashes," periods of intense warmth resulting from the dilation of skin arterioles, during menopause are associated with LH pulses. Such pulses elicit alpha adrenergic signals that stimulate thermoregulatory neurons and neurons that release GnRH.

Hormones that help to maintain pregnancy include hCG, estradiol, estriol, progesterone, relaxin, and human chorionic somatomammotropin (hCS). Certain embryonic cells produce **human chorionic gonadotropin** (ko-re-on'ik go-nad'o-tro'pin), or hCG, an LHlike hormone that causes the corpus luteum to double in size and continue to secrete estradiol and progesterone. These hormones maintain the uterine lining and prevent menstruation during pregnancy. Secretion of hCG persists for 5 to 6 weeks until the placenta can produce estrogen and progesterone. The hCG excreted early in pregnancy is used for most pregnancy tests.

During the first 6 weeks of pregnancy the corpus luteum, under the influence of hCG, produces estradiol and progesterone. As hCG production drops, the placenta starts to secrete estrogens and progesterone. Placental estrogens are about 10 percent estradiol and 90 percent estriol, a less active kind of estrogen. Secretion of these hormones increases throughout pregnancy, reaching a peak concentration 50 times that of the menstrual cycle and a peak activity 30 times that of the menstrual cycle. The lower activity is due to estriol's weaker effects. A few weeks before the onset of labor, progesterone reaches a peak 10 times that in the menstrual cycle. Typical hormone concentrations in urine during pregnancy are shown in Figure 5.20.

Estrogens and progesterone have special functions during pregnancy. Estrogens cause the uterus, breasts, and external genitalia to enlarge. Progesterone

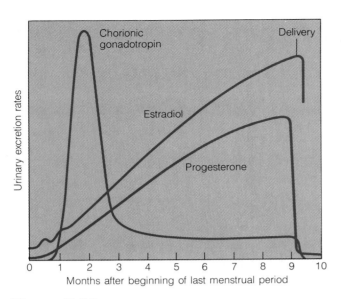

Figure 5.20

Urinary excretion of hormones that maintain pregnancy.

stimulates development of the breasts and maternal placenta and uterine tube secretions. It suppresses uterine muscle contractility.

Relaxin gradually relaxes pelvic ligaments, so that by the time of birth the pelvic outlet is larger and slightly flexible. These changes ease passage of the fetus through the birth canal.

Along with other hormones, hCS stimulates breast development and, like growth hormone, maintains positive nitrogen balance, mobilizes maternal fats for energy, and maintains maternal blood glucose concentration to meet the needs of both mother and fetus.

See Questions—Objectives 15 through 18

Endocrine Functions of Thymus, Pineal Gland, and Other Organs

The **thymus** (thi′mus) **gland,** located below the thyroid gland and between the lungs, functions in early life but atrophies (shrinks) shortly after puberty. It produces **thymosins** (thi′mo-sinz) and is essential for the development of certain lymphocytes and immunological competence (chapter 6).

Questions

Objective 15

(a) What events occur in the ovarian cycle and what is their significance?

(b) How is the ovarian cycle controlled?

Objective 16

(a) What is menarche?

(b) What events occur in the preovulatory phase of the menstrual cycle?

(c) What is the significance of preovulatory events?

Questions Continued

Objective 16 Continued

(d) What events occur during ovulation?

(e) Why is ovulation significant?

(f) What events occur in the postovulatory phase of the menstrual cycle?

(g) What is the significance of postovulatory events?

Objective 17

(a) How is the menstrual cycle regulated?

(b) How is the menstrual cycle disrupted by pregnancy?

(c) What events occur in menopause and how do they cause the menstrual cycle to cease?

Objective 18

(a) How do hormone secretions change in pregnancy?

(b) How is lactation initiated? (Refer back to the discussion of the posterior pituitary gland.)

Questions

Objective 19

(a) What other human hormones have been identified?

(b) What are their main functions?

The **pineal** (pin'e-al) **gland,** derived from neural tissue and found between the cerebral hemispheres, secretes **melatonin** (mel-ah-to'nin) in darkness and **serotonin** in light and appears to be involved in circadian (daily) rhythms. Melatonin secretion is relatively high before age 5 and gradually decreases to its low adult level by the end of puberty. Removing the pineal gland in animals allows premature sexual development and pineal gland abnormalities in male children can lead to premature puberty. These observations suggest that melatonin inhibits sexual maturation and that puberty occurs when the melatonin level fails to suppress pituitary hormones that stimulate the gonads.

At least 10 polypeptide **enteric** (en-ter'ik) **hormones** are released into the blood from various cells in the digestive tract lining. These hormones have as their target organs the liver, pancreas, or the digestive tract itself. Four of the hormones are reasonably well understood. **Gastrin** (gas'trin), from G cells in the stomach lining, stimulates other kinds of stomach cells to release hydrochloric acid and the inactive enzyme pepsinogen. **Secretin** (se-kre'tin), from the duodenum, stimulates release of fluid and bicarbonate from the pancreas. **Cholecystokinin** (ko-le-sis-to-ki'nin), also called CCK or **pancreozymin** (pan-kre-o-zi'min), comes from the duodenum. It stimulates the gallbladder to release bile and the pancreas to release digestive enzymes. **Gut glucagon,** released from the duodenum when significant amounts of carbohydrate are present, stimulates the pancreas to release insulin.

The term hormone originally described secretions of discrete endocrine glands, but most organs are now known to secrete regulatory molecules. **Erythropoietin** (e-rith''ro-poi-e'tin), derived from a blood protein, stimulates bone marrow to produce erythrocytes (chapter 6). After activation in the kidneys, **vitamin D** behaves as a hormone, stimulating synthesis of a calcium carrier protein in lining cells of the small intestine. This protein facilitates calcium absorption. **Atrial natriuretic** (a'tre-al nat-re-ur-et'ik) **hormone** (ANH), found in granules in the cells of the atria of the heart, greatly increases the rate at which fluid is filtered from the blood into the kidney tubules. It probably increases sodium excretion by blocking the transport of sodium back into blood once it enters kidney tubules. ANH relaxes smooth muscle in many blood vessels probably by releasing cGMP.

See Questions—Objective 19

Study and Review

Did you get the essentials?

The following summary contains the basic concepts from chapter 5 except for the key terms that have been omitted. Read the chapter. Then try to fill in the blanks from memory. If you cannot fill in all the blanks, review the chapter paying particular attention to boldface terms.

Endocrine structures secrete _____ into the blood where they travel to and exert their effects on _____ _____ . Compared to the nervous system, the endocrine system produces its effects more _____ but they last a _____ period of time. Hormones in inactive form are called _____ , and some hormones are transported on specific _____ in the blood. Self-regulation of most hormones involves _____ _____ .

Steroid and thyroid hormones act by the _____ _____ mechanism but most other hormones act by the _____ mechanism. In addition, calcium ions and _____ _____ proteins also regulate some cellular processes. The actions of hormones also can be affected by how tightly they bind to _____ , by how much their effect is multiplied via _____ _____ , and by how quickly they are _____ .

Special regulatory molecules called _____ are made and released as needed, often act by altering the _____ concentration, and occur in pairs with _____ actions.

The _____ pituitary gland secretes _____ , _____ , _____ , _____ , _____ , and _____ under the effects of _____ and _____ hormones from the _____ . The _____ pituitary gland is a storage area for _____ and _____ , which are synthesized in the _____ . Of the pituitary hormones those that act on other glands are called _____ hormones. The most common pituitary disorders involve excesses or deficiencies of _____ hormone.

The thyroid gland uses the element _____ and the amino acid _____ to make the hormones _____ and _____ , which control the _____ rate. Secretion of the hormones is regulated by _____ and _____ and by _____ from the hormones themselves.

Calcium metabolism is regulated by _____ from the parathyroid glands and _____ from the thyroid gland. The blood level of calcium regulates hormone secretion by _____ .

Continued on next page

Study and Review

The adrenal gland consists of an inner _____, which secretes _____ and _____, and an outer _____, which secretes _____, _____, and _____. Of these _____ provides negative feedback to the _____ and _____.

The endocrine functions of the pancreas are carried out by alpha cells secreting _____ when blood glucose is _____ and _____ cells secreting _____ when blood glucose is _____.

The ovaries secrete _____, _____, and _____, whereas the testes secrete _____ and _____. Male hormones are regulated by _____ _____.

In the ovarian cycle under the stimulation of _____, follicular cells divide to form _____ _____, one of which becomes a dominant _____ _____, while others degenerate. When mature, this follicle ruptures and the _____ is released from the ovary at _____. This leaves a _____ _____, which secretes the hormones _____ and _____. If the ovum is not fertilized this body degenerates into a _____ _____. Correlated with the ovarian cycle is the _____ cycle. During its _____ phase, menstruation is completed and the uterine lining begins to _____. During its _____, or _____, phase the _____ becomes glandular and secretes _____ and stores _____. The menstrual cycle occurs at approximately _____ intervals from _____ at about age 13 to _____ at about age 50.

Hormones that regulate pregnancy include: _____ _____ _____. _____ is the release of milk from mammary glands. The glands are prepared to secrete milk by the hormones, _____, _____, and _____. Secretion itself is initiated by the infant's _____, which stimulates _____ to send signals to the _____. These signals release _____ which stimulate synthesis of _____.

Other endocrine structures include the _____ (a gland involved in immunity), the _____ gland that probably regulates rhythms, and cells of digestive tract that secrete _____ hormones.

Blood, Gas Exchange, and Immunity

6

Blood

Blood, the fluid pumped through the blood vessels by the heart, consists of 55 to 60 percent liquid **plasma** (plaz′mah) and 40 to 45 percent **formed elements**—red blood cells called **erythrocytes** (er-ith′ro-sītz), white blood cells called **leukocytes** (lu′ko-sītz), and cell fragments called **platelets** (Table 6.1). The presence of formed elements and plasma proteins makes blood about five times as viscous as water. It has a pH of 7.35 to 7.45, depending on how much carbon dioxide it is carrying. Blood performs several important functions (Table 6.2).

Plasma is more than 90 percent water (by weight) and contains proteins, electrolytes (inorganic ions), gases, nutrients, waste products, hormones, and blood clotting factors (Table 6.3). Cholesterol and other blood lipids are transported on various blood proteins. If clotting factors are removed from plasma, the fluid that remains is called **serum** (se′rum). Each liter of plasma contains 60 to 80 grams of protein, mostly made in the liver and composed

Table 6.1
Normal Numbers of Formed Elements in Human Blood

Element	Normal Numbers per Microliter*
Erythrocytes	
Adult male	4.6 to 6.2 million
Adult female	4.2 to 5.4 million
Infant and child	4.5 to 5.0 million
Neonate	5.0 to 5.1 million
Adult at 4,000 meters	6.0 to 8.0 million
(altitude)	
Leukocytes	5,000 to 9,000
Percent of all leukocytes	
Neutrophils	50%-70%
Eosinophils	1%-4%
Basophils	0.1%
Monocytes	2%-8%
Lymphocytes	20%-40%
Platelets	250,000 to 300,000

*1 microliter = 1/1,000,000 liter = 1 cubic millimeter

Table 6.2
The Functions of Blood

Functions	Examples
Transport	
Nutrients	Glucose, amino acids, vitamins, and minerals are carried to nearly all cells.
Oxygen	Oxygen is transported from the lungs to all cells of the body by erythrocytes.
Carbon dioxide	Carbon dioxide is transported from cells to the lungs.
Waste products	Urea, uric acid, and other waste products are carried from cells to the liver, lungs, and kidneys.
Hormones	Hormones are carried from endocrine cells to target cells.
Protection	
Invasion of microorganisms	Leukocytes engulf and destroy some microorganisms. Antibodies produced by certain leukocytes react with specific antigens, such as those on microorganisms.
Damage from toxic substances	Some toxic substances are carried to the liver, where they are rendered nontoxic, and then to the kidneys, where they are excreted. Some toxic substances are inactivated by antibodies called antitoxins.
Blood loss	The clotting mechanism seals off blood vessels. Substances released from platelets initiate clotting after tissue injury.
Regulation	
Water content of cells	Centers in the hypothalamus detect changes in the blood osmotic pressure and indirectly cause the blood to gain or lose water, thereby maintaining nearly constant water content of cells.
pH of body fluids	Certain neural centers detect carbon dioxide concentration and regulate the pH of body fluids indirectly. (An excess of carbon dioxide causes acid to accumulate.) Buffers in blood help to keep blood pH nearly constant.
Body temperature	Temperature sensors in the hypothalamus detect changes in body temperature and act to increase or decrease blood flow to the skin, thereby increasing or decreasing heat loss, respectively.

Table 6.3
Constituents of Plasma

Constituent	Amount/Concentration*
Water	90% of plasma
Electrolytes (inorganic)	<1% of plasma
Na⁺	142 mEq/l (142 mmol/l)
K⁺	4 mEq/l (4mmol/l)
Ca²⁺	10 mEq/l (5mmol/l)
Mg²⁺	3 mEq/l (1.5 mmol/l)
Cl⁻	103 mEq/l (103 mmol/l)
HCO₃⁻	27 mEq/l (27 mmol/l)
HPO₄²⁻	4 mEq/l (2mmol/l)
SO₄²⁻	1 mEq/l (0.5 mmol/l)
Gases	<1% of plasma
CO₂ (mostly as HCO₃⁻)	60 ml/100 ml
O₂	0.2 ml/100 ml
N₂	0.9 ml/100 ml
Nutrients	<1% of plasma
Glucose and other carbohydrates	100 mg/100 ml
Amino acids	40 mg/100 ml
Lipids	500 mg/100 ml
Cholesterol	150–250 mg/100 ml
Vitamins	Traces
Trace elements	Traces
Waste products	<1% of plasma
Urea	<20 mg/100 ml
Creatinine	<1 mg/100 ml
Uric acid	5 mg/100 ml
Bilirubin	0.2–1.2 mg/100 ml
Proteins	6% of plasma (2.5 mmol/l)
Albumins	4.5 g/100 ml
Globulins	2.5 g/100 ml
Fibrinogen	0.3 g/100 ml
Hormones	Very small amounts
pH	7.35–7.45

*Concentrations for some substances are expressed in both milliequivalents (mEq) and millimoles (mmol). One millimole is one-thousandth of a gram molecular weight of a substance. For substances that have a valence of 1, mEq and mmol are equal; for substances that have a valence of 2, 2 mEq equals 1 mmol.

of about 55 percent albumin, 38 percent globulin, and 7 percent fibrinogen, an important protein in the blood clotting mechanism.

Albumins (al-bu′minz) are large, globular proteins negatively charged at normal blood pH. They transport calcium ions, relatively water-insoluble blood components, and drugs such as aspirin, barbiturates, and digitalis (a heart drug). Albumins exert osmotic pressure that helps hold water in the blood.

Globulins (glob′u-linz), proteins larger than albumins, vary in size and function. **Alpha globulins** include **haptoglobulin** that transports hemoglobin from erythro-cytes, **ceruloplasmin** (se-roo″lo-plaz′min) that transports copper, **prothrombin** (pro-throm′bin) and other blood clotting factors, and some lipid transport proteins. **Beta globulins** include **transferrin** (tranz-fer′in) that transports iron and some other lipid transport proteins. **Gamma globulins,** also called immunoglobulins or antibodies, are concerned with immunity. They are synthesized by white blood cells called **lymphocytes** (lim′fo-sītz).

Among formed elements, oxygen-carrying erythrocytes are by far the most abundant; they make up about 40 percent of the total blood volume in females and about 45 percent in males. Leukocytes are classified as granular if their cytoplasm contains easily stained granules or agranular if it does not.

See Questions—Objective 1

Erythrocytes, Hemoglobin, and Iron

Erythrocytes are biconcave disks with a large surface area that facilitates gas diffusion into and out of the cell. The erythrocyte plasma membrane is rendered flexible, yet durable, by a meshlike network of proteins beneath the membrane. These proteins include actin filaments, actin-binding proteins, **spectrin** (spek′trin), and **ankyrin** (an-ki′rin) that crosslinks spectrin molecules. These proteins enable erythrocytes to withstand elongation as they pass through narrow capillaries and to regain a normal shape in larger blood vessels.

Each erythrocyte contains about 280 million molecules of **hemoglobin** (hem-o-glo′bin), a pigment that is bright red when oxygenated. It gives erythrocytes the name "red blood cells." Hemoglobin (Figure 6.1) consists of the protein **globin** with associated nonprotein pigment units called **heme** (hēm). Globin contains four polypeptides—two identical alpha chains and two identical beta chains. Imbedded in each chain is a molecule of heme, a chemically complex molecule called a porphyrin. Each heme contains an iron ion that can bind an oxygen molecule (O_2), so each hemoglobin molecule can carry four O_2 molecules.

Hemoglobin in erythrocytes increases the oxygen-carrying capacity of blood to about 70 times the amount that can dissolve in plasma. Hemoglobin also transports carbon dioxide, though much carbon dioxide is transported as bicarbonate. Because carbon dioxide binds to globin, hemoglobin can carry both oxygen and carbon dioxide simultaneously. However, under normal physiological conditions erythrocytes traveling from the lungs to other tissues carry large quantities of oxygen and almost no carbon dioxide. In the tissues, they lose oxygen and acquire carbon dioxide, but they still carry more oxygen than carbon dioxide on their way back to the lungs.

Questions

Objective 1

(a) What use might be made of knowing normal values for blood constituents?

(b) What are the major functions of blood?

(c) What are the major components of plasma?

(d) How does serum differ from plasma?

(e) What kinds of formed elements are found in blood?

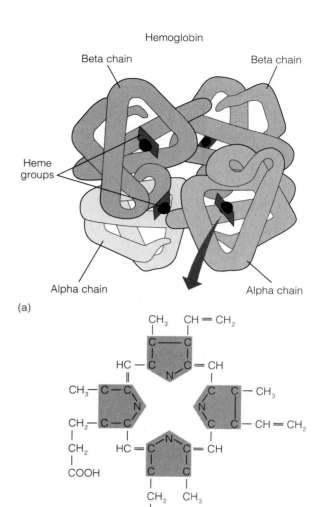

(a)

(b)

Figure 6.1

(*a*) The three-dimensional structure of a hemoglobin molecule. (*b*) The molecular structure of heme is enlarged. Note the iron atom in the center of the molecule.

Iron is essential, not only for hemoglobin synthesis, but also for synthesis of myoglobin in muscle, and cytochromes that function in oxidative metabolism. Iron metabolism begins with oxidation of Fe^{3+} in some foods to Fe^{2+} and the binding of Fe^{2+} to the protein **gastroferrin** (gas-tro-fer′in) in the stomach. Gastroferrin ferries iron to the small intestine, where it is transferred to a carrier protein in the intestinal lining cells and absorbed into the blood. Once absorbed, iron is oxidized to Fe^{3+} and combines loosely with a plasma protein **transferrin.** Of the body's 4 grams of iron, over two-thirds is in hemoglobin, a small amount in cyclochromes and myoglobin, and the rest is stored in the liver in **ferritin** (fer′it-in). Ferritin consists of a protein **apoferritin** (ap-o-fer′it-in) with iron loosely bound to it.

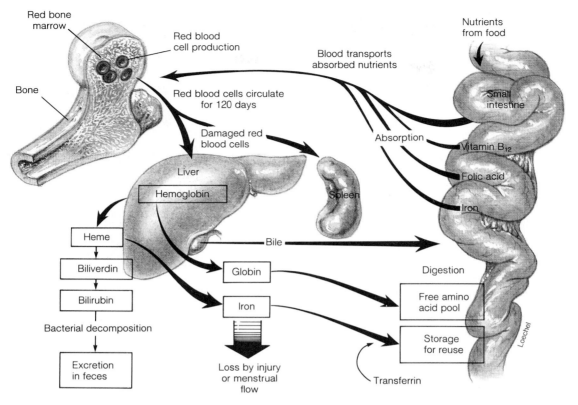

Figure 6.2

Formation and destruction of erythrocytes, including
degradation of hemoglobin.

When the plasma iron concentration drops, iron
from ferritin is released and carried by transferrin to bone
marrow and other tissues. Under normal conditions small
amounts of iron (about 0.6 mg) are lost daily through the
feces. Bleeding due to injury or excessive menstrual flow
can seriously deplete iron stores. When excess iron accumulates in the body it is deposited in cells in insoluble
granules of **hemosiderin** (hem-o-sid′er-in). Such deposits
result from many blood transfusions and not from excessive dietary iron. Dietary iron can be increased by eating
iron-enriched cereals, green leafy vegetables, and red
meats.

Iron absorption appears to be regulated mainly
by the amount of iron already available in the body, though
the details of regulatory processes are not fully understood. When apoferritin is saturated with iron, it cannot
accept iron from transferrin and transferrin, too, becomes
saturated. When transferrin is saturated, it cannot accept
iron from the intestinal mucosa and fails to be absorbed
until some transferrin becomes available.

Erythrocytes are constantly dying and being replaced at a rate of about 2.5 million per second (Figure
6.2). Tissue oxygenation is the major regulator of erythrocyte production. Disorders that reduce blood flow to the
tissues, impair blood oxygenation, and reduce erythrocytes and hemoglobin, or living at high altitudes where

the atmospheric oxygen concentration is low stimulate
erythrocyte production. The hormone **erythropoietin**
(er″ith-ro-poi-e′tin) is synthesized mainly by the kidney
and secreted when blood oxygen decreases. It induces bone
marrow cells to divide and form erythrocytes within 5 days.
Iron and vitamins such as vitamin B_{12} and folic acid are
necessary for the production of new erythrocytes.

When a person donates a pint (475 ml) of blood
about 10 percent of the total blood volume is removed.
Through a variety of homeostatic mechanisms, the body
replaces lost volume in a matter of hours, plasma proteins
in one to three days, and lost cells and platelets in three
to four weeks.

Erythrocyte destruction is less wasteful than it
might appear. Phagocytic cells in the liver, spleen, and bone
marrow digest erythrocytes with inflexible membranes,
and separate hemoglobin into heme and globin. They
digest the globin to amino acids for new proteins and the
heme into iron and porphyrin. Iron is reused in making
erythrocytes and porphyrin is degraded to the green pigment **biliverdin** (bil-e-ver′din) and then to the yellow-
orange pigment **bilirubin** (bil-e-roo′bin). Bilirubin is secreted in bile and excreted in the feces.

See Questions—Objective 2

Questions

Objective 2

(a) Describe the chemical properties of erythrocytes.

(b) How are components of erythrocytes reused?

(c) What mechanisms rid the body of nonreusable components?

(d) How is iron absorbed, transported, stored, and excreted?

(e) What factors control the number of erythrocytes in blood?

Leukocytes and Platelets

Leukocytes, the largest of blood cells, include **neutrophils** (nu'tro-filz), **basophils** (bas'o-filz), **eosinophils** (e-o-sin'o-filz), **monocytes** (mon'o-sitz), and **lymphocytes.** Hormones similar to erythropoietin are now known to stimulate leukocyte production, with development of each kind of cell being controlled by a different hormone.

Neutrophils, the most numerous of all leukocytes, are produced and released into the blood in greater numbers after injury or infection where they are key participants in the **inflammatory process.** Eosinophils, which increase in number during allergic reactions, probably detoxify foreign substances, turn off inflammatory reactions, and break down products of immune reactions. Basophils synthesize and secrete histamine and heparin, both of which participate in the body's defense mecha-

nisms. **Histamine** (his'tam-in) initiates the inflammatory response and **heparin** (hep'ar-in) inhibits blood clotting, especially where clots might occlude vessels. Monocytes are especially adept at phagocytizing large particles of debris. When they enter tissues they are transformed into macrophages, which have two important functions. They digest debris left from neutrophils that have died after ingesting smaller particles, and they stimulate lymphocytes to participate in immunologic reactions. Lymphocytes are found mainly in the lymph nodes, spleen, and tonsils. They function in immunity. Lymphocytes and monocytes enter blood along with lymph that has collected in lymph vessels.

Platelets, also called **thrombocytes** (throm'bo-sitz), are fragments of megakaryocytes only 2 to 4 μm in diameter. They are replaced every 5 to 10 days. Platelets contain enzymes in mitochondria and can oxidatively metabolize glucose for energy. Their main functions are to plug leaks in small blood vessels and to release substances essential for blood clotting.

See Questions—Objective 3

Hemostasis

The arrest of bleeding is called **hemostasis** (hem-o-sta′sis) and includes three processes—blood vessel spasm, platelet plug formation, and blood clotting. Related to hemostasis is the removal of blood clots after they have arrested bleeding.

When a blood vessel is injured, the smooth muscle in its walls contracts and sticky cells of the vessel lining adhere to one another. If a blood vessel lining is disrupted and underlying collagen is exposed, platelets adhere to the collagen and to each other to form a **platelet plug** at the site of injury. Cytoplasmic vacuoles in adhering platelets release substances such as ADP and serotonin, which increase platelet stickiness. This causes circulating platelets to adhere to those already attached to the collagen. In other words, stickiness produces more stickiness by positive feedback and clumps of platelets rapidly increase in size. Platelets also release epinephrine; it and serotonin cause further blood vessel constriction.

A pair of prostaglandins also contributes to controlling blood flow and bleeding. Platelets release a prostaglandin called thromboxane A_2 (TXA_2), which acts for only about 30 seconds to cause platelet aggregation and vasoconstriction. Arterial lining cells release an antagonistic prostaglandin called PGI_2 (prostacyclin), which acts for about 2 minutes to prevent platelet aggregation and cause vasodilation. TXA_2 may go too far and contribute to emboli (clots in blood vessels), atherosclerosis, and heart attacks. PGI_2 seems to protect against these disorders and probably reduces the likelihood of metastasis of cancer cells by preventing them from adhering to membrane surfaces.

Bleeding from small injuries and breaks in capillaries can be stopped by vessel spasms and platelet adhesion, but more extensive injuries require blood clotting. Blood clotting, which can be initiated by extrinsic or intrinsic mechanisms, involves several steps that end in blood clot formation (Figure 6.3).

Numerous blood clotting factors circulate in blood plasma at all times in inactive forms (Table 6.4). Most are proteins and they include some protease enzymes that activate other proteins. Some are cofactors, substances that promote the formation of enzyme-substrate complexes and subsequent reactions. Several factors are synthesized in the liver with the aid of vitamin K. Without vitamin K, these protein factors lack an important carboxyl group and function poorly in clotting reactions.

Questions
Objective 3

(a) What are the characteristics and relative numbers of the different kinds of leukocytes?

(b) What are the major functions of the different kinds of leukocytes?

(c) What are platelets and what are their functions?

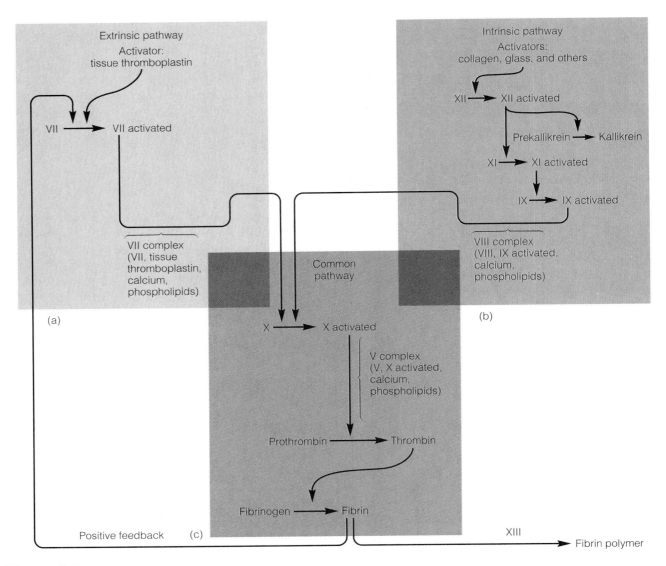

Figure 6.3

The mechanisms of blood coagulation: (*a*) extrinsic pathway,
(*b*) intrinsic pathway, and (*c*) final common pathway.

Tissue damage activates the **extrinsic mechanism** of blood clotting. In addition to causing blood vessel spasm and platelet plug formation, tissue damage causes injured cells to release a membrane lipoprotein called **tissue thromboplastin** (throm-bo-plas'tin). With calcium ions and other factors, tissue thromboplastin activates factor VII. Active factor VII in turn activates factor X and the blood clotting mechanism is underway. This mechanism is said to be extrinsic because the initiating factor comes from injured cells anywhere in the body.

Blood coming in contact with a negatively charged surface, such as collagen, within a blood vessel wall activates the **intrinsic mechanism** of blood clotting. Steps in this mechanism are as follows: Factor XII converts prekallikrein to **kallikrein** (kal-e-kre'in). Kallikrein and

kininogen from platelets accelerate factor XII activation; with factor XII they activate factor XI. Active factor XI activates factor IX; active factor IX with factor VIII and calcium ions activate factor X.

Crossovers occur between the extrinsic and the intrinsic pathways. Activated factor XII from the intrinsic system activates factor VII of the extrinsic system. Also factor VII activated in the extrinsic pathway activates factor IX of the intrinsic pathway. Such crossovers prevent uncontrolled bleeding in some clotting factor deficiencies by activating whichever clotting mechanism is functional.

Once factor X is activated by either mechanism the pathways converge. Factor X converts the blood protein **prothrombin** to the active enzyme **thrombin.** Thrombin

Table 6.4
Blood Coagulation Factors

Factor Number	Name	Nature and Origin	Function
I	Fibrinogen	Protein synthesized in liver	Precursor to fibrin
II	Prothrombin	Enzyme synthesized in liver in presence of vitamin K	Precursor to thrombin
III	Tissue thromboplastin	Lipoprotein released from tissues	Activates factor VII
IV*	Calcium ions	Inorganic ion present in plasma	Necessary for reactions in most stages of process
V	Proaccelerin or labile factor	Protein synthesized in liver	Required for extrinsic and intrinsic mechanisms
VI	Number no longer used; substance now shown to be activated factor V		
VII	Proconvertin	Enzyme synthesized in liver in presence of vitamin K	Necessary for extrinsic mechanism, activates factor X
VIII	Antihemophilic factor	Globulin synthesized in liver; absent in inherited disorder hemophilia A	Necessary for intrinsic mechanism
IX	Plasma thromboplastin (or Christmas factor)	Enzyme synthesized in liver; absent in inherited disorder hemophilia B	Necessary for intrinsic mechanism
X	Stuart-Power factor	Enzyme synthesized in liver in presence of vitamin K	Necessary for extrinsic and intrinsic mechanisms
XI	Plasma thromboplastin antecedent	Enzyme synthesized in liver	Necessary for intrinsic mechanism
XII	Hageman factor	Enzyme synthesized in liver	Necessary for intrinsic mechanism and to activate plasmin
XIII	Fibrin stabilizing factor	Enzyme found in platelets and plasma	Crosslinks filaments and makes fibrin polymer
	Prekallikrein	Enzyme activated by small amount of factor XII	Activates more factor XII; accelerates cascade
	High-molecular weight kininogen	Cofactor in plasma	Assists kallikrein in activating factor XII
Pf$_1$	Platelet accelerator	Platelets; same as factor V	Accelerates action of platelets
Pf$_2$	Thrombin accelerator	Platelets	Accelerates thrombin formation
Pf$_3$	Platelet thromboplastic factor	Platelets	Necessary for intrinsic mechanism
Pf$_4$	Platelet factor 4	Platelets	Binds heparin, a natural anticoagulant, during clotting

*IV no longer used; referred to simply as Ca^{2+}

acts on another blood protein **fibrinogen,** converting it to **fibrin.** Polymerization of fibrin occurs by the overlapping of molecules and formation of crosslinks between them. Crosslinking is established by an enzyme called **fibrin stabilizing factor,** or factor XIII. Polymerized fibrin is a fibrous material that forms a loose meshwork over the injured area. Blood cells become trapped in the meshwork, reinforce the platelet plug, and close off the opening. Thus, a blood clot consists of fibrin, blood cells, and platelets.

The sequence of reactions in blood clotting constitutes a cascade—each step in the process causes production of more molecules of the product of the next step. Also, once thrombin is formed, it activates more factor VII, thereby acting as a positive feedback mechanism to accelerate the clotting process.

In addition to initiating coagulation, mechanisms also exist to inhibit it after bleeding has been arrested and to prevent it unless a vessel has been damaged. These mechanisms include activation of a protein called **antithrombin** (ant-e-throm'bin) by heparin. Giving heparin to people prone to form thrombi makes use of this mechanism. Clotting also is limited to some degree by clotting factors being used up in reactions and by the action of

PGI₂ (prostacyclin), which inhibits platelet aggregation. The competing actions of PGI_2 and thromboxane help to regulate the clotting process.

Anticoagulants (ant-e-ko-āg′u-lantz) are substances that reduce the clotting ability of blood by interfering with reactions in the clotting mechanism. Some anticoagulants are used in the body and others are used only in the laboratory. **Heparin,** a complex polysaccharide produced by both mast cells and basophils, reduces the blood's ability to clot. It interferes with activation of thrombin and certain other clotting factors. **Coumarin** (koo′mar-in), an anticoagulant derived from sweet clover, competes with vitamin K and interferes with synthesis of factors V and VII in the liver. Similar drugs such as **dicumarol** (di-koo′mar-ōl) and **coumadin** (koo′mah-din), or **warfarin** (war-far′in), also reduce the likelihood of blood clots within blood vessels. Coumarin derivatives take up to 2 days to act but have longer lasting effects than heparin and can be taken orally.

The laboratory anticoagulants, including heparin, prevent blood samples from clotting before tests can be performed. Sodium citrate and ammonium oxalate bind to calcium ions so they cannot facilitate reactions in the clotting process.

Blood clots are temporary structures that seal off a damaged area until healing can take place. After half an hour or more, the clot retracts and becomes smaller and more dense, probably by the action of platelets trapped in the clot. As the fibrin filaments gather around the platelet aggregation, cytoplasmic processes of platelets attach to fibrin and pull the fibers closer together. This is called **clot retraction.** When it occurs in a test tube, one can observe fluid squeezed from the clot.

At the same time a clot is forming, the inactive enzyme **plasminogen** (plaz-min′o-jen) is deposited in the clot. When plasminogen is activated, it becomes **plasmin** (plaz′min) and begins to digest the clot. Several factors are known to participate in activating plasmin. Activated factor XII not only participates indirectly in the activation of thrombin, it also is broken down to form proteolytic fragments called **prekallikrein activators** (Figure 6.4).

The protein **tissue plasminogen activator** (tPA) appears to be released from the epithelial linings of blood vessels and other tissues in response to injury and other stimuli. tPA causes digestion of fibrin specifically by activating plasminogen.

In addition to the above mechanisms, thrombin also activates plasmin and plasmin breaks active factor XII into prekallikrein activators, thereby accelerating clot digestion by positive feedback. Much of the plasmin is incorporated into the clot where it degrades both fibrinogen and fibrin. Plasmin is a slow-acting proteolytic enzyme that gradually dissolves away the clot while tissue repair is taking place. The products of the action of plasmin are called **fibrin degradation products** (FDP), or fibrin split

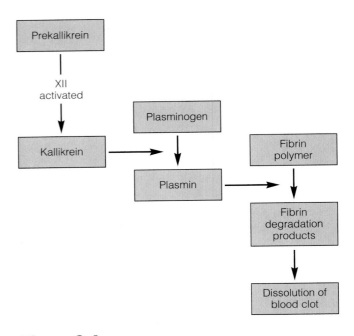

Figure 6.4

The mechanism by which blood clots are dissolved.

products. Maintaining an intact circulatory system through which blood flows freely requires a continuous balance between hemostasis and fibrinolysis.

Treatment for thrombi (fixed clots in blood vessels) and emboli (circulating clots) often involves the use of substances to dissolve the clots and thereby restores circulation to a tissue before severe ischemia and cell death result. **Streptokinase** (strep-to-ki′nās), an enzyme made by certain hemolytic streptococci, digests most any kind of protein. Given intravenously, it digests blood clots in coronary vessels especially if given within a few hours of the onset of symptoms. Tissue plasminogen activator appears to act faster and more specifically than streptokinase in dissolving blood clots. Its amino acid sequence is known, and it is made by genetically engineered bacteria.

Clotting disorders include hemophilia and disseminated intravascular coagulation. **Hemophilia** (hēm-o-fil′e-ah) refers to a group of inherited diseases, each caused by a different genetic defect that prevents the synthesis of a coagulation factor. The most common type, hemophilia A, which accounts for about 80 percent of all cases, is due to the absence of factor VIII. The next most common type, Christmas disease or hemophilia B, is due to the absence of factor IX (Christmas factor, named for the patient in whom it was identified). Both are inherited as sex-linked recessive characteristics and affect males almost exclusively. von Willebrand's disease, caused by an autosomal recessive gene, is due to low factor VIII activity. The factors most often missing in hemophilia function in the intrinsic mechanism, and symptoms such as ease of bruising and joint pain result from internal bleeding. Small breaks in blood vessels in joints where the vessels are continually subjected to stress are not repaired as they are when the intrinsic clotting mechanism is operating normally. Though the extrinsic mechanism usually is functional, it is not activated by internal vessel damage.

In **disseminated intravascular coagulation** (DIC), many small clots are disseminated throughout the body and hemorrhages at other sites occur simultaneously. It occurs in the presence of some bacterial toxins, complications of childbirth, burns, and various kinds of trauma.

Though the reasons for clot formation are not well understood, such clots block small blood vessels and impede blood flow. They also deplete the supply of clotting factors so hemorrhage occurs at some sites while clots are present at others. Ultimately, treatment should be directed toward alleviating the cause, but immediate treatment may be necessary to avert death. Whether to treat the patient for hemorrhage or clotting is a matter of some controversy.

See Questions—Objectives 4 and 5

Questions

Objective 4

(a) What is hemostasis and what factors contribute to it?

(b) Explain the similarities and differences in the extrinsic and intrinsic blood clotting mechanisms.

(c) How is clotting prevented when it is not needed?

Objective 5

(a) How are clots dissolved?

(b) What are the effects of disorders in hemostasis?

Gas Exchange

Gas exchange depends on diffusion of gases across membranes in the lungs and blood and between blood and the cells of various tissues. How much and how fast a gas diffuses depends on the partial pressures of the gases.

The earth's atmosphere is a mixture of gases—nearly four-fifths nitrogen, about one-fifth oxygen, and small amounts of carbon dioxide and other gases. Each gas in a mixture exerts a pressure, its **partial pressure,** independently of other gases. The total atmospheric pressure is the sum of the partial pressures of gases in the atmosphere. Gases dissolve in blood and tissue fluids in proportion to their partial pressures and their solubility in the liquid.

Dry atmospheric air at sea-level conditions is 20.9 percent oxygen and 0.04 percent carbon dioxide and exerts a total pressure of 760 torr. From this information, we can determine that the partial pressure of oxygen is 158 torr (20.9% of 760) and the partial pressure of carbon dioxide is 0.3 torr (0.04% of 760). In addition, nitrogen exerts a pressure of about 596 torr and water vapor about 6 torr (more in humid air). Nitrogen is breathed in and out and diffuses into and out of the blood but is not directly involved in normal respiration.

The different partial pressures of oxygen and carbon dioxide in alveoli, blood, and tissues, create diffusion gradients. Net diffusion of gases is down these gradients from higher to lower partial pressure. Diffusion itself changes partial pressures and gradients. Net diffusion is greater and faster in large gradients than in small ones.

Gas exchange occurs continuously across respiratory membranes, with alveolar gas losing oxygen to the blood and gaining carbon dioxide from it. Gas in respiratory passageways at the end of inspiration never reaches the respiratory membranes and remains unchanged. The composition of inhaled, alveolar, and exhaled gases is summarized in Table 6.5.

To follow the course of diffusion in the lungs, let us consider the average partial pressures of the gases in the alveoli and in the blood (Figure 6.5). (We refer to average pressures because pressures are constantly changing.) The average partial pressure of oxygen, P_{O_2}, is 100 torr in the alveoli and 40 torr in blood entering alveolar capillaries. Net oxygen diffusion is out of the alveoli and into the blood. The partial pressure of carbon dioxide, P_{CO_2}, is 40 torr in the alveoli and 45 torr in blood entering alveolar capillaries. Net carbon dioxide diffusion is out of the blood and into the alveoli. As diffusion proceeds, blood oxygen increases to a maximum and blood carbon dioxide decreases to a minimum as blood leaves the alveolar capillaries. Partial pressures of gases as a result of gas exchange in lungs and tissues are summarized in Table 6.6.

Table 6.5
Composition of Inhaled, Alveolar, and Exhaled Gases (expressed as partial pressures in torr)*

Gas	Inhaled	Alveolar	Exhaled
Oxygen	158.0	100.0	116.0
Carbon dioxide	0.3	40.0	26.8
Nitrogen	596.0	573.0†	570.2
Water vapor	5.7	47.0	47.0
Total	760.0	760.0	760.0

*Partial pressures are given for sea-level conditions. Above sea level, the barometric pressure decreases and so does the partial pressure of the gases.

†The decrease in the partial pressure of nitrogen is due mainly to the relative increase in water vapor and not to the removal of nitrogen in the gas exchange process.

See Questions—Objectives 6 and 7

Transport of Gases

Once oxygen enters the blood it is transported throughout the body. As oxygen is used and carbon dioxide released in cellular respiration, carbon dioxide enters the blood and is transported to the lungs. Special mechanisms are involved in the transport of both oxygen and carbon dioxide.

Oxygen entering blood first dissolves in the plasma, quickly diffuses into erythrocytes, and then combines with hemoglobin, forming **oxyhemoglobin.** As some oxygen moves to hemoglobin, more can dissolve in plasma. As much as 99 percent of blood oxygen is bound to hemoglobin in erythrocytes, but the quantity varies depending on how much hemoglobin the blood contains. Because of hemoglobin, blood typically carries about 70 times as much oxygen (200 ml/l) as can dissolve in plasma (3 ml/l). In tissue capillaries, this process is reversed. As oxygen diffuses into interstitial fluid, lowering the plasma P_{O_2}, hemoglobin continuously releases more oxygen into the plasma.

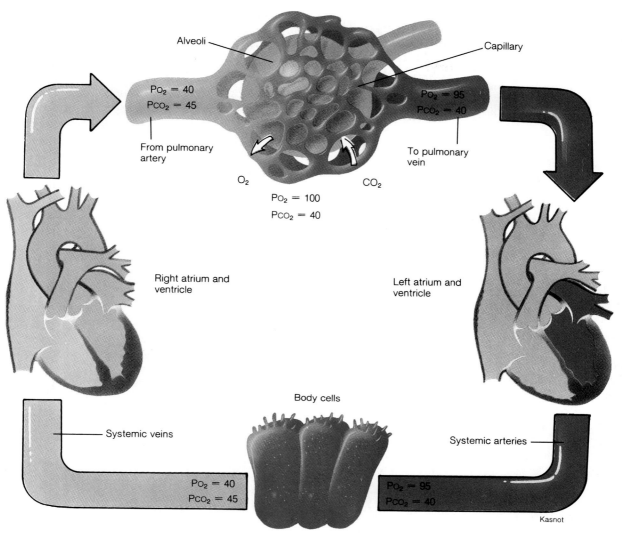

Alveoli

Capillary

$P_{O_2} = 40$
$P_{CO_2} = 45$

$P_{O_2} = 95$
$P_{CO_2} = 40$

From pulmonary
artery

To pulmonary
vein

O_2

CO_2

$P_{O_2} = 100$
$P_{CO_2} = 40$

Right atrium and
ventricle

Left atrium and
ventricle

Body cells

Systemic veins

Systemic arteries

$P_{O_2} = 40$
$P_{CO_2} = 45$

$P_{O_2} = 95$
$P_{CO_2} = 40$

Kasnot

Figure 6.5

The diffusion of gases in external and internal respiration, as
partial pressures of each gas (in torr).

Table 6.6

Average Partial Pressures (in torr)
of Oxygen and Carbon Dioxide

Gas	Atmosphere	Alveoli	Arterial Blood	Cell	Venous Blood
P_{O_2}	158	100	95	25	40
P_{CO_2}	0.3	40	45	46	45

When completely saturated, each hemoglobin
molecule carries four molecules of oxygen. Oxygen combines with and is released from hemoglobin as shown in
the following equation:

$$Hb + O_2 \rightleftarrows HbO_2$$

deoxygenated   oxygen   oxyhemoglobin
hemoglobin

(Deoxygenated hemoglobin carries H^+ instead of oxygen.)
This reaction goes to the right in the lungs during oxygen
loading and to the left in the tissues during oxygen un-
loading. When P_{O_2} is high and P_{CO_2} is low, as in the alveolar capillaries, hemoglobin readily combines with

How can partial pressures help to explain gas exchange?

Objective 7

(a) Where and how does gas exchange occur in the lungs?

(b) How does gas exchange in other tissues differ from that in the lungs?

molecules. The degree of saturation increases as the P_{O_2} increases, markedly up to P_{O_2} of 50 torr and to a lesser degree at higher partial pressures. Changes in pH and temperature change the molecular shape of hemoglobin and alter its affinity for oxygen (Figure 6.6). Lowering the pH or raising the temperature decreases affinity; conversely raising the pH or lowering the temperature increases affinity. The former occurs in metabolically active tissues and the latter in resting tissues.

Metabolically active cells such as contracting muscles produce lactic acid and give off heat. Blood passing through such tissues becomes more acidic and warmer and its hemoglobin releases more oxygen. Erythrocytes contain the enzyme carbonic anhydrase, which causes carbon dioxide entering the cells to combine with water. Some of the resulting carbonic acid ionizes:

$$CO_2 + H_2O \rightleftarrows H_2CO_3 \rightleftarrows H^+ + HCO_3^-$$

As the above reaction goes to the right, H^+ lowers the blood pH and hemoglobin loses more oxygen. All these processes increase oxygen availability in plasma near metabolically active cells. They also act locally to assure that hemoglobin releases oxygen in proportion to the metabolic activity of tissues.

Finally, 2,3-diphosphoglycerate (DPG), a metabolic product of erythrocytes, binds to hemoglobin and causes it to release oxygen. This process occurs continuously, assuring some oxygen release regardless of the local pH or temperature.

Carbon dioxide (CO_2) is transported in three ways—dissolved in plasma, on globin, and as bicarbonate (Table 6.7). CO_2, being about 20 times more water-soluble than oxygen, diffuses rapidly across interstitial fluid into blood and can be carried in plasma without a carrier as can hemoglobin. Some CO_2 enters erythrocytes and binds to globin, forming **carbaminohemoglobin** (kar-bam″in-o-hēm-o-glo′bin). Carbonic anhydrase causes the remaining CO_2 to combine with water to form carbonic acid. (See equation above.) As some carbonic acid ionizes, H^+ binds to negative charges on globin and bicarbonate ions diffuse into plasma. Some H^+ binds to plasma proteins.

Exposed amino acids of globin in desaturated (deoxygenated) hemoglobin bind both H^+ and CO_2. With some O_2 bound to heme, hemoglobin can carry both gases simultaneously.

About 70 percent of HCO_3^- formed in erythrocytes diffuses into the plasma, creating an electrical gradient across erythrocyte membranes. Electrical balance between erythrocytes and plasma is restored by **chloride shift,** the diffusion of chloride ions into erythrocytes.

When blood reaches alveolar capillaries, CO_2, HCO_3^-, and H^+ movements are reversed. CO_2 diffuses from plasma into the alveoli. Carbonic acid reforms and

available oxygen. When hemoglobin reaches the tissues, where P_{CO_2} is higher and P_{O_2} is lower, hemoglobin readily releases oxygen.

Hemoglobin is 100 percent saturated with oxygen when every hemoglobin molecule has bound four oxygen

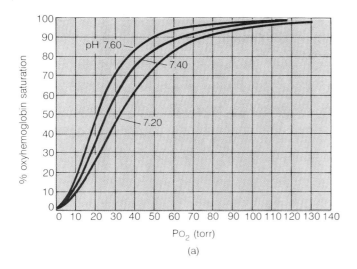

(a)

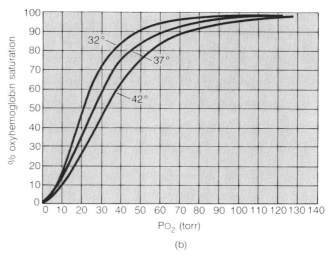

(b)

Figure 6.6

(*a*) The oxyhemoglobin dissociation curves for pH. These curves illustrate two principles: (1) the percent saturation of hemoglobin increases with P_{O_2}, rapidly up to P_{O_2} of 50 torr and more slowly at higher P_{O_2}; and (2) pH affects the percent saturation at any P_{O_2}. The curve labeled pH 7.4 is the normal dissociation curve. At pH 7.6, hemoglobin is more saturated at any given P_{O_2} and at pH 7.2, it is less saturated at any given P_{O_2}. (*b*) The oxyhemoglobin dissociation curves for temperature (Celsius). These curves illustrate the relationship between P_{O_2} and percent saturation of hemoglobin, and show the effect of temperature on hemoglobin saturation. The curve labeled 37° is the normal curve. At 42° C, hemoglobin is less saturated at any P_{O_2}, and at 32° C, hemoglobin is more saturated at any P_{O_2}. (*c*) Effects of 2,3-DPG on hemoglobin saturation.

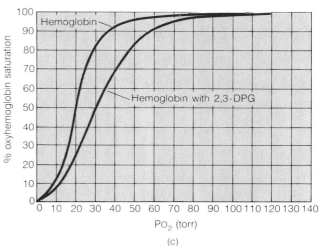

(c)

carbonic anhydrase breaks it into CO_2 and H_2O. Hemoglobin also releases H^+ and CO_2 and becomes ready to carry more oxygen. Carbon dioxide diffusion continues until the P_{CO_2} in the blood and alveoli are equal.

See Questions—Objective 8

Chemical Regulation of Respiration

When the partial pressures of CO_2 or O_2 or the concentration of H^+ in the blood change, chemical regulation of breathing can occur. **Chemoreceptors** that detect such changes are found in **carotid bodies** and in **aortic bodies** in arterial walls. Chemoreceptors are stimulated by an increase in P_{CO_2} or H^+ in arterial blood, and under some conditions by a decrease in P_{O_2}. When stimulated, they send signals to inspiratory neurons and increase the fre-

Table 6.7
Transport of Carbon Dioxide

Mechanism	Percent
Dissolved gas	5–7
Bicarbonate ions	70–72
Associated with hemoglobin and plasma proteins	22–23
Total	100

quency of breathing. This increases *both* the rate at which O_2 enters the blood and the rate at which CO_2 leaves it.

Chemical regulation of respiration also occurs in a **chemosensitive area** near the ventral surface of the medulla. This area responds directly and quickly to increased H^+ in cerebrospinal fluid. It also responds to increased

Questions

Objective 8

(a) What factors affect the transport of oxygen in the blood?

(b) How is oxygen released from the blood?

(c) By what mechanisms is carbon dioxide transported in the blood?

Questions

Objective 9

How are chemical signals involved in regulating breathing?

Basic Concepts of Body Defense

In a general sense, **immunity** (*immune,* safe or free of burden) refers to all physiological mechanisms that defend against infectious agents. **Nonspecific defenses** involve natural barriers such as the skin and its antimicrobial secretions, inflammation, interferon, and complement. **Specific defenses** respond to an **antigen** (an'te-jen), a foreign molecule, by producing specific **antibodies** (an'te-bod-e) that inactivate the antigen or cytotoxic reactions that kill the microorganism or other cell with that antigen.

See Questions—Objective 10

Nonspecific Defenses

Inflammation is the first step in combating infection. When bacteria or other foreign materials enter the skin, they are engulfed by leukocytes that release lysosomal enzymes. The enzymes destroy microbes and activate complement, which, in turn, causes mast cells to release **histamine.** Histamine dilates blood vessels increasing blood flow and making vessel walls more permeable, thus allowing clotting factors and nutrients to reach an injured area and wastes to leave it. Histamine also causes hay fever symptoms and itching of insect bites.

Tissue injury activates small peptides called **kinins,** found in blood in inactive form, and clotting factors if there is bleeding. Kinins further increase blood flow and vessel permeability and attract phagocytes to injured

blood P_{CO_2} as carbon dioxide easily crosses the blood-brain barrier and enters the chemosensitive area, where it combines with water to form carbonic acid. Some carbonic acid then ionizes and releases H^+, which directly stimulates the chemosensitive area. CO_2 may have a greater effect in cerebrospinal fluid than in blood because cerebrospinal fluid lacks the buffers (substances that resist pH change) found in blood. CO_2 that reaches cerebrospinal fluid quickly ionizes and H^+ stimulates the chemosensitive area, whereas H^+ remaining in blood may be rendered innocuous by buffers.

See Questions—Objective 9

Questions

Objective 10

How are specific and nonspecific defenses alike and how are they different?

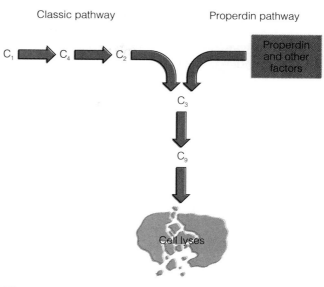

Figure 6.7

Pathways of complement activation.

tissue. Clotting factors stop bleeding by forming a clot where blood vessels are severed. The clot dries and becomes a scab. Inflamed tissues release **leukocytosis-promoting factor** (LP factor), which causes leukocytosis, an increase in the number of blood leukocytes. Many leukocytes leave capillaries by **diapedesis** (di-ah-ped-e'sis), or passing between cells in capillary walls, and move toward chemical attractants in injured tissues. Avidly phagocytic neutrophils congregate in tissue fluids in this manner.

Pain after tissue injury may be due to activation of **bradykinin** and other kinins, which also dilate blood vessels. How bradykinin stimulates pain receptors in the skin is unknown, but **prostaglandins** seem to intensify its effect. Aspirin relieves pain by interfering with prostaglandin synthesis, but no pain relievers have been found to interfere with the bradykinin release or action.

Phagocytes move by **chemotaxis** along a concentration gradient toward a chemical attractant (such as might be produced by an invading microorganism). The phagocyte's membrane forms a vacuole around the object and its lysosomes fuse with the membrane and release enzymes into the vacuole. The enzymes digest microbes into small molecules (amino acids, sugars, and fatty acids) the phagocyte can use for energy to find and digest more microbes. Phagocytes use large amounts of oxygen to form microbe-killing hydrogen peroxide, superoxide (O_2^-), and hypochlorite (the active ingredient in household bleach).

During the entire inflammatory reaction, healing is stimulated by a variety of polypeptide growth factors including growth hormone, insulin, and factors that cause epidermis and nerve to grow. Vitamins also are important in healing—vitamin A for cell division, vitamin C for collagen synthesis, and vitamin K for blood clotting. Vitamin E also may promote healing and prevent formation of ex-

cessively large, disfiguring scars. Growth hormone stimulates cell division and causes macrophages to produce more than twice as much bacteria-killing superoxide.

Fever is a nonspecific defense against microorganisms. It increases the rate of chemical reactions including those that inactivate or kill disease-causing microbes. Treating viral infections in children with aspirin may lead to the often fatal **Reye's syndrome,** in which blood ammonia rises, and vomiting, liver damage, deep coma, and cerebral edema occur.

The body's defense against viruses involves a small protein molecule called **interferon** (in-ter-fe'ron) produced by cells infected with viruses. Several interferons among different species and different tissues in the same species have been identified. Human leukocytes make **alpha interferon,** fibroblasts make **beta interferon,** and lymphocytes (T cells) make **gamma interferon.** Once released, interferon binds to surface receptors on adjacent cells, where a single molecule stimulates a cell to make many molecules of **antiviral protein.** When viruses enter such cells, antiviral protein prevents virus replication.

The **complement system,** or **complement,** is a set of over 20 proteins that circulate in plasma in inactive form and account for about 10 percent (by weight) of plasma proteins. Complement, so named because it complements, or completes, some immunologic reactions, also enhances phagocytosis, produces inflammation, and breaks down microorganisms. It acts long before specific immune reactions can occur. Two complement reaction pathways produce the same results (Figure 6.7). Complement proteins C1 through C9 (C stands for complement) participate in the **classic pathway.** Properdin and other factors replace C1, C2, and C4 in the **properdin** (pro-per′din) **pathway.** Several factors are common to both pathways, so their effects are the same. Certain factors increase blood vessel permeability, stimulate chemotaxis, and cause mast cells to release histamine. Some bind to microorganisms already coated with antibodies from specific immune reactions, causing them to be phagocytized. The C5 through C9 sequence, called the **membrane attack complex,** causes **immune cytolysis** (si-tol′is-is), or cell membrane rupture.

See Questions—Objective 11

Specific Defenses—Immunity

Immune functions are carried out by **B lymphocytes (B cells)** and **T lymphocytes (T cells),** characterized in Table 6.8. Differentiation of B lymphocytes was first studied in the **bursa of Fabricius** (fab-ris′e-us) of birds. Humans have no such bursa, but they do have B cells and bursal-equivalent tissue—probably liver or lymphoid tissues. B cells are found in all lymphoid tissues and in blood. T lymphocytes differentiate in the **thymus gland** from birth to puberty. In addition to releasing T cells into the blood, the thymus secretes **thymosin,** which stimulates lymph tissues to make lymphocytes, and other hormones that cause lymphocytes to become T cells. After puberty T cells are thought to form in bone marrow. T cells further differentiate into (1) cytotoxic (killer) T cells, (2) delayed-type hypersensitivity T cells, (3) helper T cells, and (4) suppressor T cells. All are found in lymph tissues and blood.

Specific immunity can be humoral or cell-mediated, and many foreign substances trigger both responses. **Humoral** (hu′mor-al) **immunity** results from antibodies mainly in blood. When stimulated by an antigen, B lymphocytes divide to form plasma cells, which in turn releases antibodies. Humoral immunity destroys bacterial toxins, bacteria, and viruses (outside cells). **Cell-mediated immunity** results from T cell actions at the cellular level, especially where antigens are inaccessible to antibodies. It kills virus infected cells, some fungi and parasites, cancer cells, and foreign tissues, including transplanted organs.

Immunity is specific because of the nature of the molecules that participate in it. These molecules include antigens, major histocompatibility complex proteins, and antibodies. Immunity develops in response to **antigens,** substances the immune system identifies as foreign. Most antigens are large, complex proteins with molecular weights greater than 10,000. Some are polysaccharides and a few are glycoproteins (carbohydrate and protein) or nucleic acids. Every antigen has at least one **antigenic determinant,** a site to which antibodies can bind, and large

(a) How is inflammation initiated?

(b) How do phagocytes contribute to inflammation?

(c) What is interferon and how does it defend the body?

(d) What is complement and how does it defend the body?

Table 6.8
Characteristics of B Cells and T Cells

	B cells	T cells
Site of production	Bursal-equivalent tissues	Thymus or elsewhere under the influence of thymic hormones
Type of immunity	Humoral	Cell-mediated and assist humoral
Subpopulations	Plasma cells and memory cells	Cytotoxic, helper, suppressor, delayed hypersensitivity, and memory cells
Antigen receptors	Yes	Yes

antigens can have several such sites. Certain small molecules called **haptens** (hap′tenz) bind to a large protein molecule where they become antigenic determinants. Neither the hapten nor the protein alone is antigenic, but together they are. When penicillin molecules (haptens) bind to protein molecules, they can act as antigens. The body's response to such antigens can be a severe allergic reaction.

Cells have many naturally occurring surface proteins that can behave as antigens. Only identical twins have identical cell surface proteins, but human cells have proteins not found in other species and family members have more common proteins than unrelated individuals. Of these proteins, **major histocompatibility** (his″to-kompat-ib-il′it-e) **complex** (MHC) **proteins** account for differences in surface antigens. MHC proteins on cell surfaces are important in immune reactions. Certain cells, especially those called macrophages, can present antigens to B or T cells. Such a cell is called an **antigen presenting cell** (APC). When APCs ingest and digest microbes, they process microbial antigens, combine them with MHC proteins, and insert the MHC-microbial antigen complex in their membranes. APCs can display Class I or Class II proteins. Cytotoxic T cells also have Class I MHC proteins, and helper T cells have class II MHC proteins. Immune reactions occur when proteins bind as an APC presents an antigen.

One way the immune system responds to a foreign substance is by making **antibodies,** proteins that can bind specifically to the antigenic determinant of an antigen. Antibodies also are called **immunoglobulins** (im-u-no-glob′u-linz), or Ig. An antibody consists of two pairs of polypeptide chains—short **light** (L) **chains** and longer **heavy** (H) **chains** (Figure 6.8)—held together by disulfide bonds. Each has a pair of identical variable regions, or antigen-binding sites, at the adjacent ends of the L and H chains. The variable regions of one kind of antibody are

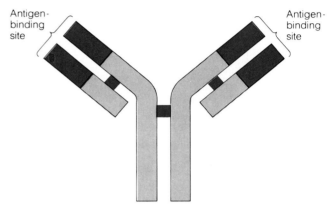

Figure 6.8

Antibody structure.

different from those of any other kind of antibody. H chain tails have another binding site that can attach to a cell membrane.

How a cell has enough genetic information to make billions of different antibodies, each with a unique variable region, has puzzled biologists. One way might be that many short pieces of DNA each specify small segments of a variable region, while a few longer pieces specify nonvariable regions. Then antibodies might be assembled from different assortments of segments.

Five immunoglobulin classes are defined according to their H chains as IgG, IgA, IgE, IgM, and IgD (Table 6.9). The most important blood immunoglobulin, **IgG,** makes up about 20 percent of plasma proteins. Variable regions of IgGs attach to microbial antigens and tissue-binding sites attach to phagocytes. Most IgGs activate complement and cross the placenta, where they provide some neonatal immunity. **IgA** is found in tears, milk, saliva, and mucus, and on all membranes that line body

Table 6.9
Properties of Immunoglobulins

Properties	Classes of Immunoglobulins				
	IgG	**IgA**	**IgE**	**IgD**	**IgM**
Number of units	1	1 or 2	1	1	5
Percent of total blood antibodies	75–85	5–15	0.5	0.2	5–10
Activation of complement	Yes	Yes	No	Yes	Yes
Crosses placenta	Yes	No	No	No	No
Binds to phagocytes	Yes	Yes	No	?	No
Binds to lymphocytes	Yes	Yes	Yes	?	Yes
Binds to mast cells and basophils	No	No	Yes	?	No

tracts open to the environment. IgA, which resists proteolytic enzyme action, binds microbial antigens before the microbes invade tissues. It activates complement but does not cross the placenta. Being abundant in **colostrum** (kolos′trum), the first fluid secreted by mammary glands, IgA provides an infant with some immunity.

Among other Igs, **IgE** binds to receptors on basophils and mast cells in body fluids and skin. Simultaneous binding of IgE to antigens such as pollens, drugs, or foods, causes mast cells to release histamine and other substances that elicit allergy symptoms. Made by both B cells and plasma cells, **IgM** consists of 5 units connected by H chains with 10 antigen-binding sites. It binds antigens and clumps microbes early in infections and activates complement. **IgD** is found mainly on B cell membranes, is rarely secreted, and has no known function.

General attributes of immunity are recognition of self versus nonself, specificity, heterogeneity, and memory (Table 6.10). **Recognition of self versus nonself** is now believed to be a lifelong process carried out by many kinds of APCs and by T cells "educated" in the thymus to recognize an antigen. Such T cells learn to destroy foreign antigens and to ignore body proteins. APCs regularly insert MHC proteins and antigen fragments on their surfaces. Some antigen fragments are derived from normal cellular proteins whereas others are foreign. T cells with the same MHC protein as the APC temporarily bind to it. They combine sequentially and randomly with many different APCs, creating immune surveillance in which a cell-mediated reaction occurs only when a foreign antigen is found.

By age 2 or 3 when the immune system is mature, it displays **specificity,** the ability to react to each antigen in a particular way. B cells recognize antigens by shape, and T cells recognize them by chemical properties of antigen fragments. By one mechanism or the other, the

Table 6.10
Attributes of Specific Immunity

Attribute	Description
Recognition of self and nonself	Ability to distinguish between substances naturally present in the body and substances foreign to the body
Specificity	Ability to react differently to each foreign substance
Heterogeneity	Presence of a large number of different lymphocytes, each capable of recognizing and responding to a different antigen
Memory	Ability to recognize a previously encountered antigen

immune system can rid the body of foreign substances without destroying the body's own proteins.

Whereas specificity refers to the immune system's ability to attack particular antigens, **heterogeneity** refers to a diverse assortment of B and T cells each capable of responding to a different antigen. Given such a lymphocyte population, it is likely that one or more will attack an antigen regardless of whether it has ever been present in the body.

Immune system **memory,** the ability to recognize a previously identified antigen, allows a rapid and specific immune response to a microbe that has caused a previous infection. Memory cells or their progeny persist for decades ever ready to react.

See Questions—Objective 12

Questions

Objective 12

(a) What cells and tissues comprise the immune system?

(b) Why is specific immunity said to have a dual role?

(c) What are the properties of antigens and MHC proteins?

(d) What are the properties of immunoglobulins and how do the types differ?

(e) How is the immune system believed to distinguish between self and nonself?

Continued on next page

Questions Continued

Objective 12

(f) Define (as they apply to the immune system): specificity	heterogeneity memory

Study and Review

Did you get the essentials?

The following summary contains the basic concepts from chapter 6 except for the key terms that have been omitted. Read the chapter. Then try to fill in the blanks from memory. If you cannot fill in all the blanks, review the chapter paying particular attention to boldface terms.

Blood consists of formed _____ such as _____ , _____ , and _____ suspended in a liquid called _____ . Leukocytes include the most numerous _____ , _____ , _____ , _____ , and _____ . _____ (cells that lack nuclei) live only about _____ days. They contain _____ , which is broken down to the element _____ , the pigment _____ , and the amino acids from the protein _____ . Cell fragments called _____ are important for _____ of blood.

Blood clotting by the _____ mechanism is initiated by _____ release from injured tissue, and blood clotting by the _____ mechanism is initiated by exposed _____ in a damaged blood vessel. The final steps: prothrombin to _____ , which converts _____ to _____ (the substance of a clot), are common to both mechanisms. _____ is a group of inherited disorders in which blood fails to clot. A clot lodged within a vessel is a _____ and one that moves is a _____ .

The behavior of gases in gas exchange is determined by _____ their partial _____ , and their ability to _____ in body fluids. In the lungs the partial pressure of oxygen is higher in _____ than in _____ , so oxygen diffuses _____ _____ . The partial pressure of carbon dioxide is higher in _____ than in _____ , so it diffuses _____ _____ . Conversely, in other tissues pressure _____ are reversed and oxygen diffuses _____ _____ , while carbon dioxide diffuses _____ _____ .

Oxygen is only _____ soluble in plasma and most of it is carried on _____. Lowering the _____ and increasing the _____ of blood facilitates release of _____. Carbon dioxide is carried in three ways, dissolved in _____, as _____ ion, and on _____. Movement of _____ ions to _____ creates an excess _____ charge that is balanced by _____ _____. Chemoreceptors in _____ and _____ bodies detect changes in concentrations of _____ and _____ ions in blood.

_____ immunity defends the body against any agent, whereas _____ immunity defends against a single agent called an _____, often by producing _____.

Inflamed tissues release _____-_____ _____, which increases the numbers of _____. Some of these cells leave the blood by _____ and move to injured tissues. In such tissues, they move to dead cells and debris by _____ and engulf them.

Cells infected with _____ produce a protein called _____, which diffuses to other cells and causes them to make _____ _____ and thereby resist infection.

The _____ system, or _____, is a nonspecific defense that attacks cell _____ in a process called _____ _____.

In specific immunity, _____ _____ are activated in _____-equivalent tissue while _____ _____ are activated in the _____ gland. The latter further differentiate into cells called _____, _____, _____, and _____ cells. The immune system produces two kinds of _____. They are _____ immunity, which involves making specific _____, and _____-_____ immunity, which involves direct attacks on foreign _____.

All _____ have _____ determinants, which can consist of small molecules called _____ bound to proteins. Antibodies, or _____, consist of 4 proteins—2 _____ and 2 _____ chains, each of which has a _____ region that reacts specifically with a particular _____. _____ are the major blood antibodies, _____ are found in secretions, _____ account for allergic reactions, _____ function early in infections and activate complement, and _____ have no known function.

Four properties of immunity are (1) _____ of _____ vs _____, (2) _____ (ability to respond to particular antigens), (3) _____ (ability to respond to many different antigens), and (4) _____ (ability to recognize and respond to previously encountered antigens).

Digestion, Absorption, and Regulation of Metabolism

7

General Properties of the Digestive System

Three major chemical processes—secretion, digestion, and absorption—occur in the digestive tract. **Secretion** includes the release of mucus, enzymes, and other digestive secretions. Mucus lubricates and protects. Enzymes and other digestive secretions are normally released in response to the presence of food in the tract, with the kind and amount of food determining the secretions. **Digestion,** or breaking of large food molecules into smaller ones, occurs in the lumen of the digestive tract—mostly located in the small intestine. Enzymes break molecules, usually by hydrolysis, into their components—amino acids, monosaccharides, fatty acids, and other substances. **Absorption,** the process of moving nutrients across the mucosa (digestive tract lining) into blood or lymph, takes place mainly in the small intestine. Structural modifications there increase the surface area and facilitate the process. A fourth process, **motility,** involves muscle contractions in the gut wall that propel partially digested foodstuffs through the tract.

See Questions—Objective 1

Mouth

Chemical processing of food in the **mouth** involves mainly the salivary glands, which secrete over 1 liter of saliva daily. Saliva contains the starch-digesting enzyme **salivary amylase** (am'il-ās) and **mucin** (mu'sin), a glycoprotein that lubricates membranes and binds food particles together. Amylase and mucin account for only 0.5 percent of the saliva volume, the rest being mainly water and electrolytes. Electrolytes in saliva are subsequently reabsorbed in the intestine and are continuously recycled. Saliva also contains a small amount fat-digesting **lingual lipase** (li'pās) and thiocyanate (SCN^-), which may destroy some bacteria in food.

See Questions—Objective 2

Stomach

The stomach mucosa consists of epithelium interrupted by millions of **gastric** (gas'trik) **pits** that open into the gastric glands. Nearest to the surface are **mucous neck cells,** which with mucosal goblet cells, synthesize and secrete enough mucus to coat the entire gastric mucosa. Deeper in the gland are the **chief cells, parietal cells,** and **argentaffin** (ar-jen'taf-in) **cells.** Chief cells secrete **pepsinogen** (pep-sin'o-jen), an inactive form of the enzyme **pepsin** (pep'sin) and small amounts of **gastric amylase** and **gas-**

Questions

Objective 1

What are the general functions of the digestive system?

Questions

Objective 2

What digestive functions occur in the mouth?

tric lipase. The amylase is inactive because of acidity and the lipase acts only on butterfat. Parietal cells secrete **hydrochloric** (hi-dro-klo'rik) **acid** (HCl) and **intrinsic factor,** a polysaccharide that binds to vitamin B_{12} and facilitates its absorption in the small intestine. Argentaffin cells secrete serotonin and histamine, but the digestive function of these substances is unknown. The pylorus contains **G cells** that secrete the hormone **gastrin** into the blood and pyloric glands that secrete a particularly viscous alkaline mucus.

In the stomach, acid and enzyme secretion create conditions that might allow digestion of the stomach itself if not properly regulated. HCl synthesis and secretion probably involves the following steps (Figure 7.1). Water in the parietal cells ionizes to form H^+ and OH^-, and Cl^- and CO_2 diffuse from interstitial fluids into the parietal cells. Carbonic anhydrase causes CO_2 to combine with H_2O to form carbonic acid (H_2CO_3). This acid partially ionizes and releases bicarbonate (HCO_3^-) and H^+. The HCO_3^- diffuses into the interstitial fluid, where it replaces negative charges lost by Cl^- entering cells. H^+ from carbonic acid combines with OH^- from water, reforming some water molecules. The Cl^- and the remaining H^+ are actively transported separately across parietal cell membranes to small passageways called canaliculi (little canals) made of folds in the membrane.

Gastrin stimulates the release of H^+ and Cl^- from the canaliculi into the lumen of the stomach. Only a small

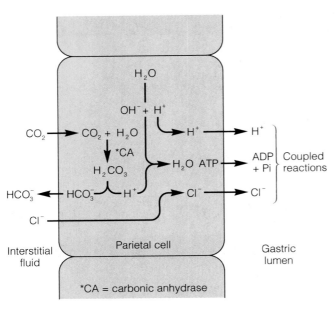

Figure 7.1

The formation of hydrochloric acid in the parietal cells of the gastric mucosa.

quantity of water is released with the ionized HCl, so the acid reaching the lumen has a very low pH of 0.8. Once it enters the lumen, HCl lowers the pH of stomach contents to about 2. This activates pepsinogen to pepsin and kills microorganisms. Pepsin, a proteolytic (protein-digesting) enzyme, also can be activated by pepsin itself. Stomach mucus near the entrance to the intestine is alkaline, thick, and sticky. It provides special protection where the most acid chyme (semifluid material ready to enter the small intestine) accumulates.

Gastric secretions are regulated by neural and chemical factors. Chemicals such as partially digested proteins, alcohol, and caffeine are **secretagogues** (se-kre'to-gogz); they stimulate secretion and act locally and specifically to release gastrin. Gastrin, being a hormone, is secreted into the blood, through which it is carried to gastric glands. Gastrin stimulates HCl secretion, until stomach acidity below pH 2 inhibits gastrin secretion by negative feedback.

In the stomach, both digestion and absorption are limited. Hydrolysis of proteins to shorter chains called polypeptides by pepsin and HCl is the main digestive action. Absorption is severely limited by the thick mucus that prevents HCl and pepsin from digesting the mucosa. Only alcohol and some lipid-soluble drugs such as aspirin can cross the mucous barrier and be absorbed in the stomach.

See Questions—Objective 3

Questions

Objective 3

What events occurring in the stomach contribute to
(a) secretion?

(b) digestion?

(c) absorption?

(d) How are gastric secretions regulated?

Liver and Pancreas

The main digestive function of the liver is to produce bile, which contains bile acids, cholesterol, bile pigments, other lipids, and electrolytes. The liver uses cholesterol to make the primary **bile acids, cholic** (ko′lik) **acid** and **chenodeoxycholic** (kēn-o-de-ox-e-ko′lik) **acid. Bile salts** are bile acids that have lost a hydrogen ion and gained a sodium or potassium ion, and the terms can be used interchangeably. Bile acids combine with either the amino acid **glycine** (gli-sēn) or an amino acid derivative **taurine** (taw′rēn) to form **conjugated bile acids,** which travel in bile to the small intestine where they emulsify (break up) fat globules. In the intestine, most bile salts are deconjugated (undergo loss of glycine or taurine) and recycled to the liver for reuse. Along with aiding fat digestion, the liver performs many other functions (Table 7.1).

Acinar cells of the pancreas secrete three proteolytic enzymes, **trypsin** (trip′sin), **chymotrypsin** (ki-mo-trip′sin), and **carboxypeptidase** (kar-box-e-pep′tid-ās) as inactive **trypsinogen** (trip-sin′o-jen), **chymotrypsinogen** (ki-mo-trip-sin′o-jen), and **procarboxypeptidase** (pro″kar-box-e-pep′tid-ās), respectively. Activation of these enzymes in the small intestine as they mix with chyme prevents them from digesting pancreatic or intestinal tissues. Acinar cells also secrete **pancreatic amylase, pancreatic lipase, cholesteryl esterase** (ko-les′ter-il es′ter-ās), **ribonuclease** (ri-bo-nu′kle-ās), and **deoxyribonuclease** (de-ox″e-ri-bo-nu′kle-ās). All pancreatic enzymes act in the small intestine. Bicarbonate secreted by pancreatic tubule cells buffers the acid in chyme and water dilutes hyperosmotic chyme.

> **See Questions—Objectives 4 and 5**

Secretion and Digestion in the Small Intestine

The **small intestine** receives chyme from the stomach and secretions from the liver and pancreas. **Intestinal glands,** or **crypts of Lieberkuhn** (kriptz of lē′ber-koon), in the mucosa secrete digestive enzymes. The small intestine has circular folds, fingerlike projections of the mucosa called **villi** (vil′e), and folds in cell membranes called **microvilli** (mi-kro-vil′e), all of which increase its absorptive surface. Within each villus are blood capillaries and a lymph vessel called a **lacteal** (lak′te-al). The close proximity of blood and lymph vessels to the epithelium facilitates absorption. Most nutrients travel from the digestive tract to the blood, but fats enter lacteals and later drain into blood.

The liver, pancreas, and the intestinal epithelium secrete substances into the intestinal lumen. Bile from the liver (via the gallbladder) and enzymes, bicarbonate, and water from the pancreas reach the duodenum via ducts.

Table 7.1
Functions of the Liver

Carbohydrate Metabolism

Glycogenesis—the removal of excess glucose from the blood and the storage of glucose in glycogen

Glycogenolysis—the conversion of glycogen back to glucose for release into the blood when the glucose concentration drops

Gluconeogenesis—the synthesis of glucose from noncarbohydrate nutrients as occurs when the glucose and glycogen supplies are depleted

Fat Metabolism

Synthesis of cholesterol, the precursor of all steroids in the body

Synthesis of phospholipids used in growth and repair of cells

Synthesis of lipoproteins, carrier molecules by which lipids are transported to other cells for energy and to adipose cells for storage

Protein Metabolism

Synthesis of plasma proteins

Deamination of amino acids and the production of ammonia

Decomposition of hemoglobin from worn-out erythrocytes

Other Synthetic Functions

Synthesis of urea, a nitrogenous waste product that is formed, in part, from ammonia released by deamination

Synthesis of several factors that are essential for the clotting of blood

Detoxification of toxic substances, usually by the addition of some chemical substance rendering the toxic substances less harmful

Storage Functions

Storage of a year's supply of vitamin A, several months' supply of vitamin D, and several years' supply of vitamin B_{12}

Combination of iron with the liver protein **apoferritin** for later use in the synthesis of hemoglobin

---

Cells in the crypts of Lieberkuhn secrete various enzymes and about 2 liters of fluid per day—more if chyme is highly acidic or hypertonic. This fluid, which has a neutral pH and is similar to interstitial fluid, dilutes chyme and provides a watery medium for digestion and absorption. Because most of the fluid is reabsorbed, net fluid movement is from the lumen toward the blood and lymph.

Epithelial cells in the mucosa and in the crypts of Lieberkuhn synthesize several enzymes. Some enzymes within the cells act as nutrients are absorbed and others act at membrane surfaces or in the lumen as mucosal cells

Questions

Objective 4

How does the liver contribute to digestion?

Objective 5

(a) How does the pancreas contribute to digestion?

(b) What would happen if proteolytic enzymes were released in active form?

creatic enzymes with the assistance of bile salts digest most nutrients. Before digestion can begin in the small intestine, acidic chyme must be buffered by bicarbonate ions from the pancreas because enzymes that act here require a nearly neutral pH. This stops the action of pepsin, but allows many other enzymes to act.

Dietary carbohydrates enter the small intestine as starch, partially digested starch and disaccharides, undigestible cellulose, and as other kinds of fiber. Pancreatic amylase completes starch digestion to the disaccharide maltose. Maltase hydrolyzes maltose to 2 molecules of glucose. Lactase hydrolyzes lactose (milk sugar) to glucose and galactose, and sucrase hydrolyzes sucrose (table sugar) to glucose and fructose. Cellulose and other kinds of fiber remain undigested because the human body has no enzymes that can attack them. Fiber has no nutrient value but provides roughage to facilitate the movement of unabsorbed material through the large intestine. The digestion of carbohydrates produces monosaccharides (Figure 7.2).

After partial digestion by pepsin in the stomach, proteins enter the small intestine as polypeptides. Alkaline conditions stop the action of pepsin, and pancreatic proteolytic enzymes (Figure 7.3) take over. Trypsinogen is activated to trypsin by intestinal enterokinase, after which trypsin activates other proteolytic enzymes. This chain of events helps to assure that enzymes are mixed with chyme before they are activated and minimizes the risk that they will digest pancreatic or intestinal tissue. Trypsin and chymotrypsin digest large polypeptides to smaller ones by breaking peptide bonds at various locations along the chain. They release individual amino acids only if they happen to act at the end of a chain. Carboxypeptidase releases single amino acids from the carboxyl end of a peptide.

Mucosal enzymes also digest proteins. Dipeptidase breaks dipeptides into amino acids and aminopeptidase releases single amino acids from the amino end of a chain. Proteins are completely digested to amino acids (Figure 7.4).

Lipids enter the small intestine undigested, except for some butterfat digested in the stomach. These lipids include triacylglycerols, cholesterol, and cholesteryl esters (cholesterol with a fatty acid attached to its alcohol group). Before digestion can begin, lipids must be emulsified. **Emulsification** (e-mul″se-fi-ka′shun), the mixing of lipids with watery digestive juices, is accomplished by bile salts, which have water-soluble and fat-soluble parts. The fat-soluble part combines with a fat globule, and the water-soluble part projects into the watery medium. This reduces surface tension on the fat globule and keeps it suspended in the water. Agitation by gentle mixing contractions then breaks large fat globules into smaller ones. Lecithins in bile also help to break up fat globules. Fragmenting large globules into many tiny ones greatly in-

are sloughed. **Enterokinase** (en-ter-o-ki′nās) converts trypsinogen to active trypsin. **Dipeptidase** (di-pep′tid-ās) and **aminopeptidase** (am′in-o-pep′tid-ās) break peptide bonds, and **maltase** (mawl′tās), **lactase** (lak′tās), and **sucrase** (su′krās) break glycosidic bonds.

Secretion in the small intestine is regulated mainly by chyme, which stimulates Brunner's glands (submucosal glands in the duodenum) to secrete mucus and the crypts of Lieberkuhn to secrete fluid. The hormone VIP (vasoactive intestinal peptide) stimulates secretion of electrolytes, which causes water to enter the lumen by osmosis. The water helps dissolve nutrients.

More digestion takes place in the small intestine than in all other parts of the digestive tract together. Pan-

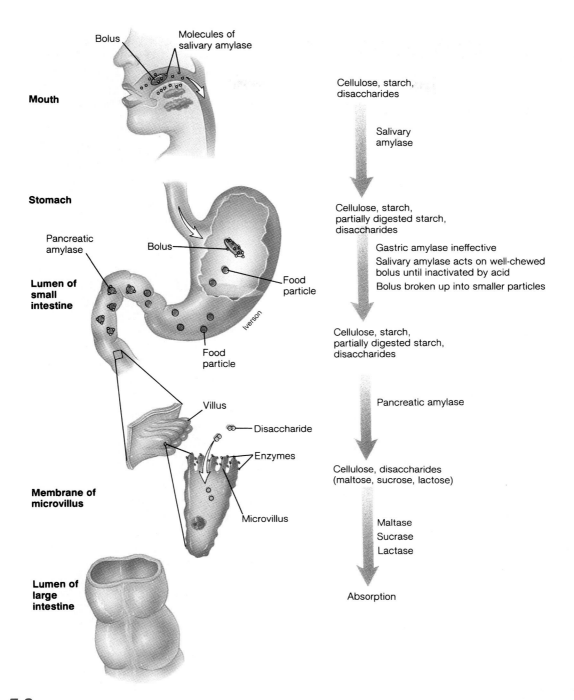

Figure 7.2

Digestion of carbohydrates.

creases the total surface area exposed to digestive enzymes. Pancreatic lipase, with the assistance of the protein **coli-pase** (ko-li′pāz), breaks ester linkages in triacylglycerols and yields glycerol, free fatty acids, and mono- and diacylglycerols.

The resulting small particles called **micelles** (mis-elz) contain bile salts, cholesterol, triacylglycerols, and products of digestion such as mono- and diacylglycerols,

free fatty acids, and glycerol. Polar molecules with their polar ends toward the watery medium keep micelles suspended as they migrate to microvilli. Fatty acids, glycerol, and acylglycerols are absorbed and bile salts are released for reuse. Dietary cholesterol and cholesteryl esters are emulsified along with other lipids. The esters are hydrolyzed by the pancreatic enzyme cholesteryl esterase. Both

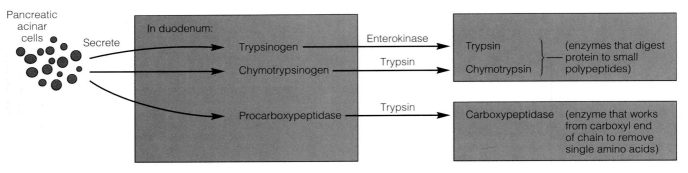

Figure 7.3

Pancreatic proteolytic enzymes.

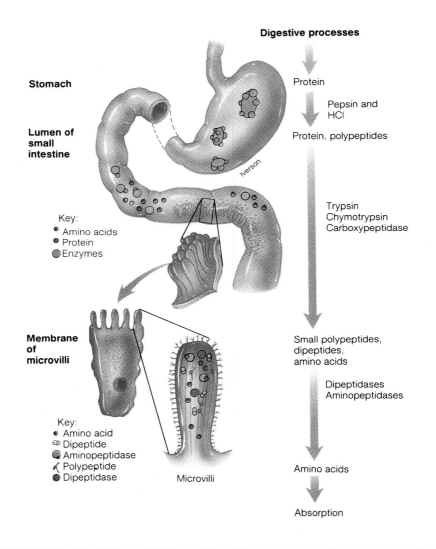

Figure 7.4

Digestion of proteins.

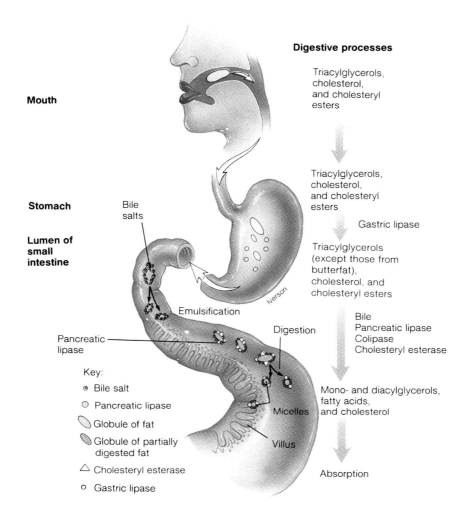

Digestive processes

Triacylglycerols, cholesterol, and cholesteryl esters

Triacylglycerols, cholesterol, and cholesteryl esters

Gastric lipase

Triacylglycerols (except those from butterfat), cholesterol, and cholesteryl esters

Bile
Pancreatic lipase
Colipase
Cholesteryl esterase

Mono- and diacylglycerols, fatty acids, and cholesterol

Absorption

Mouth

Stomach

Lumen of small intestine

Bile salts

Emulsification

Pancreatic lipase

Digestion

Micelles

Villus

Key:
- Bile salt
- Pancreatic lipase
- Globule of fat
- Globule of partially digested fat
- Cholesteryl esterase
- Gastric lipase

Figure 7.5

Digestion of lipids.

cholesterol and fatty acids are ferried in micelles to microvilli where they are absorbed. Lipid digestion is summarized in Figure 7.5.

Foods, which consist mostly of previously living cells, contain nucleic acids that are digested along with other nutrients. Pancreatic ribonuclease and deoxyribonuclease break RNA and DNA, respectively, into nucleotides. Intestinal phosphatases split phosphate from nucleotides, forming nucleosides. Intestinal nucleosidases split nucleosides into purines, pyrimidines, ribose, and deoxyribose, most of which are absorbed and used by cells to make new nucleotides.

In an average sized adult about 1 kilogram of food and 7.5 liters of gastrointestinal secretions enter the digestive tract daily. Most of this volume is absorbed in the small intestine—300 g to 400 g of monosaccharides, 50 g to 100 g of amino acids, about 100 g of lipids, 50 g of various ions, and 8 liters of water.

See Questions—Objectives 6 and 7

Absorption in the Small Intestine

Absorption takes place across mucosal cells of the intestinal villi. Adjacent to the mucosal cell membranes is a thin **unstirred water layer** that substances diffuse through to reach the mucosal membrane. Substances pass through the membrane by a variety of mechanisms—simple diffusion, facilitated diffusion, and active transport. After entering mucosal cells, they also must pass through the inner cell membrane, the interstitial fluid, and the single layer of cells in the walls of blood or lymph capillaries.

Digested carbohydrates and sugars from nucleic acids are absorbed as monosaccharides. Absorption of 5-carbon-monosaccharides probably occurs by simple diffusion, but absorption of 6-carbon-sugars is more complex. Glucose moves across the mucosal surface membrane by **cotransport** on a carrier molecule that also carries sodium ions (Figure 7.6a). Energy from ATP actively

Questions

Objective 6

(a) What secretions are produced in the small intestine?

(b) What are the actions of various secretions found in the small intestine?

Objective 7

Summarize the steps in digestion that occur in the small intestine.

---

transports Na^+ and a glucose molecule attached to the carrier moves with it. Because of cotransport, the glucose transport rate is proportional to Na^+ concentration in the intestinal lumen, at least until the carrier molecules become saturated. Glucose moves by diffusion out the basal side of the cell, across the interstitial fluid, and into a blood capillary. Na^+ moves to the interstitial fluid by the sodium-potassium pump. Galactose is similarly transported on the same carrier. Fructose moves by facilitated diffusion and needs neither ATP nor sodium ions. It is phosphorylated and converted to glucose in mucosal cells, so intestinal capillary blood contains little fructose.

Of digested proteins, which are absorbed as individual amino acid molecules, half come from dietary proteins and half from the breakdown of proteins within

the digestive tract. The latter consists of about equal amounts of enzymes and proteins from sloughed epithelial cells. Amino acids are transported across mucosal surface membranes on different carrier molecules that also transport Na^+ and require energy from ATP. The carrier molecule does not move Na^+ unless it also has an amino acid attached. Each of four kinds of carrier molecules transports a specific group of amino acids. Amino acids move out of mucosal cells to the blood in the same way as glucose molecules. Amino acid transport is summarized in Figure 7.6b.

Products of lipid digestion—free fatty acids, cholesterol, glycerol, and mono- and diacylglycerols—travel in micelles across the unstirred water layer to the mucosal membrane. Micelles probably disintegrate near the mem-

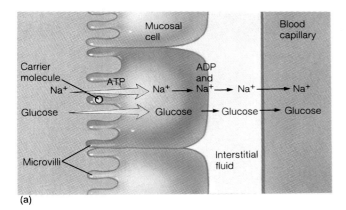

(a)

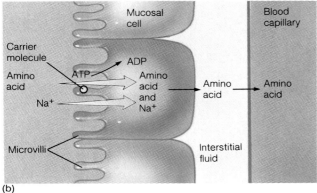

(b)

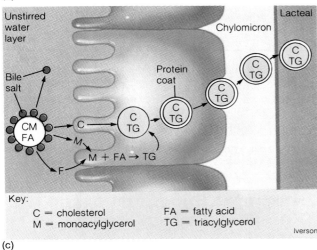

Key:

C = cholesterol   FA = fatty acid
M = monoacylglycerol   TG = triacylglycerol

Iverson

(c)

Figure 7.6

(*a*) The cotransport of glucose and sodium ions during absorption. (*b*) The transport of amino acids during absorption. (*c*) The absorption of lipids.

brane releasing a collection of lipid-soluble molecules, which then diffuse through membrane lipids and enter mucosal cells. Some micelles may actually fuse with mucosal cell membranes and enter the cell intact. Inside the mucosal cells, free fatty acids recombine with mono- and diacylglycerols to re-form triacylglycerols. A small globule of triacylglycerols and cholesterol is coated with protein

to make a particle called a **chylomicron** (ki-lo-mi′kron). Chylomicrons are extruded from the mucosal cells into the interstitial fluids and then to lacteals. Chylomicrons circulate through lymphatic vessels and later enter the blood. This process is summarized in Figure 7.6c.

Bile salts released when micelles disintegrate remain in the intestinal lumen and emulsify more fat globules. As bile salts are reused, they are swept along the intestine, and most are actively transported into mucosal cells and diffuse into the blood. Less than 5 percent are excreted with the feces. Bile salts reaching the liver are reexcreted into bile. This **enterohepatic** (en-ter-o-he-pat′ik) **circulation** is so efficient that some bile salt molecules make the trip 5 times a day.

Nucleic acid digestion releases 5-carbon-sugars (ribose and deoxyribose), purines, pyrimidines, and phosphates. The sugars and phosphates probably are absorbed by simple diffusion and the purines and pyrimidines by active transport.

Large quantities of water from ingested fluids and digestive juices pass through the small intestine daily; 90 percent of this water is absorbed. Water moves freely in both directions across the mucosal membrane in the small and large intestines, but the net movement is out of the intestine into the blood.

Some sodium and potassium ions diffuse in both directions across the mucosal membrane. Sodium also is actively transported as we have seen, and potassium is actively secreted into the intestinal lumen in the jejunum, ileum, and colon.

Calcium and iron are actively absorbed in the small intestine. Calcium ions are absorbed on a carrier protein, and the availability of the carrier molecule is enhanced by vitamin D and parathyroid hormone. Iron ions are absorbed on gastroferrin. The absorption rate is regulated by the amount of iron already stored in the ferritin in intestinal mucosal cells and certain other tissues. Transport and storage proteins help to assure adequate supplies of certain key elements—calcium for maintaining bone strength and normal muscle and nerve function and iron for making new erythrocytes.

Chloride ions move to equalize charges. They passively follow active absorption of sodium ions or they can be actively transported into the lumen of the ileum and the colon, when H^+ from bacterial action accumulates in the lumen.

Most vitamins released by digestion are absorbed in the upper part of the small intestine. Typically, water-soluble vitamins (B complex and C) are rapidly absorbed, but vitamin B_{12} binds to intrinsic factor in the stomach and the complex is absorbed across the mucosa of the ileum. Fat-soluble vitamins (A, D, E, and K) are absorbed with other fats provided pancreatic lipase and bile salts are available.

See Questions—Objective 8

Questions

Objective 8

(a) How are carbohydrates absorbed?

(b) How are lipids absorbed?

(c) How are amino acids absorbed?

(d) How are other nutrients absorbed?

Large Intestine and Rectum

In the large intestine, secretions are limited to mucus from goblet cells in the epithelium and crypts of Lieberkuhn and bicarbonate from the mucosal epithelium. The bicarbonate helps to neutralize acids produced by bacterial action. No enzymes are secreted in the large intestine.

Except in newborns and people who have undergone prolonged antibiotic therapy, the large intestine contains numerous bacteria called intestinal flora. They enter the digestive tract early in life with food and maintain themselves by metabolizing nutrients that remain in the chyme. Any digestion in the large intestine is performed by the bacteria, which release byproducts of their metabolic processes, especially amino acids and the vitamins K, B$_{12}$, thiamine, and riboflavin. Small amounts of these substances are absorbed in the large intestine.

From the liter of chyme entering the large intestine daily, 100 ml to 200 ml of water, a few electrolytes, and some vitamins are absorbed. When water is present, the large intestine can absorb up to 5 liters per day. It also efficiently absorbs sodium ions without cotransport if stimulated by aldosterone. Because of the absorptive capacities of the colon, fluids and medications are frequently instilled rectally, especially in children.

See Questions—Objective 9

Integrated Functions of the Digestive System

Food entering the stomach initiates secretion of mucus, pepsinogen, and gastrin; gastrin stimulates secretion of HCl. As digestion proceeds, polypeptides suppress HCl secretion by negative feedback. Absorption is limited to molecules that can diffuse through mucus on the stomach lining. When acid chyme reaches the small intestine, it initiates neural and hormonal signals that reduce stomach motility and secretions (Figure 7.7).

Functions of the liver and pancreas are regulated by neural and hormonal stimuli. The same parasympathetic stimuli that act on the stomach also stimulate enzyme synthesis in the pancreas. However, the hormones secretin and CCK are responsible for releasing bile and

Questions

Objective 9

How does the large intestine contribute to
(a) secretion?

(c) absorption?

(b) digestion?

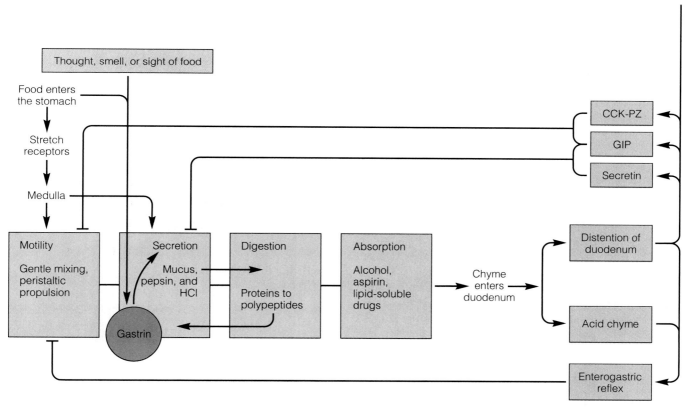

Figure 7.7

Processes that occur in the stomach and their regulation.

Questions

Objective 10

How are digestive processes coordinated?

Objective 11

What are the properties and actions of enteric hormones?

---

pancreatic secretions. Both hormones are secreted by the duodenal epithelium in response to the presence of chyme, and they reach the liver and pancreas via the blood.

Secretin stimulates the liver to produce and secrete bile and the pancreatic tubular cells to secrete water and bicarbonate ions. Acidity is a particularly strong stimulator of secretin release—the more acidic the chyme, the more secretin is released. As bicarbonate reaches the duodenum, it neutralizes the acidity, and as acidity is reduced so is secretin release.

CCK has two important effects, independently discovered and named. Later, it was learned that the same substance produces both effects. As cholecystokinin, the hormone stimulates bile release from the gallbladder. (The name aptly describes this action: "chol" means bile, "cyst" means bladder, and "kinin" means to move.) When it reaches the gallbladder, the hormone causes the muscular wall and duct to contract rhythmically. These peristaltic waves push bile along the bile duct and open the sphincter between the duct and the duodenum. As pancreozymin, the hormone stimulates the secretion of pancreatic enzymes. (Again the name aptly describes the action: "pancreo" refers to the pancreas and "zymin" means enzyme producer.) Both secretin and CCK potentiate or augment the other's effects, and their combined effects are greater than those of either alone.

Acid chyme entering the small intestine initiates both motility and release of secretin and CCK. The hormones, in turn, cause the pancreas to release bicarbonate ions, fluid, and enzymes and the gallbladder to release bile. Bicarbonate ions neutralize the chyme, bile emulsifies fats, and the enzymes digest various nutrients to small molecules. Intestinal enzymes complete the digestive process.

Amino acids, monosaccharides, glycerol, fatty acids, mono- and diacylglycerols, cholesterol, nucleic acid components, vitamins, minerals, electrolytes, and water are made available for absorption. Motility of villi and microvilli facilitate absorption. Most substances are absorbed by active transport or diffusion and enter the blood, but products of fat digestion are packed in chylomicrons and enter lacteals.

Processes that occur in the liver, pancreas, and small intestine are summarized in Figure 7.8. Actions of all substances involved in digestion are summarized in Table 7.2. The current understanding of enteric hormones is summarized in Table 7.3.

See Questions—Objectives 10 and 11

Overview of Metabolic Regulation

As we have seen, many metabolic processes occur at the cellular level and depend on delivery of nutrients to cells. In cells, nutrients are catabolized for energy or used to synthesize proteins and other substances cells need for maintenance, growth, and repair. Regulatory processes supply nutrients to cells according to their metabolic needs by (1) coordinating nutrient processing with absorption, (2) stimulating protein synthesis as needed for tissue maintenance, growth, and repair, and (3) maintaining a metabolic rate that will support these processes.

See Questions—Objective 12

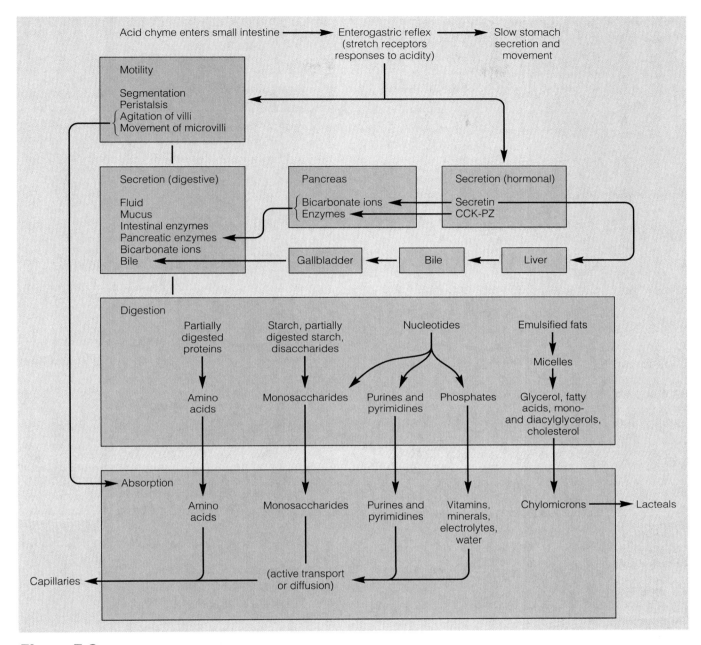

Figure 7.8

Processes that occur in the liver, pancreas, and small intestine and their regulation.

Questions

Objective 12

What are the major processes involved in regulating metabolism?

Table 7.2
Summary of Substances Involved in Digestion

Site of Action	Substance	Nature of Substance	Source	Action
Mouth	Salivary amylase	Enzyme	Salivary glands	Digests starch to intermediate products (and maltose, if time permits)
Stomach	Hydrochloric acid	Mineral acid	Stomach	Activates pepsin, kills bacteria
	Pepsin	Enzyme	Stomach	Digests proteins to polypeptides
	Rennin	Enzyme	Child's stomach	Coagulates milk proteins
	Gastric lipase	Enzyme	Stomach	Digests small amounts of butterfat to glycerol and fatty acids
Small Intestine	Sodium bicarbonate	Salt	Pancreas	Neutralizes hydrochloric acid
	Pancreatic amylase	Enzyme	Pancreas	Digests starch and intermediate products to maltose
	Pancreatic lipases	Enzymes	Pancreas	Digests lipids to fatty acids, glycerol, and cholesterol
	Trypsin	Enzyme	Pancreas	Digests proteins and polypeptides to small polypeptides
	Chymotrypsin	Enzyme	Pancreas	Digests proteins and polypeptides to small polypeptides
	Carboxypeptidases	Enzymes	Pancreas	Remove amino acids from the carboxyl end of polypeptides
	Nucleases	Enzymes	Pancreas	Break down DNA and RNA
	Bile salts	Steroids	Liver	Help to form emulsions and micelles
	Enterokinase	Enzyme	Intestinal epithelium	Activates trypsin
	Dipeptidases	Enzymes	Intestinal epithelium	Break dipeptides into amino acids
	Aminopeptidase	Enzyme	Intestinal epithelium	Remove amino acids from the amino end of polypeptides
	Maltase	Enzyme	Intestinal epithelium	Digests maltose to glucose
	Lactase	Enzyme	Intestinal epithelium	Digests lactose to glucose and galactose
	Sucrase	Enzyme	Intestinal epithelium	Digests sucrose to glucose and fructose
	Nucleotide-digesting enzymes	Enzymes	Intestinal epithelium	Break DNA and RNA into purines, pyrimidines, phosphate, ribose, and deoxyribose

Absorptive vs. Postabsorptive Metabolism

Almost as soon as food enters the small intestine absorption begins. **Absorptive metabolism** refers to metabolic events during absorption (for about 4 hours after a meal). **Postabsorptive metabolism** refers to metabolic events after absorption is complete (for about 2 hours before a meal and during sleep).

In the absorptive state, monosaccharides and amino acids travel via the portal vein directly to the liver. Products of fat digestion travel via lymph vessels to blood and then to the liver. Blood entering the liver, therefore, is nutrient laden.

During absorption all cells generally receive sufficient nutrients, and excess nutrients are processed in a variety of ways (Figure 7.9). Lipoprotein lipase digests triacylglycerols in chylomicrons, and glycerol and fatty acids enter adipose cells, where they reform triacylglycerols and are stored. Such storage conserves excess nutrients during absorptive metabolism when most cells obtain energy from glucose.

During absorption blood glucose usually exceeds that needed to maintain a normal blood glucose concentration of 70 mg/dl to 100 mg/dl. Excess glucose is normally converted to other metabolites—polymerized into glycogen or used to make triacylglycerols in liver and adipose cells. Most amino acids are delivered to cells and are used to make proteins, and only limited quantities remain in cellular amino acid pools. Unused amino acids are deaminated mainly in the liver and used for energy or to synthesize fatty acids. Amino groups are excreted in urea. In summary, absorptive metabolism supplies cells with nutrients for energy and synthesis, maintains a normal blood glucose concentration, and conserves energy by storing glycogen and triacylglycerols.

Table 7.3
Current Understanding of Enteric Hormones

Hormone	Site of Production	Main Site of Action	Main Functions
Gastrin	G cells of the pyloric portion of stomach	Stomach	Stimulates secretion of acid and motility of upper part of stomach.
Secretin	Mucosal glands of upper small intestine	Pancreas, stomach	Increases pancreatic secretion of bicarbonate; decreases gastric section; augments action of CCK.
Cholecystokinin-pancreozymin (CCK)	Mucosal glands of upper small intestine	Pancreas, gallbladder	Causes contraction of gallbladder; stimulates secretion of pancreatic enzymes; augments action of secretin; inhibits gastric emptying.
Glucagon	Duodenum	Pancreas	Stimulates release of insulin from pancreas when glucose concentration is high in intestine.
Gastric inhibitory peptide (GIP)	Duodenum and jejunum	Stomach, pancreas	Inhibits gastric secretion and motility; stimulates insulin secretion.
Vasoactive intestinal peptide (VIP)	Small intestine	Small intestine, stomach, blood vessels	Stimulates intestinal secretion of electrolytes (and thus water); inhibits acid secretion; dilates peripheral blood vessels.
Motilin	Duodenum	Stomach	Stimulates acid secretion, increases motility.
Substance P	Neurons and endocrine-type cells of gastrointestinal tract	Small intestine	Increases motility of small intestine (not proven to enter blood); may increase mucus secretion.
Bombesin (gastrin-releasing peptide)	Neurons? of gastrointestinal tract?	Stomach, small intestine, gallbladder	Increases gastrin secretion; increases motility of small intestine and gallbladder.
Somatostatin (growth-hormone-inhibiting hormone)	Pyloric mucosa?	?	Same hormone from hypothalamus has several effects on gastrointestinal tract, but effects of locally produced hormone not yet clear; probably inhibits gastrin secretion.
Chymodenin	Duodenum?	Pancreas	Stimulates chymotrypsin release from pancreas.
Bulbogastrone	Duodenum?	Stomach	Inhibits HCl secretion.
Urogastrone	Various	Stomach	Inhibits HCl secretion.
Villikinin	Intestinal mucosa	Villi	Stimulates movement of villi and increases lymph flow.
Enkephalins	Various	Cholinergic neurons	Suppresses release of acetylcholine?
Neurotensin	N cells of villi	?	Modulates neural signals?

The postabsorptive state is characterized by the controlled release of nutrients from glycogen and triacylglycerols and, if necessary, the eventual use of proteins for energy (Figure 7.10).

Postabsorptive metabolism maintains the blood glucose concentration in the normal range. Liver glycogen formed only a few minutes to a few hours earlier is degraded as needed. Triacylglycerols in fat deposits can be broken down. Fatty acids are metabolized for energy; glycerol goes mainly to the liver where it is used to synthesize glucose. Metabolism of fatty acids by cells that can use them, called **glucose sparing,** conserves glucose for neurons and erythrocytes. In starvation, fatty acid metabolism increases and numerous ketone bodies are produced. Under such conditions neurons can metabolize ketone bodies. Some cellular proteins, especially from muscle cells, are degraded to amino acids. After deamination, their carbon skeletons can be used for energy or to make glucose. In summary, postabsorptive metabolism supplies cells with nutrients for energy and synthesis, maintains a normal blood glucose concentration, and uses energy from molecules stored in absorptive metabolism.

Figure 7.9

Metabolic events in the absorptive state.
(*a*) Triacylglycerols are digested to glycerol and fatty acids. They are used in adipose tissues to reform triacylglycerols and are stored in fat deposits. (*b*) In addition to being used to maintain blood glucose concentration, glucose is converted to glycogen mainly in the liver, but also in muscle cells. When glucose is present in great abundance, it can be used to synthesize triacylglycerols in the liver and adipose tissues. (*c*) Most amino acids are used by cells to synthesize proteins. Excess amino acids cannot be stored directly, but they can be deaminated, and their carbon skeletons converted to fatty acids, and eventually incorporated into triacylglycerols.

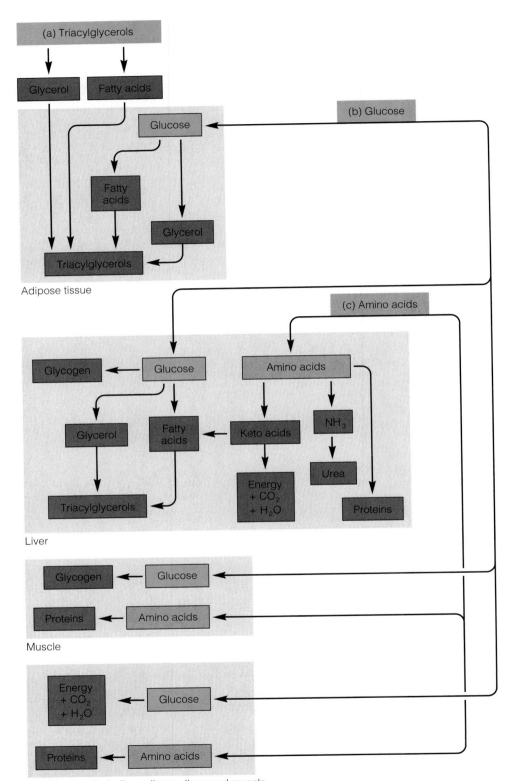

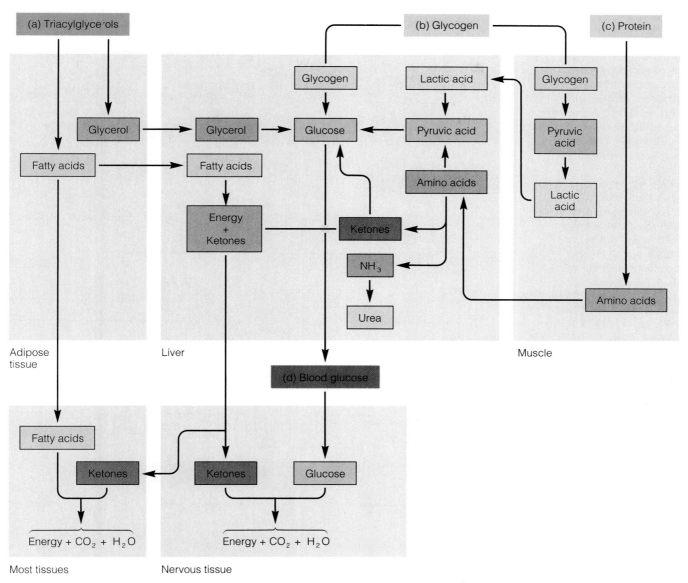

Figure 7.10

Metabolic events in the postabsorptive state.
(*a*) Triacyglycerols are broken down to fatty acids, which are metabolized for energy, and glycerol, which is converted to glucose in the liver and used for energy. (*b*) Glycogen from the liver and muscles is metabolized to lactic acid or pyruvic acid. These molecules are converted to glucose in the liver.
(*c*) When the need for glucose is great, proteins are broken down to amino acids and transported to the liver, where they are deaminated and their keto-acid derivatives used to synthesize glucose. (*d*) The focal event in the postabsorptive state is to maintain a normal blood glucose concentration and a constant supply of glucose for nervous tissue, even though nervous tissue eventually metabolizes ketones under starvation conditions.

Complex mechanisms simultaneously regulate absorptive and postabsorptive metabolism and control shifts from one to the other. Blood glucose is regulated by three negative feedback mechanisms—insulin-glucagon, epinephrine-hypothalamic, and growth hormone-hypothalamic. Blood glucose normally rises to 120 mg/dl to 140 mg/dl as glucose is being absorbed for 1 to 2 hours after a meal. As these mechanisms exert their effects, it returns to normal within 2 hours after glucose absorption ceases.

In the **insulin-glucagon mechanism** (Figure 7.11a), insulin and glucagon are released from the pancreas according to negative feedback from the blood glucose concentration. High blood glucose (during absorptive

Figure 7.11

(*a*) The insulin-glucagon mechanism. When the blood glucose concentration increases, insulin is produced, and its action reduces the blood glucose concentration. When the blood glucose concentration is reduced, glucagon (instead of insulin) is produced, and the action of glucagon increases the blood glucose concentration. This system of negative feedback shown by the "figure eight" in the diagram represents one of the homeostatic mechanisms that helps to regulate the blood glucose concentration. (*b*) The epinephrine-hypothalamic mechanism. A decrease in blood glucose stimulates glucose receptors in the hypothalamus to send nerve signals to adipose tissue and to the adrenal gland. Nerve signals resulting from stress have the same effect. Once stimulated, the adrenal gland releases epinephrine, which stimulates the breakdown of glycogen to glucose in muscles. Fatty acids released from adipose tissue provide an alternative energy source and supplement glucose, thereby increasing the blood glucose concentration. (*c*) The growth hormone-hypothalamic mechanism. When the blood glucose decreases or the body is under stress, the hypothalamus releases GHRH, which, in turn, causes the anterior pituitary to release growth hormone. Growth hormone decreases glucose uptake and increases fatty acid release, thereby increasing the blood glucose concentration. The increase in glycogen storage preserves a supply of glycogen, which can be released later to maintain a stable blood glucose concentration over a longer period of time.

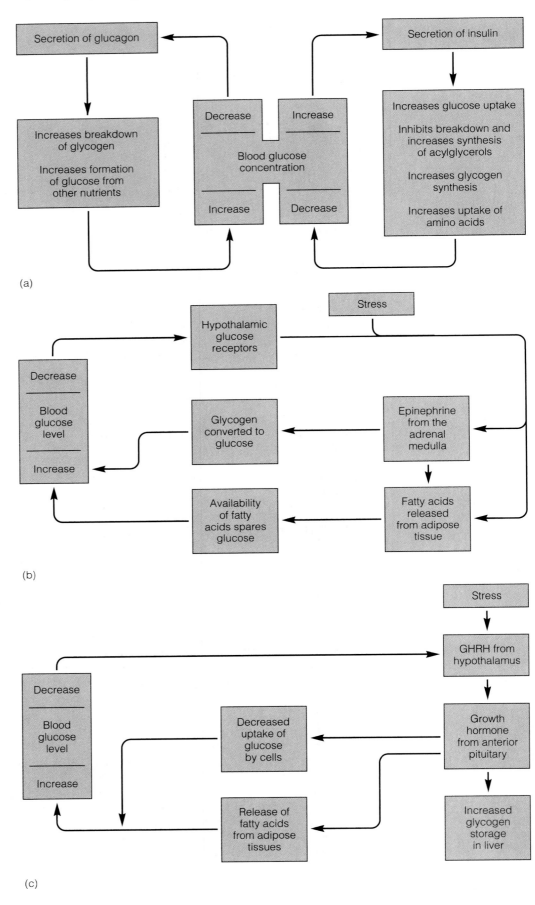

metabolism) initiates insulin secretion. Insulin causes glucose and amino acids to enter most cells, inhibits breakdown and fosters synthesis of triacylglycerols, and increases glycogen synthesis, especially in the liver. As a result of these actions of insulin, the blood glucose becomes sufficiently low, so it no longer stimulates insulin release. A 2-hour postprandial (after a meal) blood glucose test is used to assess insulin activity and to screen for diabetes mellitus.

As the blood glucose drops during postabsorptive metabolism, the lack of glucose initiates glucagon secretion. Glucagon increases the breakdown of glycogen and triacylglycerols and increase glucose synthesis by gluconeogenesis. These effects elevate blood glucose by making glucose available and by supplying glucose-sparing fatty acids. More glucagon, and cortisol, which amplifies the effects of glucagon, are secreted during strenuous exercise, starvation, and other stresses.

Glucagon secretion seems to be regulated by the amount of glucose entering glucagon-secreting cells of the pancreas. Insulin facilitates entry of glucose into those cells and suppresses glucagon secretion. When insulin is not being released, glucose fails to enter the cells and glucagon is secreted. In individuals unable to produce insulin, glucagon is secreted even when blood glucose is high. Without insulin the glucagon-secreting cells fail to get the message that glucose is present and glucagon is not needed.

Blood amino acids also affect the insulin-glucagon mechanism. High concentrations of blood amino acids stimulate the release of insulin *and* glucagon. Having both hormones in the blood tends to stabilize blood glucose, so a high protein, low carbohydrate diet can damp oscillations in the blood glucose concentration. Even without a high amino acid level, both hormones are secreted continuously at a basal level and are present in small amounts in the blood. They exert regulatory effects when they are secreted above the basal rate.

Under normal circumstances the insulin-glucagon mechanism lowers blood glucose in the absorptive state and elevates it in the postabsorptive state. During any kind of exertion, neural signals to the adrenal medulla initiate epinephrine secretion.

The hormone epinephrine and sympathetic signals initiated by the hypothalamus comprise the **epinephrine-hypothalamic mechanism** (Figure 7.11b). This mechanism can be activated by stress or by a decrease in the blood glucose level, which stimulates hypothalamic glucose receptors to relay signals to the adrenal medulla. Stimulation of the adrenal medulla initiates epinephrine secretion. With respect to glucose metabolism, epinephrine behaves much like glucagon. It causes breakdown of glycogen and triacylglycerols and increases gluconeogenesis. This mechanism operates in the postabsorptive state and during stress and is turned off in the absorptive state and in nonstressful situations by negative feedback that shows an adequate blood glucose level.

Once thought to act only during growth, it is now known that growth hormone affects metabolism throughout life. Stress, starvation, or excitement create postabsorptive-like conditions and activate the **growth hormone-hypothalamic mechanism** (Figure 7.11c). The hypothalamus secretes GHRH, which stimulates the anterior pituitary gland to secrete growth hormone. Growth hormone increases liver glycogen, decreases glucose uptake by cells, increases amino acid uptake and protein synthesis, and releases fatty acids from adipose tissue. These effects increase glycogen *and* blood glucose directly and by glucose sparing.

Prolonged hypersecretion of growth hormone accelerates insulin secretion as homeostatic mechanisms attempt to counteract the excess of growth hormone. This can lead to diabetes mellitus by exhausting the pancreas' capacity to produce insulin.

See Questions—Objectives 13 and 14

Regulation of Protein Metabolism

Regulation of absorptive and postabsorptive metabolism helps to regulate protein metabolism. Absorptive processes foster amino acid uptake by cells, whereas postabsorptive processes allow protein breakdown and the use of amino acids to make glucose if needed. Steroid and thyroid hormones control synthesis of specific proteins. They supply enzymes, plasma proteins, and other proteins as needed, and they maintain nitrogen balance while fostering growth and tissue maintenance and repair.

Negative nitrogen balance can be severely debilitating. Cells cannot replace enzymes or make other proteins they need. A lack of blood clotting proteins leads to excessive bleeding after injuries. Too little globin reduces the numbers of new erythrocytes and leads to anemia. Lymphocytes cannot make enough antibodies, so disease resistance is lowered. Tissue injuries are slow to heal. Growth is impaired in children. Finally, phagocytic cells ingest and degrade plasma proteins. This process makes amino acids available to repair essential cellular structures, but it further depletes plasma proteins. Blood oncotic pressure, which holds fluid in blood, decreases and tissue edema results. Such edema accounts for the pot-bellies of starving children.

Normal to positive nitrogen balance provides adequate supplies of amino acids. As newly absorbed amino acids reach the liver, many are used to make plasma proteins. The liver also synthesizes nonessential amino acids by transamination and degrades them by deamination. Cells use amino acids to synthesize enzymes and maintain cellular structures. Certain cells make special proteins such as antibodies and hormones.

Questions

Objective 13

(a) What events occur in absorptive metabolism?

(b) What events occur in postabsorptive metabolism?

(c) How do absorptive and postabsorptive metabolism differ?

Objective 14

How do the following mechanisms regulate shifts between absorptive and postabsorptive metabolism?
(a) insulin-glucagon mechanism

(b) epinephrine-hypothalamic mechanism

(c) growth hormone-hypothalamic mechanism

Unlike carbohydrates deposited in glycogen and lipids deposited in fatty tissue, the body has no protein deposits. Yet, when dietary protein is abundant, most cells, and liver cells in particular, synthesize more protein than is essential for normal function. Then, if dietary protein becomes inadequate, the excess protein can be degraded and the amino acids reused or released into the blood for uptake by cells that need them. This mechanism temporarily avoids the effects of dietary protein deficiency. Protein homeostasis is summarized in Figure 7.12.

The various processes of protein metabolism just described occur continuously to maintain body tissues. In wound healing and growth, protein synthesis accelerates and more dietary protein is needed to sustain them. Growth hormone stimulates protein synthesis by facilitating amino acid transport into cells and by accelerating certain steps in the transcription and translation processes of protein synthesis. Mobilized fatty acids provide energy for protein synthesis and raw materials and energy for lipid synthesis. Growth occurs by synthesis of protein and lipid

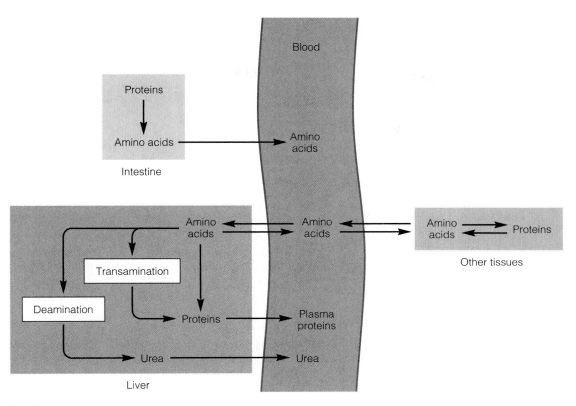

Figure 7.12

Protein homeostasis. Amino acids enter the blood from the intestine. The blood carries amino acids to the liver and to all other tissues. The liver carries out a variety of processes in protein metabolism. Cells in most other tissues use amino acids mainly to synthesize proteins. Small excesses of proteins can be degraded and their amino acids used to make needed proteins when the diet is temporarily deficient in protein.

components of new cells. Growth hormone itself is regulated by GHRH and GHIH from the hypothalamus (Figure 7.13). What releases GHRH is not entirely clear, but once released it causes the anterior pituitary gland to release growth hormone.

Growth also is stimulated by insulin, androgens, and thyroid hormones. Insulin increases protein synthesis by accelerating amino acid transport and transcription-translation and by making glucose, and therefore energy, available to cells. Androgens act with growth hormone to stimulate protein synthesis in muscles where they produce male musculature. Synthetic androgens (anabolic steroids) are often abused for this purpose. Thyroid hormones increase the overall metabolic rate, thereby increasing both protein synthesis and protein breakdown. They must be present for growth hormone to act and for normal brain development and function at all ages.

In contrast to growth stimulating hormones, glucocorticoids in pharmacological concentrations tend to inhibit growth. They mobilize amino acids from tissues except the liver and increase the uptake of amino acids in liver cells. The liver uses some amino acids for protein synthesis, but it deaminates many for use in gluconeo-

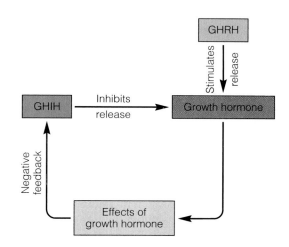

Figure 7.13

Growth hormone secretion is regulated by negative feedback.

genesis. Prostaglandins also slow growth, probably by interfering with the effect of growth hormone on fat deposits.

See Questions—Objective 15

Questions

Objective 15

What factors regulate protein metabolism?

Regulation of Overall Metabolism

Overall metabolism is regulated so that sufficient energy is available to keep all physiological processes operating, to maintain body temperature, and to allow physical activity. The overall **metabolic rate,** how fast the body uses energy, is proportional to the rate at which it releases heat. For example, about 40 percent of the energy in food is transferred to ATP during metabolism, and the remainder is released as heat. The metabolic rate can be determined by measuring heat released from the body in a body-sized chamber, but measuring oxygen use is less cumbersome. Using data from studies that relate heat release to oxygen use and an estimate of body surface area, heat release can be calculated and the metabolic rate determined.

Energy changes in chemical reactions are measured in units called calories. One **calorie** (c) is the amount of energy required to raise the temperature of 1 gram of water 1 degree C (from 14° C to 15° C). A **kilocalorie** (kcal), which equals 1,000 calories, is more convenient to describe human energy metabolism.

Humans use energy for metabolic processes and voluntary activities. Energy used for metabolic processes such as breathing, heartbeat, and maintaining body temperature is called **basal metabolism** and the rate at which such energy is expended is the **basal metabolic rate** (BMR). The total metabolic rate is the sum of BMR and energy used for voluntary physical activities.

A major factor in determining the basal metabolic rate is lean body mass—total body mass, less body fat. BMR can be estimated from body surface area (from which heat is lost). Measured in kilocalories per square meter of body surface per hour, BMR is higher in males than females and higher in the young than in the elderly.

Some people gain weight as they age because they do not reduce food intake as their metabolic rate decreases.

The main factor that increases the metabolic rate above the basal rate is physical activity. Maximal muscular activity for even a short time elevates heat production as much as 100 times that under resting conditions. Swinging a 2-pound weight in each hand while walking more than doubles the calories used. In general, physical activity increases the total daily caloric requirements by 40 to 50 percent for a sedentary person (typist, seamstress), 60 to 70 percent for a moderately active person (salesperson, nurse), to 80 to 100 percent for a very active person (mail carrier, bricklayer).

Certain hormones and sympathetic signals can alter the BMR. Excesses of glucocorticoid, thyroid, or growth hormones increase the basal metabolic rate by directly increasing the rate of cellular metabolism. Sympathetic signals also increase the basal metabolic rate, mainly by increasing glycogenolysis in liver and muscle. In newborn infants, such signals liberate heat from **brown fat** around internal organs. Because oxidation and phosphorylation are uncoupled in brown fat metabolism, little energy is captured in ATP and much is released as heat.

Sleep can decrease the metabolic rate 10 to 15 percent below the BMR measured while awake but resting. This change is thought to be caused by decreased sympathetic activity and decreased skeletal muscle tone.

The ratio of carbon dioxide released to oxygen consumed during metabolism is the **respiratory quotient** (R). R varies depending on what is being metabolized. For carbohydrate R is 1.00 because for every O_2 molecule used, a molecule of CO_2 is released. For triacylglycerols, which contain more hydrogen, R is 0.71 because O_2 is used faster than CO_2 is given off. For amino acids R is 0.83. During the metabolism of a mixed nutrient meal, at first R is very close to 1.00 as carbohydrate molecules reach cells. It approaches 0.71 when cells metabolize fats, as in starvation or untreated diabetes. R can be used to estimate proportions of nutrients being metabolized at any given time.

See Questions—Objectives 16 and 17

Metabolic Disorders

Sugar diabetes, or **diabetes mellitus,** is characterized by high blood glucose and occurs in two forms. In **insulin-dependent diabetes mellitus** (IDDM), the B cells of the pancreas produce little or no insulin. In **noninsulin-dependent diabetes mellitus** (NIDDM), cells have reduced insulin sensitivity, probably because of altered receptors. Other characteristics of these diseases are summarized in Table 7.4. Over 11 million Americans have diabetes, but only 6.5 million know they have the disease and less than 1 million have IDDM and take insulin daily.

Questions

Objective 16

(a) How does basal metabolism differ from total metabolism?

(b) What factors affect metabolic rates?

Objective 17

What is the significance of the respiratory quotient?

Table 7.4

Differences between Insulin-dependent and Noninsulin-dependent Diabetes

Characteristic	Insulin-dependent (Juvenile-onset)	Noninsulin-dependent (Maturity-onset)
Age at onset	Usually under 20	Usually over 40
Percent of all diabetics	About 10%	About 90%
Time of year of onset	Fall and winter	No seasonal trend
Appearance of symptoms	Sudden and usually acute	Gradual and slow to develop
Metabolic acidosis	Frequent	Rare
Obesity at onset	Uncommon	Common
B cells	Decreased	Variable
Insulin	Decreased	Variable
Inflammation of pancreas	Present at onset	Absent
Family history of diabetes	Less common	More common
Association with HLA antigens	Yes	No

The major effects of diabetes are decreased entrance of glucose into most cells, increased fat metabolism, and depletion of tissue proteins. (Insulin is not required for glucose to enter liver or brain cells, except for certain hypothalamic cells.) Diabetic symptoms are closely related to the metabolic disorders it causes (Figure 7.14).

Obesity is defined as a body weight more than 20 percent above the ideal weight due to excess fat. By laboratory calculations, weight should increase at the rate of 1 pound for each excess of 3,500 kcal eaten and decrease similarly for excess calories used. Any nutrients can be metabolized to acetyl-CoA and used to synthesize fat. Stored fat remains as long as energy from food equals energy used.

The causes of obesity are not well understood, but behavioral, physiological, genetic, and developmental factors are being studied. Behavioral factors include feelings, learned behaviors, and exercise habits. Some people overeat when they are anxious and others when they are bored. Children learn to overeat to satisfy oversolicitous parents. Many people become obese when they fail to reduce food intake as their metabolic rate decreases. Sedentary people may gain weight because they get too little exercise.

Malfunction of hunger-satiety centers is an important physiological factor, and can be caused by a hypothalamic tumor, a deficiency in CCK, or an abnormally high operating set point for body weight. People with a high set point have difficulty losing weight and must maintain stringent long-term control of eating behavior to keep from regaining it. Conversely, people with a low set point must constantly force themselves to eat to gain weight and to maintain weight gain.

Other physiological factors include decreased activity of sodium-potassium pumps and depressed autonomic function. People whose cells have fewer than normal sodium-potassium pumps have a lower basal metabolic rate. They gain on a diet that merely maintains weight in a person with a normal number of sodium-potassium pumps. Finally, depressed sympathetic function leads to lower energy use in stressful situations and depressed parasympathetic activity conserves energy by slowing the heart rate and other physiological processes.

In a rare, genetic obesity an enzyme needed to break down stored fat is absent or deficient. Any nutrients converted to fat are stored and cannot be broken down easily.

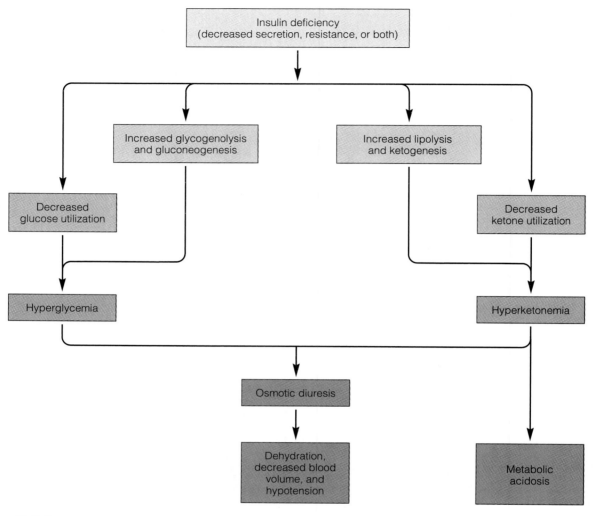

Figure 7.14

Physical effects of diabetes.

The metabolism of obese people may differ from that of lean people though the fat in their bodies is qualitatively no different. Some obese people metabolize glucose more slowly than normal. On reducing diets, they apparently metabolize fats more slowly and completely because their blood has fewer ketone bodies than that of normal people on the same diet. Many obese people are insulin resistant and predisposed toward diabetes, hypercholesterolemia, gallstones, and hypertension. Fat deposits increase the effort required for breathing, and carbon dioxide retention can lead to sleepiness and reduced activity. Treatment of obesity is largely unsuccessful and likely to remain so until causes are better understood. Amphetamines should not be used because they are addictive and suppress appetite for only a few weeks.

Usually seen in adolescent and young adult females, **anorexia nervosa** (an-o-rek′se-ah ner-vo′sah) is a persistent desire to lose weight even when weight is already below normal. Patients have a faulty body image and see themselves as fat no matter how much weight they lose. Symptoms are those of starvation—emaciation, hair loss, and cessation of menstruation in women—but the starvation is self-induced. **Bulimia** (bu-lim′e-ah), or binge eating, is followed by self-induced vomiting. Patients with bulimia often fail to feel full after a meal, and they have only about half the blood CCK of normal individuals. Bulimia sometimes precedes anorexia, but the relationship between the disorders is not clear.

Victims of anorexia nervosa become dehydrated from excreting an unusually high volume of urine. Administering saline solution fails to increase blood levels of ADH as it normally does, so water is not reabsorbed even when blood osmotic pressure is high. Yet, ADH in cerebrospinal fluid of anorexics is high. These ADH abnormalities may relate to the development of anorexia nervosa.

What causes anorexia nervosa is unknown, but psychological factors and hypothalamic disorders are being studied. Current treatment makes use of psychotherapy and gavage (stomach tube) feeding. A typical patient receives 3 liters per day of fluid containing carbohydrate, hydrolyzed protein, fat, electrolytes, and vitamins, which will cause a weight gain of about 1 pound. Temporary gavage feeding quickly increases weight and restores nutritional balance, but unless the patient's body image and eating habits can be altered, anorexia nervosa can be fatal.

Physiological, psychological, or social impairment directly associated with persistent and excessive use of alcohol is called **alcoholism.** Both genetic and environmental factors appear to contribute to alcoholism, but exactly how is unclear. In the United States, where 10 percent of the drinkers consume 50 percent of the alcohol, over 10 million people are severely alcoholic and another 7 million are alcohol abusers. Alcoholism is the third leading cause of death in the 25-to-65 age group.

Two types of alcoholism may exist. Type 1, which includes most alcoholics, develops after age 25 usually in stressful environments. Type II occurs in sons of men who became alcoholics near or before age 25 and who also display aggressive behavior. Such individuals have a distinctive brain wave that persists even during abstinence. A genetic factor in alcoholism may be adjacent to a gene for MN blood antigens and the enzyme alcohol dehydrogenase, which oxidizes alcohol.

Studies of volunteers with a family history of alcoholism and control subjects lacking such a history show differences in alcohol metabolism. In one study, acetaldehyde, which is produced by the action of alcohol dehydrogenase on alcohol, rose much more sharply in the subjects with a family history of alcoholism than in the control subjects. Acetaldehyde production may be genetically determined and suitable for screening people at risk of alcoholism. Family and other environmental stresses contribute to alcoholism in genetically susceptible individuals. Habitual drinking exacerbates the stresses and creates a vicious cycle.

Blood alcohol concentrations in excess of 100 mg/dl (defined as intoxication by law in most states) affects brain centers and produces diminished coordination and judgment. Higher levels lead to speech impairment and stupor. At blood levels above 400 mg/dl, alcohol appears to disrupt cell membranes. Coma and hypothermia occur and can lead to death. Because alcohol dilates peripheral blood vessels and allows heat loss, it lowers body temperature and, contrary to a popular notion, does not help a person to keep warm while hunting or skiing. The major effects of excessive alcohol consumption are summarized in Figure 7.15 .

See Questions—Objective 18

Questions
Objective 18

How is regulation disturbed in
(a) noninsulin-dependent diabetes?

(b) insulin-dependent diabetes?

(c) obesity?

(d) anorexia nervosa?

(e) alcoholism?

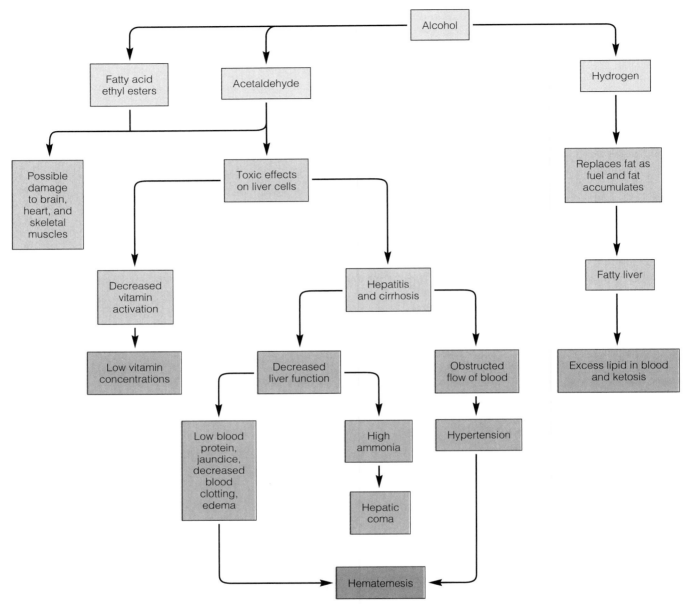

Figure 7.15

Some of the effects of excessive alcohol consumption.

Detoxification—First Defense Against Foreign Substances

Rendering poisonous substances harmless can be a matter of life or death. Substances foreign to our bodies enter when we breathe polluted air, consume food containing pesticides or additives, use tobacco, alcohol, or drugs. Any foreign substance is potentially toxic (poisonous); its effect depends on how much is present and how well the body can detoxify it.

Most chemical reactions that **detoxify,** or render foreign substances harmless, are carried out by liver cell enzymes. The enzymes ordinarily inactivate hormones and other normal body substances, but they can detoxify many foreign substances, too. Detoxification reactions include oxidation, reduction, hydrolysis, and conjugation (Figure 7.16). Such reactions usually convert fat-soluble substances to water-soluble substances that can be excreted in the kidneys. Many such reactions are inducible, that is, the presence of the potentially toxic substance induces activity of the enzymes that detoxify it.

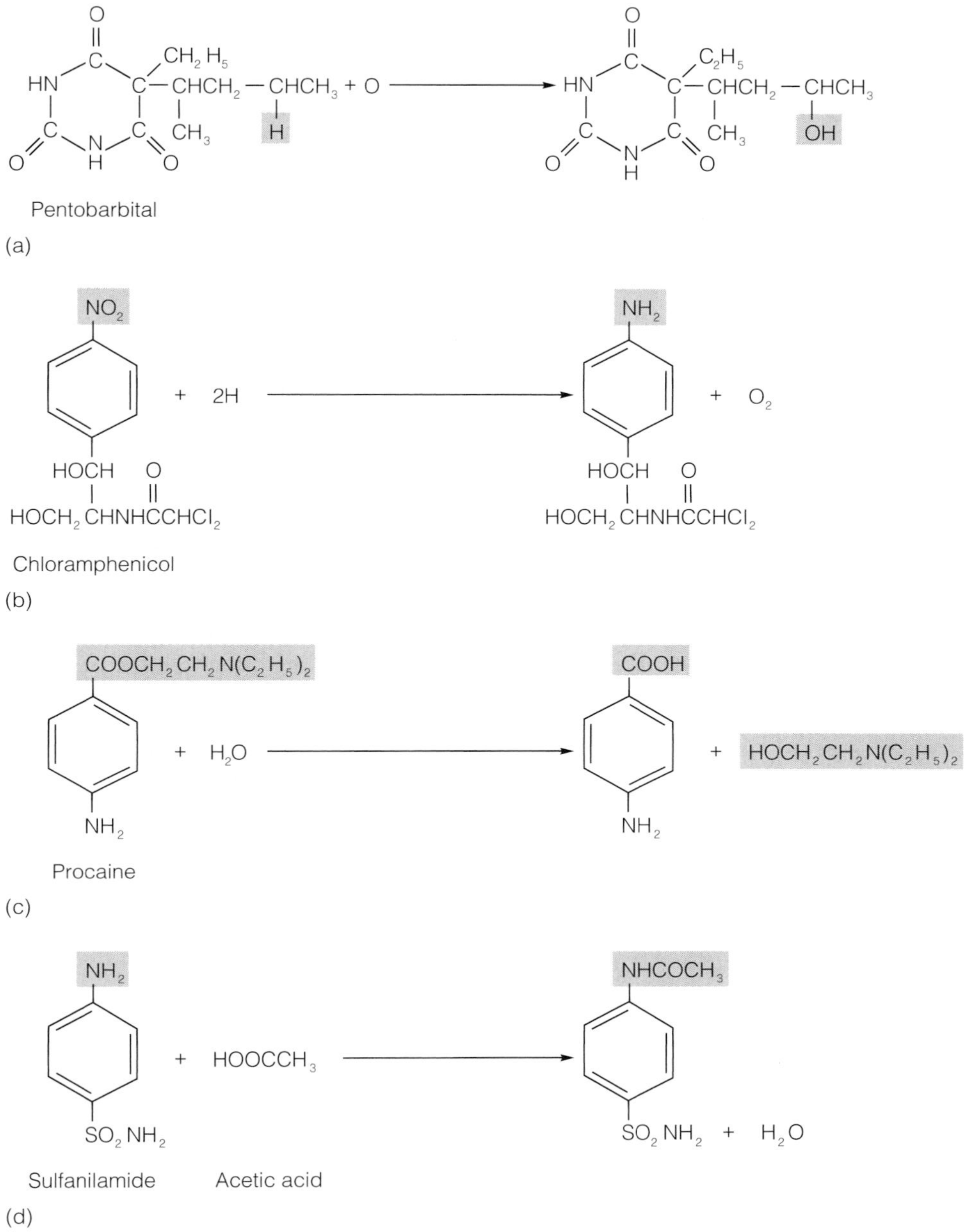

Figure 7.16

The inactivation of foreign substances: (*a*) oxidation of pentobarbital, (*b*) reduction of chloramphenicol, (*c*) hydrolysis of procaine, and (*d*) conjugation of sulfanilamide.

Combinations of drugs can have unpredictable and undesirable results when they are detoxified by the same enzyme. The drug tolbutamide, used to increase insulin activity, is metabolized and inactivated more slowly when the anti-inflammatory agent phenylbutazone, coumarin anticoagulants, or the antibiotic chloramphenicol are present. In addition to drugs, insecticides, dyes, and other foreign substances can be detoxified in the liver.

When the liver is subjected to large amounts of toxic substances, as in an alcoholic, it produces more detoxifying enzymes—at least until the liver itself becomes impaired. When an alcoholic suddenly abstains from alcohol, the liver makes an overabundance of detoxifying enzymes and sometimes destroys medications before they can have their effect. Individual differences exist in what can be detoxified, and good general health and nutritional status maximize this capability. In the elderly, detoxification can be impaired because the body, including the liver, is less efficient. Infants are extremely sensitive to foreign substances because their immature livers fail to make sufficient quantities of detoxifying enzymes.

Fetuses are particularly vulnerable to toxic substances that cross the placenta. The mother may suffer no adverse effects, but the fetus may be deformed, as was tragically demonstrated in the 1950s by the absence of limbs in infants whose mothers took the tranquilizer and anti-nausea drug thalidomide. When the mother uses alcohol or narcotics, the fetus becomes dependent and suf-

fers withdrawal symptoms at birth and usually has other birth defects including mental retardation. Large numbers of infants are hospitalized in intensive-care nurseries for alcohol and drug withdrawal. Many also are infected with AIDS.

Even though our livers detoxify foreign substances, it is unwise to expose ourselves unnecessarily to them. New drugs, food additives, insecticides, and other manufactured substances should be evaluated carefully before they are marketed. Strong laws preventing environmental pollution should be enforced.

See Questions—Objective 19

Questions

Objective 19

(a) What is detoxification?

(b) How does the body detoxify substances?

(c) Why is it important for a physician to know all medications a patient is taking?

(d) Why are the very young and very old most susceptible to damage from toxic substances?

Study and Review

Did you get the essentials?

The following summary contains the basic concepts from chapter 7 except for the key terms that have been omitted. Read the chapter. Then try to fill in the blanks from memory. If you cannot fill in all the blanks, review the chapter paying particular attention to boldface terms.

The three chemical processes carried out by the digestive tract are _____ , _____ , and _____ .

_____ glands in the mouth secrete the enzyme _____ _____ . The stomach glands have cells differentiated to secrete various digestive substances—_____ and _____ factor from parietal cells, enzymes from _____ cells, mucus from _____ _____ cells, and the hormone _____ from _____ cells. Secretions are regulated by neural factors and by _____ . Action of _____ _____ is stopped by _____ in the stomach, but the enzyme _____ at pH _____ starts digestion of _____ .

The liver secretes _____ , which is stored in the _____ , and released into the small intestine to _____ _____ . Liver functions include _____ storage, _____ synthesis, and _____ of foreign substances. The pancreas releases _____ from clusters of cells called _____ and _____ ions from _____ cells. Proteolytic enzymes are released in inactive forms called _____ , _____ , and _____ . Other enzymes from the pancreas include _____ .

The small _____ receives secretions from the _____ and _____ . Structures that increase the surface area include _____ _____ , _____ , and _____ . Enzymes include _____ that activates _____ ; _____ and _____ that break peptide bonds and _____ , _____ , and _____ that break glycosidic bonds. Most absorption takes place in the _____ intestine. The sugar, _____ , sodium, and _____ acids are absorbed by _____ _____ . Products of _____ digestion _____ across the mucosa into _____ and packaged in _____ for transport. _____ _____ , which initiate lipid digestion by causing formation of _____ , are reusable because of _____ _____ . In the large _____ absorption is limited mainly to _____ .

Enteric hormones include _____ that stimulates _____ secretion in the stomach, _____ that stimulates _____ secretion from the pancreas, and _____ that stimulates secretion of both _____ from the liver and _____ _____ from the _____ .

Continued on next page

Study and Review

_____ metabolism refers to events that take place while nutrients are being absorbed. _____ metabolism refers to events that take place when no nutrients are being absorbed. In the former, _____ and _____ are stored, and in the latter they are used for _____ . When glycogen becomes depleted, some cells use _____ _____ for energy. This is called _____ _____ because it conserves _____ for cells that cannot use _____ _____ . Mechanisms of regulating metabolism include the _____ - _____ mechanism centered in the pancreas, the _____ - _____ mechanism that uses the adrenal medulla and a part of the brain, and the _____ _____ - _____ mechanism that

The _____ rate is measured in _____ (units of heat), and is proportional to the rate at which the body uses energy. A _____ (used to express the quantity of energy in foods) is 1,000 times a _____ .

The _____ _____ rate is the rate at which the body uses energy to maintain itself in an awake-resting state. This quantity plus the energy used for _____ equals the total _____ rate. Factors that alter metabolic rates include _____ .

The _____ _____ is the ratio of _____ _____ released to _____ consumed. It is especially low for _____ because they contain so much _____ .

The metabolic disorder _____ _____ is often _____ dependent in young patients whose _____ cells have been destroyed, but it is usually _____ _____ dependent in older patients who cells _____ to respond to _____ . The former is treated with _____ , the latter with diet and _____ _____ .

_____ is excess _____ _____ due to excess _____ . Treatment will remain difficult until _____ are better understood. _____ _____ is a persistent desire to lose weight even after the body becomes _____ . _____ is binge eating usually followed by self-induced _____ .

_____ , excessive intake of alcohol, leads to physical, economic, and social problems. Alcohol is metabolized to _____ , which causes liver damage.

Rendering a substance harmless is called _____ . Chemical reactions that can do this include _____ , _____ , _____ , and _____ .

Nutrition

8

Principles of Nutrition

The science of **nutrition** deals with providing all nutrients—carbohydrates, proteins, fats, vitamins, minerals, and water—for maintaining health through food ingestion. Carbohydrates and fats provide calories to meet the body's energy demands. Carbohydrates can be converted to glycogen and stored in the body, and some are high in fiber (roughage), which helps move foodstuffs through the digestive tract. Fats also serve as energy storage molecules, and some fats are essential for synthesis of cell membranes, prostaglandins, and other substances. Dietary protein provides amino acids for the synthesis of new proteins and other cell components, and is important during growth, repair, pregnancy, and lactation. Vitamins, minerals, and water also are essential to maintain health. Most vitamins and minerals assist enzymes in catalyzing important reactions. Water, which makes up about 60 percent of the body weight, participates in nearly every chemical reaction as reactant, product, or reaction medium.

See Questions—Objective 1

Questions

Objective 1

(a) Define nutrition and name six kinds of nutrients.

(b) What are the principles of good nutrition?

General Nutritional Requirements

Nutritional requirements are determined at the cellular and molecular levels according to energy expenditure, protein synthesis, enzyme activity, and electrolyte balance.

No specific carbohydrates are required in the human diet, but at least 20 percent of total calories should come from carbohydrate, and 50 percent is recommended. Some carbohydrate is needed to provide Krebs cycle intermediates (oxaloacetate, for example) even when the body is obtaining most of its energy from fats. Metabolism of fat without carbohydrate as occurs in untreated diabetes can lead to ketosis and acidosis.

Many Americans eat too little complex carbohydrate (polysaccharide, such as starch), too little fiber, and too much refined sugar. Microbes metabolize such sugar and deposit dental plaque in which they become entrapped. And they produce acid that attacks tooth surfaces causing caries.

Dietary fiber is any plant material that humans cannot digest. Two major kinds are insoluble cellulose and soluble pectins. Each comes from grains, fruits, and vegetables. Cellulose, found in wheat, celery, and many other vegetables, increases fecal bulk and decreases the time required for material to traverse the intestine. Some is digested by microorganisms, which release metabolic products—short chain fatty acids, water, carbon dioxide, and other gases. Soluble fiber, found in fruits, oat bran, and peas and beans, delays gastric emptying, slows glucose absorption (making blood sugar easier to control in diabetics), and lowers serum cholesterol and triglycerides.

Either too little or too much fiber can be detrimental. A lack of fiber can contribute to diverticulitis, inflammation of diverticuli (little pouches) in the colon, and colon cancer. Fiber helps to prevent diverticulitis by keeping the colon distended so diverticuli do not form and by reducing colon transit time. Reducing transit time also helps prevent colon cancer, probably by preventing accumulation of carcinogenic products of bacterial metabolism. Fiber also lowers blood pressure in some people. Excess fiber can decrease the absorption of elements such as calcium, iron, copper, and zinc. Many nutritionists are currently studying fiber, but which and how much fiber humans need have not been established.

Dietary protein is digested to amino acids and used to make new proteins and other nitrogenous molecules—porphyrins for heme in hemoglobin and nucleotides such as DNA, RNA, and ATP. Of the 20 amino acids used in these synthetic reactions, the **essential amino acids** must be present in the diet because the cells cannot synthesize them in adequate amounts. Of the essential amino acids, 8 are needed throughout life, and 2 are needed during periods of rapid growth (Table 8.1). Though cells

Table 8.1	
Essential Amino Acids	
Amino Acids Required throughout Life	
Leucine	Phenylalanine
Isoleucine	Threonine
Lysine	Tryptophan
Methionine	Valine
Amino Acids Required during Periods of Rapid Growth	
Arginine	Histidine

can make small quantities of arginine and histidine, they cannot make these amino acids fast enough to support rapid protein synthesis.

Adequate amounts of all amino acids used in protein synthesis must be present at the same time for synthesis to occur. As mRNA codons are read, the appropriate amino acid attached to its tRNA must be immediately available for a polypeptide chain to be lengthened. Dietary proteins have high **biological value** when they supply amino acids in the proportions needed for the synthesis of human proteins. The biological value of egg albumin is 100 (the highest possible value) because its amino acid content is nearly the same as the average for human cells. Most meat proteins have a biological value of about 70 and those in cereals and grain have a biological value of about 40. Some plant proteins are low and others high in a few human essential amino acids. Eating plant foods in appropriate combinations in the same meal increases the biological value of their proteins.

Proteins normally eaten by humans have an average biological value of about 70, a fact used in setting the recommended daily protein allowance of about 0.8 g/kg body weight. Males require slightly more protein per kilogram body weight than females because their bodies contain relatively more muscle mass. Exercise increases the need for calories, protein, and vitamin B_6, which is required for protein metabolism.

The **essential fatty acid** linoleic acid is the only specific lipid required in the diet. An essential fatty acid is one that must be present in the diet because the body cannot synthesize it. If cells have sufficient linoleic acid, they can synthesize other unsaturated fatty acids to maintain cell membranes and make prostaglandins and leukotrienes. Because linoleic and other unsaturated fatty acids are abundant in most vegetable fats, a tablespoon of margarine daily satisfies this nutritional need.

Omega-3 fatty acids are highly unsaturated, long chain fatty acids with a double bond starting at the third carbon from the methyl end. Abundant in fish fats and green leafy vegetables, they may have important functions not yet recognized as essential nutrients. Omega-3 fatty acids may improve sharpness of vision, lower blood cholesterol and triglycerides, and retard platelet aggregation in atherosclerotic arteries. Fish fats may reduce the risk of atherosclerosis in another way. They contain free radicals (atoms with highly excited electrons) that inhibit a growth factor needed for arterial smooth muscle proliferation.

Most Americans eat too much rather than too little lipid. Excesses of animal fats are particularly detrimental because they contain large quantities of saturated fats and cholesterol, which increase the risk of heart disease as is discussed later.

Vitamins are small organic molecules required in the diet because human cells fail to produce them or produce too small amounts of them to maintain health. Many vitamins are necessary for synthesis of coenzymes such as NAD and FAD or other essential molecules such as collagen and visual pigments. When first discovered, vitamins were designated by letters, A, B, and so forth. As multiple factors were found in the original vitamin B, they were given numerical subscripts. Though vitamins now have names based on their chemical properties, letters are still used. B vitamins and vitamin C are water-soluble, whereas vitamins A, D, E, and K are fat-soluble. **Minerals** are inorganic substances, usually ions. They act as enzyme cofactors, maintain membrane excitability, and form part of substances such as bone, heme, and thyroid hormones.

Water maintains adequate hydration of body cells and normal volumes of extracellular fluids. Most metabolic reactions take place in water, and many actually involve water molecules. Dietary water comes from beverages and foods. Watermelon, fruits used to make juices, squash, cucumbers, and leafy vegetables contain much water. In addition, **metabolic water** is released in cells when hydrogen is transferred to oxygen in oxidation. The body needs enough water, dietary and metabolic, to maintain **water balance** in which intake equals output.

Cells can catabolize most nutrients for energy, but they capture only about 40 percent of that energy in ATP and release the remainder as heat. Even so, cellular energy capture is fairly efficient compared to engines that can use only 35 percent of the energy in fossil fuels they burn. Foods differ in energy content as can be measured by various laboratory techniques and expressed in units such as kilocalories (kcal). Cells derive 4 kcal per gram of protein or carbohydrate and about 9 kcal per gram of fat ingested. Therefore, adding fat raises the energy content of a diet much more than adding carbohydrates or proteins.

Questions

Objective 2

(a) What nutritional needs do dietary carbohydrates meet?

(b) Why are essential amino acids needed?

(c) What needs do lipids, including linoleic acid, meet?

Objective 3

(a) Why, in general, are vitamins and minerals required in the human diet?

(b) What functions does water serve?

Objective 4

(a) What factors determine a person's energy requirements?

(b) How is the specific dynamic action of food related to energy requirements?

Laboratory measurements of the energy in foods are somewhat higher than that obtained by the body for two reasons:

1. Some food does not reach cells because it is not digested or not absorbed. People vary considerably in digestive system efficiency from low in those who can "eat anything and not gain a pound," to high in those in whom "everything they eat turns to fat." A high-fiber, low-fat diet moves through the digestive tract faster than a high-fat, low-fiber diet. Faster transit usually means that a smaller proportion of nutrients are absorbed.

2. On most ordinary diets, the body releases about 10 percent more heat during digestion and metabolism than at other times. This heat, called **specific dynamic action,** is highest for high protein diets because more energy is needed to digest proteins than other foods.

To maintain a constant body weight, energy obtained from food should equal energy expended for basal metabolism and physical activity. If energy intake exceeds energy use, weight gain occurs. If energy use exceeds energy intake, weight loss will occur. Ideal or desirable weight can be defined as the weight for a given height associated with the lowest mortality (death) rate.

See Questions—Objectives 2, 3, and 4

Table 8.2

The Four Basic Food Groups and Their Contributions to Nutrition

Group	Recommended Daily Amounts	Nutritional Needs Satisfied
Milk group	Serving: 8 oz milk or 1½–2 oz cheese 2 servings for adults 3 servings for children and pregnant women 4 servings for lactating mothers	Each serving provides 8 g protein, 3 g fat, and about 140 calories. Contains vitamins A and riboflavin, and provides an especially good supply of calcium, phosphorus, potassium, and magnesium. Low-fat dairy products reduce saturated fat and cholesterol in the diet.
Meat group	Serving: 3 oz lean meat, fish, or poultry, or 1 c cooked legume, 2 eggs, or 4 T peanut butter 2 servings	Each serving provides 15–25 g protein, 10–40 g fat, and 150–400 calories. Contains phosphorus, iron, sodium, potassium, riboflavin, and other B vitamins. Legumes are very low in fat. Peanut butter is high in fat but also high in niacin. Eggs provide some vitamin A but are high in cholesterol.
Group	**Recommended Daily Amounts**	**Nutritional Needs Satisfied**
Fruit and vegetable group	Serving: 1/2 cup 4 servings, including one good source of vitamin A and one of vitamin C	Foods that provide vitamin A: spinach and other green, leafy vegetables; carrots and other yellow vegetables; and fruits. Foods that provide vitamin C: citrus fruits and juices, and tomatoes. Fruits and vegetables also provide some B vitamins and vitamin K, as well as some minerals. Raw foods in this group have larger amounts of vitamins than cooked foods. All fruits and vegetables provide fiber.
Bread and cereal group	Serving: 1 slice of bread or 1 oz of prepared cereal or 1/2 cup of cooked cereal, rice, pasta, or noodles 4 servings	Each serving provides energy, some protein, and, if made of whole grain, a good supply of B vitamins. Eating food from this group with a legume increases the protein value of both foods. All breads and cereals provide fiber.

Recommended Dietary Allowances and Dietary Guidelines

Guidelines help people to select a nutritious diet without needing to know the physiological actions of nutrients. Such guidelines include the basic four food groups (Table 8.2) and Recommended Dietary Allowances (Table 8.3). Companies that process and package foods provide label information showing what percentage of recommended dietary allowances a food contains. At present, **recommended dietary allowances (RDAs),** the amount of a nutrient thought to satisfy the needs of 97.5 percent of the human population, are the best available standards for human nutritional needs. Unfortunately, RDAs have not been established for all known essential nutrients, and a diet that satisfies all RDAs may not be nutritious. For example, a processed cereal with vitamins added to meet RDAs may contain too much refined sugar and other additives. In spite of these limitations, RDAs remain the *only* available scientifically determined measures of human nutritional needs. Special RDAs are provided for infants and children and pregnant and lactating women.

The typical diet in the United States has changed significantly since 1900. By the 1970s people were eating more meat, poultry, fish, dairy products, fats and oils, sugars, and salt and less grain products, eggs, fruits, and vegetables. Because of these changes and their health implications, in 1977 the U. S. Senate published *Dietary*

Table 8.3

Food and Nutrition Board, National Academy of Sciences—National Research Council Recommended Dietary Allowances,[a] Revised 1989
Designed for the maintenance of good nutrition of practically all healthy people in the United States

Category or Condition	Age (years)	Weight[b] (kg)	(lb)	Height[b] (cm)	(in)	Protein (g)	Fat-Soluble Vitamins Vitamin A (μg RE)[c]	Vitamin D (μg)[d]	Vitamin E (mg α-TE)[e]	Vitamin K (μg)	Water-Soluble Vitamins Vitamin C (mg)	Thiamine (mg)
Infants	0.0–0.5	6	13	60	24	13	375	7.5	3	5	30	0.3
	0.5–1.0	9	20	71	28	14	375	10	4	10	35	0.4
Children	1–3	13	29	90	35	16	400	10	6	15	40	0.7
	4–6	20	44	112	44	24	500	10	7	20	45	0.9
	7–10	28	62	132	52	28	700	10	7	30	45	1.0
Males	11–14	45	99	157	62	45	1000	10	10	45	50	1.3
	15–18	66	145	176	69	59	1000	10	10	65	60	1.5
	19–24	72	160	177	70	58	1000	10	10	70	60	1.5
	25–50	79	174	176	70	63	1000	5	10	80	60	1.5
	51 +	77	170	173	68	63	1000	5	10	80	60	1.2
Females	11–14	46	101	157	62	46	800	10	8	45	50	1.1
	15–18	55	120	163	64	44	800	10	8	55	60	1.1
	19–24	58	128	164	65	46	800	10	8	60	60	1.1
	25–50	63	138	163	64	50	800	5	8	65	60	1.1
	51 +	65	143	160	63	50	800	5	8	65	60	1.0
Pregnant						60	800	10	10	65	70	1.5
Lactating	1st 6 months					65	1300	10	12	65	95	1.6
	2nd 6 months					62	1200	10	11	65	90	1.6

Reprinted with permission from *Recommended Dietary Allowances*, c. 1989, by the National Academy of Sciences. Published by National Academy Press, Washington, DC.

[a] The allowances, expressed as average daily intakes over time, are intended to provide for individual variations among most normal persons as they live in the United States under usual environmental stresses. Diets should be based on a variety of common foods in order to provide other nutrients for which human requirements have been less well defined. See text for detailed discussion of allowances and nutrients not tabulated.

[b] Weights and heights of Reference Adults are actual medians for the U.S. population of the designated age, as reported by NHANES II. The median weights and heights of those under 19 years of age were taken from Hamill et al., (1979) (see pages 16–17). The use of these figures does not imply that the height-to-weight ratios are ideal.

Goals for the United States. The typical American diet at the time of the report is contrasted with the diet that would satisfy the goals in Figure 8.1. These dietary goals attempt to lower the incidence of heart disease, stroke, atherosclerosis, hypertension, some cancers, obesity, diabetes, osteoporosis, tooth decay, and diverticulitis. The National Research Council found that in 1988 the American diet averaged 36 percent fat, 45 percent carbohydrate, and 19 percent protein. Men averaged 1.75 times and women 1.4 times the protein RDA, and men averaged 435 mg and women 304 mg cholesterol daily. Dietary guidelines also have been prepared by the American Cancer Society, the American Heart Association, and the U. S. Department of Agriculture (listed here).

1. Eat a variety of foods.
2. Maintain desirable weight.
3. Avoid too much fat, saturated fat, and cholesterol.
4. Eat foods with adequate starch and fiber.
5. Avoid too much sugar.
6. Avoid too much sodium.
7. If you drink alcoholic beverages, do so in moderation.

More specific recommendations from the National Research Council and the Surgeon General are as follows: Eat fish and skinless poultry instead of meat several days per week. To minimize saturated fats and cholesterol, trim fat from red meats when they are used, use low-fat dairy products, and reduce egg yolks in diet. Have 5 or more servings of fruits and vegetables (including peas and beans) and 6 or more servings of starches (including whole-grain foods) per day. Avoid alcoholic beverages if you are, or plan to become, pregnant. To lose weight or prevent weight gain, limit fats, sugars, and other high calorie foods and engage in regular sustained physical activity.

Water-Soluble Vitamins					Minerals						
Riboflavin (mg)	Niacin (mg NE)[f]	Vitamin B_6 (mg)	Folate (μg)	Vitamin B_{12} (μg)	Calcium (mg)	Phosphorus (mg)	Magnesium (mg)	Iron (mg)	Zinc (mg)	Iodine (μg)	Selenium (μg)
0.4	5	0.3	25	0.3	400	300	40	6	5	40	10
0.5	6	0.6	35	0.5	600	500	60	10	5	50	15
0.8	9	1.0	50	0.7	800	800	80	10	10	70	20
1.1	12	1.1	75	1.0	800	800	120	10	10	90	20
1.2	13	1.4	100	1.4	800	800	170	10	10	120	30
1.5	17	1.7	150	2.0	1200	1200	270	12	15	150	40
1.8	20	2.0	200	2.0	1200	1200	400	12	15	150	50
1.7	19	2.0	200	2.0	1200	1200	350	10	15	150	70
1.7	19	2.0	200	2.0	800	800	350	10	15	150	70
1.4	15	2.0	200	2.0	800	800	350	10	15	150	70
1.3	15	1.4	150	2.0	1200	1200	280	15	12	150	45
1.3	15	1.5	180	2.0	1200	1200	300	15	12	150	50
1.3	15	1.6	180	2.0	1200	1200	280	15	12	150	55
1.3	15	1.6	180	2.0	800	800	280	15	12	150	55
1.2	13	1.6	180	2.0	800	800	280	10	12	150	55
1.6	17	2.2	400	2.2	1200	1200	320	30	15	175	65
1.8	20	2.1	280	2.6	1200	1200	355	15	19	200	75
1.7	20	2.1	260	2.6	1200	1200	340	15	16	200	75

[c]Retinol equivalents. 1 retinol equivalent = 1 μm retinol or 6 μg β-carotene. See text for calculation of vitamin A activity of diets as retinol equivalents.

[d]As cholecalciferol. 10 μg cholecalciferol = 400 IU of vitamin D.

[e]α-Tocopherol equivalents. 1 mg d-α tocopherol = 1 α-TE. See text for variation in allowances and calculation of vitamin E activity of the diet as a α-tocopherol equivalents.

[f]1 NE (niacin equivalent) is equal to 1 mg of niacin or 60 mg of dietary tryptophan.

The Surgeon General's Report also makes recommendations for people with special needs. Fluoride should be obtained from public water supply or dental care products. People especially vulnerable to dental caries should severely limit use of high sugar foods. Adolescent girls and women should increase their intake of foods high in calcium, especially low-fat dairy products. Children, adolescents, and women of childbearing age should eat lean meats, fish, and iron-enriched cereals to satisfy needs for iron. Finally, people with disorders such as lactose intolerance, phenylketonuria, or food allergies that prevent their following guidelines should make a special effort to substitute nutritious foods for the foods they cannot eat.

See Questions—Objective 5

Questions

Objective 5

What are the advantages and disadvantages of guidelines for planning a nutritious diet?

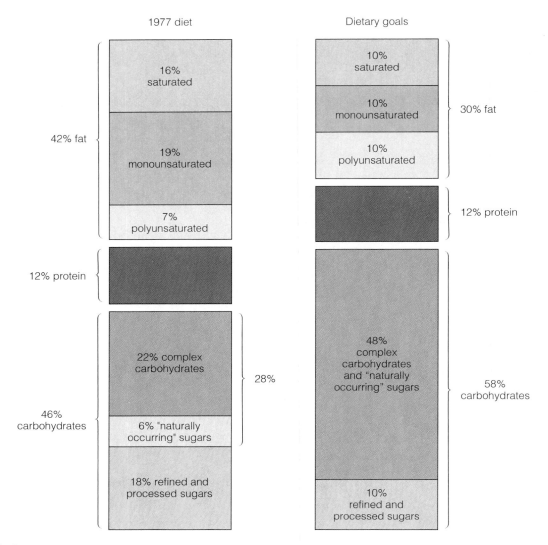

1977 diet

16% saturated

19% monounsaturated

42% fat

7% polyunsaturated

12% protein

22% complex carbohydrates

28%

46% carbohydrates

6% "naturally occurring" sugars

18% refined and processed sugars

Dietary goals

10% saturated

10% monounsaturated

30% fat

10% polyunsaturated

12% protein

48% complex carbohydrates and "naturally occurring" sugars

58% carbohydrates

10% refined and processed sugars

Figure 8.1

A typical American diet prior to establishment of dietary goals is compared with a diet consistent with those goals.

(Source: *Dietary Goals for the United States,* U.S. Government Printing Office, Washington, DC)

What Vitamins and Minerals Do in Cells

Actions, sources, and effects of excesses and deficiencies of vitamins and minerals are summarized in Table 8.4. **Water-soluble vitamins** dissolve in body fluids, and the kidneys excrete excesses. **Fat-soluble vitamins** are absorbed with other lipids and are stored mainly in the liver. Because they are stored, excesses can be toxic.

Among the B vitamins, **thiamine** (thi'am-in) is water soluble and destroyed by temperatures above 100° C. Boiling, frying, or pressure cooking foods reduces their thiamine content. Activated thiamine (with a pyrophosphate group) is a coenzyme for enzymes that oxidize and remove carbon dioxide from a molecule. The enzyme that converts pyruvate to acetyl-CoA is an example.

Riboflavin (ri-bo-fla'vin) is slightly water soluble and not destroyed by heat at most cooking temperatures. It consists of a 3-ring structure (flavin) with an attached ribose. Riboflavin is an essential component of the hydrogen carrier flavin adenine dinucleotide (FAD).

Water-soluble, heat-resistant **niacin** (ni'as-in), or nicotinic acid, is readily converted to nicotinamide and

Table 8.4
Vitamins and Minerals

	Biochemical Actions	Dietary Sources	Effects of Excess	Effects of Deficiency
Water-Soluble Vitamins				
B₁ (thiamine)	Part of decarboxylation enzyme in Krebs cycle	Organ meats, whole grains, and legumes	None known	Arrests metabolism at pyruvate; beriberi (neurological impairment, heart failure)
B₂ (riboflavin)	Part of FAD and FMN	Dairy products, eggs, and whole grains	None known	Sensitivity to light, eye lesions, and cracks in corners of mouth
Niacin	Part of NAD and NADP	Whole grains, meats and legumes; can be formed from tryptophan	Flushing of skin in sensitive people	Pellagra (skin and digestive lesions, mental disorders)
Pantothenic acid	Part of coenzyme A	Widely distributed	None known	Fatigue, nervous and motor impairment (rare in humans)
B₆ (pyridoxine)	Coenzyme for metabolism of amino acids and fats	Whole grains, meats, and vegetables	None known	Dermatitis, nervous disorders, kidney stones, impaired immunity
Folacin (folic acid)	Coenzyme for metabolism of amino acids and nucleic acids	Meats, legumes, green vegetables, and wheat	None known	Impairs production of erythrocytes, intestinal disturbances
B₁₂ (cyanocobalamin)	Coenzyme in nucleic acid metabolism	Meat, eggs, and dairy products	None known	Pernicious anemia, nervous disorders
Biotin	Coenzyme in fat and glycogen synthesis, amino acid metabolism	Egg yolks, legumes, vegetables, and meats	None known	Dermatitis, muscle pains, weakness, and depression
Choline (may not be a vitamin)	Forms phospholipids and acetylcholine	Egg yolk, liver, grains, and legumes	None known	None known in humans
C (ascorbic acid)	Acts in synthesis of collagen and matrix of connective tissue	Citrus fruits	Possibly kidney stones	Scurvy (degeneration of teeth, skin, and blood vessels)
Fat-Soluble Vitamins				
A (carotene)	Forms visual pigments; maintains epithelia	Green and yellow vegetables, fruits, milk, and egg yolks	Headache, loss of appetite, elevated blood calcium, and peeling of skin	Night blindness, excess keratin in tissues of eye
D (calciferol)	Acts in absorption of calcium and bone growth	Fish oils, liver, and fortified dairy products	Kidney damage, vomiting, diarrhea, and weight loss	Bone softness and deformity (rickets in children and osteomalacia in adults)
E (tocopherol)	Maintains integrity of erythrocytes	Green, leafy vegetables; seeds; and oils	None known	Anemia because of fragility of erythrocytes
K (phylloquinone)	Acts in synthesis of prothrombin	Liver; green, leafy vegetables; gut bacteria	May cause jaundice in high doses	Failure of blood coagulation, hemorrhage

Table 8.4
Continued

	Biochemical Actions	Dietary Sources	Effects of Excess	Effects of Deficiency
Minerals				
Calcium	Bone formation, muscle contraction, nerve impulse transmission, blood clotting	Eggs, fish, dairy products, and legumes	Renal damage	Tetany, softening of bones, hemorrhage
Phosphorus	Bone formation, buffers, phosphates in lipids and nucleotides	Dairy products, legumes, meats, and grains	Hypocalcemia	Probably loss of minerals in bones
Magnesium	Cofactor for enzymes, regulates nerve and muscle function	Green vegetables, meat, and milk	Respiratory depression if renal excretion depressed	Tetany
Sodium	Excitability of cells, maintenance of ionic and osmotic balance	Table salt and most foods	Hypertension and edema	Dehydration, renal failure, and cramps in muscles
Potassium	Excitability of membranes in nerve and muscle function	Most foods	Heart arrhythmias	Alteration in muscle contraction and ECG patterns
Sulfur	Part of certain amino acids and other important compounds	Foods containing proteins	None known	None known; deficiency may never have been observed
Chlorine	Osmotic and acid-base balance	Table salt and most food	Edema	Cramps in muscles and alkalosis
Iron	Part of heme of hemoglobin and cytochromes	Liver, eggs, nuts, legumes, and raisins	Hemochromatosis	Lack of hemoglobin in erythrocytes, anemia
Copper	Acts in hemoglobin formation, mitochondrial function, and melanin synthesis	Liver and meats	Tachycardia, hypertension, and coma	Anemia
Cobalt	Acts in hemoglobin formation	Meats	Cardiomyopathy	Anemia
Iodine	Part of thyroxine	Fish, iodized salt	None	Cretinism, goiter
Manganese	Cofactor in enzymes of oxidative and mucopolysaccharide metabolism	Leafy vegetables and whole grains	Muscle weakness, nervous disturbances	Decrease in rate of cellular respiration
Zinc	Part of insulin and some enzymes in nucleic acid metabolism	Many foods	None known	Growth inhibition, testicular atrophy, skin lesions
Fluorine	Suppresses action of oral bacteria, component of bones and tooth enamel	Milk, dentrifices	Mottling of teeth	Dental caries
Molybdenum	Purine metabolism, oxidation of aldehydes	Most foods and water	None	Goutlike symptoms
Chromium	Mediates insulin effects at cell membranes	Unknown	None	Impaired glucose metabolism
Selenium	Part of glutathione peroxidase, prevents peroxide formation	Most foods and water	Toxic in animals	Muscular pain, cardiomyopathy

used to make coenzymes such as nicotinamide adenine dinucleotide (NAD). NAD accepts hydrogen removed from glyceraldehyde-3-phosphate in glycolysis, pyruvate, and several Krebs cycle intermediates. Cells synthesize some niacin from tryptophan if they have pyridoxine (vitamin B$_6$), so the need for dietary niacin depends on how much tryptophan and pyridoxine food contains. Niacin deficiency causes pellagra.

Water-soluble, heat-stable **pantothenic** (pan-to-then'ik) **acid** is required for coenzyme A synthesis. Coenzyme A, in turn, is needed for fatty acid oxidation, to allow acetate to enter the Krebs cycle, and to make acetylcholine.

Substances with vitamin B$_6$ activity include **pyridoxine** (pi-rid-ox'in), **pyridoxal** (pi-rid-ox'al), and **pyridoxamine** (pi-rid-ox'am-ēn), all of which are water soluble and heat stable. Already phosphorylated in food or phosphorylated in cells, these molecules are required for transamination, decarboxylation, the transfer of sulfur between amino acids, and the synthesis of prophyrins and various neurotransmitters. They also help convert the amino acid tryptophan to niacin.

The vitamin **folacin** (fo'la-sin), or folic acid, and certain related compounds are slightly water soluble and are rapidly destroyed by heat. Folacins transfer single-carbon groups (such as $-CH_3$, $-CH_2-$, and $-CHO$) in several metabolic processes, including the synthesis of purines, choline, the amino acid serine, and methionine tRNA. Methionine tRNA is needed to start a polypeptide chain in protein synthesis. Because folacin participates in several processes associated with cell division, folacin deficiency interferes with such processes as red blood cell production, normal growth, and division of malignant cells. The discovery that aminopterin and methotrexate competitively inhibit folacin has led to their use as chemotherapeutic agents.

Vitamin B$_{12}$, or **cyanocobalamin** (si''an-o-ko-bal'am-in), is water soluble and heat stable. The most chemically complex vitamin, it contains pyrrole rings (similar to those in heme) and an atom of cobalt. Cyanocobalamin transfers hydrogen atoms or methyl groups in various reactions, including one that activates folacin. It is obtained mainly from foods derived from animals, so strict vegetarians can develop a cyanocobalamin deficiency. Intestinal microorganisms make the vitamin, but in insufficient quantities to meet human needs. For absorption, cyanocobalamin must be bound to a transport protein called intrinsic factor, without which pernicious anemia results. Like folacin, cyanocobalamin is essential for normal cell division.

Sulfur-containing **biotin** (bi'o-tin) is heat stable and slightly water soluble. It helps to transfer carbon dioxide to make malonyl-CoA in fatty acid synthesis or oxaloacetate from pyruvate. Biotin is inactivated by avidin, an egg white protein. Because biotin is present in many foods, a deficiency is unlikely, unless the diet contains large quantities of raw egg white.

The vitamin-like molecules **choline** (ko'lēn), **inositol** (in-os'it-al), and **lipoic** (lip-o'ik) **acid** serve as coenzymes in various reactions. They are not true vitamins because they can be made in human cells. For example, when folacin, cyanocobalamin, and methyl groups are available, cells make enough choline to support metabolism. If any of those substances is deficient, choline is likewise deficient. Choline is used to make the neurotransmitter acetylcholine and as a component of the lipid portion of cell membranes. Both inositol and choline form part of phosphoglycerides that are essential parts of cell membranes. Lipoic acid is a coenzyme in the reaction that converts pyruvate to acetyl-CoA. Deficiencies of any of these substances are unlikely because they are found in many foods.

Humans require **ascorbic** (as-kor'bik) **acid,** or vitamin C, because they lack an enzyme for its synthesis. It facilitates adding hydroxyl ($-OH$) groups to certain amino acids and steroids. Because hydroxylated proline and lysine are required for collagen synthesis, an ascorbic acid deficiency leads to defective collagen in bone and causes slow wound healing and capillary fragility. Being an antioxidant, vitamin C counteracts oxidation of molecules within cells. It also promotes iron absorption and is required for the synthesis of the hormone thyroxin. Factors that increase a person's need for vitamin C include stress, oral contraceptives, smoking, and diabetes.

Large doses of ascorbic acid have been prescribed to prevent colds, but without evidence for effectiveness or a proposed mechanism of action. Large excesses of ascorbic acid can cause kidney stones and mobilize calcium from bones. After taking large doses of vitamin C, some people suffer symptoms of vitamin C deficiency (scurvy) when they stop.

The fat-soluble vitamins A, D, E, and K are found in the lipid part of foods. They are absorbed, transported in lymph and blood, and enter cells like other lipids.

Several different **carotene** (kar-o-tēn) molecules have **vitamin A** activity when converted to retinol in the liver. Retinol is used to synthesize visual pigments, rhodopsin in rods, and similar pigments in the cones. Vitamin A also helps to maintain epithelial cells in skin and other organs and the structure and permeability of cell membranes. How the vitamin is involved in these processes is unclear. Smoking may increase a person's need for vitamin A. Deficiency of vitamin A leads to night blindness.

Steroids called **calciferols** (kal-sif'er-olz) have **vitamin D** activity. Most human vitamin D comes from activation of **cholecalciferol** (ko-le-kal-sif'er-ol), or vitamin D$_3$, by ultraviolet light striking the skin. **Ergocalciferol** (er-go-kal-sif'er-ol), or vitamin D$_2$, also is active in humans. Spending time in bright sunlight reduces the need for vitamin D in the diet. Vitamin D fosters calcium and phosphorus absorption and helps to deposit them in bone. Fortifying milk with vitamin D helps to assure normal bone growth in children. Childhood vitamin D deficiency leads to rickets.

Several **tocopherols** (tok-of'er-olz) have **vitamin E** activity. Being fat-soluble antioxidants, vitamin E molecules function in the cell membrane to prevent oxidation of polyunsaturated fatty acids in the membrane itself and to protect various substances inside the cell, including vitamin A, from oxidation. The trace element selenium and sulfur-containing amino acids appear to work with vitamin E in preventing oxidation. When vitamin E is deficient, unsaturated fatty acids can be oxidized to peroxides that damage cells and lyse red blood cells. Exposure to smog and a diet high in polyunsaturated fats can increase a person's need for vitamin E.

Three molecules have vitamin K activity in humans. **Phylloquinone** (fi-lo-kwin'ōn), or **vitamin K₁,** is found in green leafy vegetables. **Farniquinone** (far-ne-kwin'ōn), or **vitamin K₂,** is produced by intestinal microorganisms. And **menadione** (men-ad-i'ōn), or **vitamin K₃,** is made in the laboratory. The liver needs vitamin K to make prothrombin and other blood clotting factors. Vitamin K antagonists slows blood clotting in patients subject to thrombi formation.

Inorganic body components called **minerals** are found mainly in the skeleton and as electrolytes in intracellular and extracellular fluids. **Calcium** and **phosphorus** account for as much as 1 percent of the body weight. The National Institutes of Health have recently recommended that women of any age consume 1,000 mg of calcium per day and 1,500 mg after menopause to lower the risk of osteoporosis. Milk and milk products contain the highest amounts of calcium of all foods and 4 to 6 servings of low-fat products are needed to meet these requirements. People who cannot tolerate milk products will find tofu (soybean curd) and lime-treated corn tortillas as good sources of calcium. Consuming glucose with calcium increases calcium absorption. Smoking, large amounts of alcohol, more than twice the protein needed, or more than 35 grams of fiber can decrease calcium absorption. High caffeine intake—more than 5 cups of coffee—may increase calcium excretion.

Other minerals present in significant quantities and essential to the diet are **sodium, potassium, magnesium, sulfur, chloride,** and **iron.** Minerals present in limited quantities, the so-called **trace elements,** include copper, cobalt, iodine, manganese, zinc, fluorine, molybdenum, chromium, and selenium. **Copper** has recently been reported to be 100 times as effective in lowering plasma cholesterol as clofibrate, a drug commonly used for that purpose. **Magnesium** activates more than 300 enzymes and

a severe deficiency can be life-threatening. Patients taking glycosides for heart disorders, some antibiotics, and some diuretics are especially susceptible to magnesium deficiency. Such a deficiency can be treated with an antacid that contains magnesium. **Selenium** (se-le'ne-um) and vitamin E protect erythrocytes against hemolysis (rupture) as long as glucose is present. **Selenosis** (se-len-o'sis) caused by too much selenium occurs in animals grazing on plants grown in high-selenium soil, but human selenosis is rare. This trace element also may help to prevent cancer, heart disease, and sudden infant death syndrome. More selenium and other trace elements may be needed by people subjected to smog or large amounts of polyunsaturated fats and by the elderly.

Other trace elements—nickel, silicon, tin, and vanadium—may be needed for normal metabolism, but their functions are not well understood. Finally, lead, mercury, barium, aluminum, boron, cadmium, and strontium have been identified in human ash. These elements, which probably accumulate as pollutants, have no known functions and some have proven harmful. Heavy metals such as lead and mercury denature proteins and interfere with oxidative metabolism. Aluminum may contribute to the development of anemia and Alzheimer's disease.

See Questions—Objective 6

Special Nutritional Problems

Nutritional problems can occur from excesses or deficiencies of certain nutrients and by accidental or intentional ingestion of nonnutrient substances. Most such problems disturb metabolism or interfere with other physiological processes.

Twenty million Americans, nearly 10 percent of the population, are still hungry, whereas 34 million are obese (15 percent above their normal weight). Though malnutrition (bad nutrition) includes overeating, we will limit this discussion to problems with insufficient nutrients to maintain normal metabolism. Two types of **malnutrition** are recognized, and both are common among children: (1) protein deficiency called **kwashiorkor** (kwash-e-or'kor) and (2) near starvation called **marasmus** (mar-az'mus). Protein deficiency may be far less important in malnutrition than was once thought. Though infants and young children need more protein per kilogram body weight than adults, they do not need a high-protein diet. Less than 7 percent of calories in breast milk come from protein, and many cereals contain more protein than breast milk.

Diets where malnutrition is common usually are not protein deficient, but they may lack essential amino acids. Kwashiorkor, "the disease that comes when a second

Questions

Objective 6

What vitamin and mineral deficiencies might lead to the following disorders and what would be the biochemical consequences:

(a) neurological disorders?

(b) skin disorders?

(c) reduced wound healing?

(d) anemia?

(e) impaired blood clotting?

(f) functional abnormalities in membranes?

(g) impaired vision?

Table 8.5
Vitamin-Responsive Inherited Metabolic Disorders

Vitamin	Biochemical Defect	Disorder	Manner of Inheritance	Therapeutic Dose/Day
B_1 (thiamine)	Pyruvate decarboxylase deficiency	Pyruvicacidemia	Unknown	5–20 mg
B_6 (pyridoxine)	Glutamate decarboxylase deficiency	Infantile convulsions	Autosomal recessive	10–50 mg
	Unknown	Hypochromic anemia	Sex-linked recessive	>10 mg
B_{12} (cyanocobalamin)	Intrinsic factor deficiency, inactive intrinsic factor, transport deficiency	Megaloblastic anemia	All autosomal recessives	5–100 μg
Folic acid	Impaired intestinal folate absorption	Megaloblastic anemia	Unknown	<0.05 mg
	A folate reductase enzyme deficiency	Mental retardation, schizophrenic psychosis	Autosomal recessive	>10 mg
Biotin	Propionyl-CoA carboxylase deficiency	Propionicacidemia ketoacidosis and retardation	Autosomal recessive	10 mg
Niacin	Impaired intestinal and renal transport of tryptophan	Hartnup disease, cerebellar ataxia	Autosomal recessive	>40 mg
D-calciferol	Unknown	Hypophosphatemic rickets	Sex-linked dominant	>100,00 units (>4 g cholecalciferol)
	Deficiency of an enzyme in calciferol synthesis	Vitamin D-dependent rickets	Autosomal recessive	>25,000 units (>1 g cholecalciferol)

child is born," is such a deficiency. Recently weaned children obtain barely sufficient protein from the adult diet when they are well. If they get sick and eat less food, they can develop protein deficiency.

Dietary protein deficiency decreases serum proteins, which reduces blood osmotic pressure. Fluid fails to return to the blood and edema and **ascites** (as-si'tēz), fluid accumulation in the abdominal cavity, develop. Ascites accounts for the "potbelly" of children suffering from kwashiorkor. Other proteins—blood clotting factors, enzymes to make pigments—are in short supply, and erythrocytes, skin cells, and other cells that normally divide rapidly become depleted. Blood is slow to clot, wounds are slow to heal, hair becomes gray, and anemia and skin lesions develop. If protein deficiency occurs in early childhood, protein synthesis and nerve cell division are impaired, causing slow growth and mental retardation.

Marasmus in children and adults is caused by inadequate caloric intake. All essential nutrients—essential amino acids, linoleic acid, vitamins, and minerals—are likely to be deficient. Victims of marasmus suffer from all the symptoms of protein deficiency, especially muscle wasting, anemia, and susceptibility to microbial and parasitic infections. Infections further decrease the amount

of nutrients available to body cells. Some people with nutritional deficiencies, especially iron or zinc deficiency, develop a curious condition called **pica** (pi'kah), a craving for nonfood substances. Women and children living in poverty are most susceptible to pica. They sometimes eat wall plaster, paste, or large quantities of ice, though these substances do not alleviate the deficiency.

Certain inherited metabolic disorders are caused by an absent or defective gene. They respond to excessively large therapeutic doses of specific vitamins (Table 8.5), possibly by stimulating synthesis of an enzyme or coenzyme. An excessively large dose of a vitamin might be therapeutic if it improves digestion, absorption, transport, or activation of a vitamin.

Certain effects of alcoholism and cystic fibrosis are to some extent vitamin-responsive. Increasing vitamin intake in alcoholics overcomes dietary deficiencies and allows greater vitamin activation by the damaged liver. Increasing vitamin intake in cystic fibrosis overcomes the deficiencies, especially of fat-soluble vitamins that are not released from foods because of pancreatic enzyme deficiencies.

Restricted diets can cause malnutrition and vitamin deficiencies at any age and any economic level.

Vegetarianism leads to vitamin B_{12} deficiency. The so-called macrobiotic diet (consisting mainly of unpolished rice) can lead to protein deficiency and to various vitamin deficiencies. Liquid protein diets or diets limited to only a few foods also lead to deficiencies. Such deficiencies can be prevented by following dietary guidelines.

High sodium intake has been implicated in hypertension, and moderate potassium or calcium offers some protection against it. However, excessive potassium can dangerously alter heart contractility. People with normal blood pressure can consume up to 6 grams of salt with no effect. Many physicians recommend that hypertensive patients consume no more than 2 grams of salt per day—about the lowest possible sodium intake on a nonhospital diet. They believe that low sodium intake reduces the amount of medication needed to control blood pressure. Most people in nonindustrialized areas eat natural non-processed foods (low in sodium and high in potassium). They have a low incidence of hypertension, maintain physical fitness by working hard to make a living, and typically lose rather than gain weight with age. Severe hypertension is related to the ability of blood vessels to constrict and their inability to dilate. Dietary fish oils rich in omega-3 fatty acids have improved blood vessel dilation in some patients with hypertension.

Researchers have related lipoproteins to coronary artery disease. LDLs contain large amounts of cholesterol that can be deposited in arteries, whereas HDLs carry cholesterol back to the liver and seem to protect against blood vessel diseases. It now appears that dietary saturated fatty acids contribute more to elevated blood cholesterol than dietary cholesterol itself. Oat bran, legumes, and other foods with water-soluble fiber reduce blood cholesterol. Such fiber binds to bile acids in the small intestine. This decreases emulsification and absorption of fats and causes bile salts—and therefore cholesterol—to be excreted. As colon bacteria digest soluble fiber, they release short chain fatty acids, which may be absorbed and act in the liver to decrease cholesterol synthesis.

A study comparing nibbling (17 snacks/day) with the same food in three meals, indicated that the nibblers had lower total and LDL cholesterol and lower serum insulin. Insulin stimulates lipid synthesis and proliferation of smooth muscle in blood vessels. Having many small meals minimizes insulin secretion, and the lower insulin reduces the risk of blood vessel disease from both lipid synthesis and smooth muscle proliferation.

The National Cholesterol Education Program makes many of the same dietary recommendations as other guidelines—total fat less than 30 percent, saturated fat less than 10 percent of calories, and cholesterol less than 300 mg/day. If this Step-One Diet fails to lower cholesterol, the Step-Two Diet, which reduces saturated fat to less than 7 percent of calories and cholesterol to less than 200 mg/day, is recommended.

Certain commonly used foods and beverages contain nonnutritive substances, such as caffeine, food additives, and pollutants. "Junk foods" and alcohol satisfy the appetite but have little nutritional value. Purines called **xanthines** (zan'thinz) occur as **caffeine** in coffee, tea, and some carbonated beverages, **theophylline** (the-o-fi'lin) in tea, and **theobromine** (the-o-bro'min) in cocoa. Xanthines inhibit the breakdown of cyclic-AMP, so any metabolic processes stimulated by cAMP are accelerated and prolonged. Glycogenesis increases, elevating the blood glucose concentration. Lipolysis increases, elevating the free fatty acid concentration in blood. Xanthines increase sympathetic stimulation, which increases insulin secretion, thereby reducing glycogenolysis. Xanthines relax smooth muscle in bronchioles and pulmonary arterioles, increasing both airflow and blood flow. They stimulate myocardial contractility and relax arterial smooth muscle, increasing cardiac output.

Caffeine has both good and bad attributes. The good attributes appear when a person drinks 2 or 3 cups of coffee per day, and more is not better. Caffeine elevates mood, probably by binding to the same brain receptors as adenosine. Adenosine tells the brain to "slow down," and caffeine overrides this effect. Caffeine also improves mental performance, counteracts asthma symptoms, and accelerates calorie use by as much as 10 percent, especially in obese people. It also disrupts sleep, reduces fertility, and may contribute to a heart attack by increasing platelet stickiness.

Candy, carbonated beverages, potato chips, and any other snack foods that contain mainly refined sugar or fat and few vitamins and minerals are commonly known as **junk foods.** They can lead to nutritional deficiencies and undesirable weight gain. Foods and beverages sweetened with saccharine or aspartame pose other hazards. Saccharine may be carcinogenic (at least in some animals) and aspartame may interfere with neurotransmitter actions in the brain. In a study on a small number of people, those receiving aspartame-sweetened foods lost more weight than those receiving unsweetened foods. In another small study, people receiving sucrose- or aspartame-sweetened foods reported the same degree of satiety regardless of which sweetener was used.

At least 2,600 different chemical substances are added to processed foods sold in the United States today. A food additive is something intentionally added to food, usually to improve its appearance, it flavor, its nutritional value, or the length of time it can remain edible on a grocery shelf. Nutritional supplements can enrich or fortify foods. **Enrichment** is the addition of nutrients to foods, such as breads and cereals, to return them to the same nutritional value they had before processing. **Fortification** is the addition of nutrients to foods, such as milk and other dairy products, that cause them to have greater than normal nutritional value. Food additives and their properties are summarized in Table 8.6.

Table 8.6
Food Additives

Type	Function	Examples	Typical Uses
Preservatives	To retard spoilage by preventing action of microorganisms	Heating and sealing, dehydration, freezing, pasteurization, salt, sugar, sodium nitrate and nitrite, sodium and calcium propionates, sodium benzoate, sulfur dioxide	Breads, cheeses, meats, jellies, syrups, fruits, and vegetables
Antioxidants	To retard spoilage by preventing oxidation	Lecithin, butylated hydroxyanisole (BHA), butylated hydroxytoluene (BHT)	Oils, cereals, potato chips, and other snack foods
Flavorings and flavor enhancers	To add or enhance flavor	Herbs, spices, synthetic flavorings, monosodium glutamate (over 1,100 substances)	Nearly all processed foods
Dyes	To add color appeal or mask undesirable colors	Natural dyes or synthetic coal tar dyes	Soft drinks, cereals, cheeses, ice cream, nearly all other processed foods
Acids and alkalies	To add or mask tartness; to mask undesirable flavors	Phosphoric and citric acids, sodium bicarbonate	Soft drinks, juices, wines, olives, salad dressings
Emulsifiers and stabilizers	To prevent separation of components and provide smooth texture	Lecithin, propylene glycol, acylglycerols, polysorbates, vegetable gums, cellulose, gelatin, seaweed extracts	Soft drinks, ice cream, salad dressings, cheese spreads, cake mixes, margarine, candies
Chelating agents	To tie up trace metals that might cause oxidation, to prevent clouding, and to improve texture	Ethylenediaminetetraacetic acid, sodium phosphate, chlorophyll	Artificial fruit drinks, salad dressings, soft drinks, beer, cheeses, canned and frozen foods

Source: Data from G. T. Miller, Jr., *Living in the Environment.*

Contaminants that accidentally enter foods—pesticide residues, heavy metals, microorganisms and their toxins, and insects—are common food pollutants. Pesticides like DDT (dichlorophenyl-tricholoroethane) and PCBs (polychlorinated biphenyls) have been used to kill crop-damaging insects. Another pesticide, EDB (ethylenedibromide), used to control insects in grain stores, has been found in grain products. Years after their use has ceased, their residues remain in the environment because living things lack enzymes to degrade them. When animals or humans eat crops treated with pesticides or grown in soil containing pesticides, they ingest pesticide residues too.

Pesticides undergo **biological magnification**—they increase in concentration as they pass along the food chain. As animals eat fat-soluble pesticides on contaminated plants, fatty tissues accumulate pesticide residues. Humans ingest more of pesticide residues from eating animals that have eaten pesticide-treated plants. This hazard can be reduced by removing fats from meats and by eating only low-fat dairy products. How pesticides damage tissues is not fully understood, but they can cause neurological disorders, liver damage, and cancer.

Heavy metals such as mercury also are found as food contaminants, especially in ocean fish and sometimes in grains. Methylmercury, a particularly toxic metal, accumulates in the tissues of swordfish, tuna, and other ocean fishes. This mercury comes mainly from natural ocean sediments and not from human activities. Methylmercury also is used to prevent fungal growth in grain. Ordinarily applied to seed grain, it is ingested only when humans eat the grain instead of planting it. Mercury poisoning causes severe neurological damage and even death if sufficient quantities are ingested. Cadmium, another heavy metal, usually enters the body from water contaminated with industrial wastes. Large amounts can cause skeletal disintegration, but small amounts—0.1 part per million—contribute to heart disease and hypertension.

Finally, many food products are contaminated with microorganisms and insect body parts—the very things additives are intended to control. The microorganisms pose a threat to health only if they can cause human infections and are virulent enough to do so, or if they have produced toxins that affect humans. Bacteria that cause food poisoning and viruses that cause hepatitis are examples. *Gonyaulax,* an alga responsible for "red tides," produces **saxitoxin** (sax-e-tox'in), which blocks sodium channels and causes severe neurologic damage. Insect parts, though aesthetically unattractive, may pose a less serious health hazard than most additives used to control them. Health food stores frequently tout a particular dietary supplement with little or no scientific evidence of its nutritional value. Such practices can be harmful as in the case of the naturally occuring amino acid l-tryptophan. Used as a nutritional supplement, it causes a rare blood disease called **eosinophilia.**

Numerous substances in plants—safrole in root beer, piperine in black pepper, hydrazines in mushrooms, solanine in potatoes, quinones in rhubarb, and theobromine in cocoa—have been reported to cause cancer in laboratory animals. Other carcinogens include alcohol, aflatoxin from molds sometimes found in peanut butter, and nitrosamines formed in the body from natural nitrates and nitrites and those added to foods such as bacon and hot dogs.

Some carcinogens act by forming highly reactive oxygen radicals that damage DNA. These radicals are counteracted by antioxidants such as vitamins A, C, and E and selenium in foods. Some researchers believe intakes of 2,500 μ of vitamin A, 1 g of vitamin C, and 200 IU (international units) of vitamin E daily may reduce the risk of cancer, but they warn that excesses of these nutrients have severe toxic effects.

Some cancers may be related to a kind of biological time warp—a conflict between our bodies that function as they have for centuries and our twentieth century diet and sedentary life-style. To improve the situation, the National Research Council recommended the following dietary changes to reduce the risk of cancer: limiting dietary fats to 30 percent of calories, reducing pickled, salt-cured, and smoked foods, increasing fruits and vegetables, and limiting alcohol. Vegetables from the cabbage family—broccoli, brussels sprouts, cauliflower—and other high fiber foods are recommended to reduce the risk of cancer.

High-fat diets are correlated with gastrointestinal, breast, and prostate cancers. Fats should be reduced to minimize the cancer risk from excess bile acids and steroids, which are linked to malignant tumors in laboratory animals. Pickled, salt-cured, and smoked foods should be limited because they can become contaminated during processing with chemicals known to cause cancer in animals. Fruits and vegetables, which contain vitamins A and C, seem to protect against cancer. Vitamin A seems to protect epithelial cells, especially in the lungs. Vitamin C may protect stomach cells by blocking the formation of nitrosamine, a known carcinogen. Alcohol should be limited because of the high incidence of mouth, laryngeal, and esophageal cancer among users, especially those who smoke. Cabbage and similar vegetables probably contain substances that enhance the activity of enzymes that destroy carcinogens. Insoluble fiber decreases transit time of feces and neutralizes soluble fiber-bile acid combinations.

Relationships among melatonin, serotonin, and the pineal gland may affect cyclic behaviors that display carbohydrate craving, weight gain, depression, and excessive sleeping. **Carbohydrate craving obesity** (CCO) is related to daily rhythms that induce late-afternoon and evening carbohydrate craving. **Premenstrual syndrome** (PMS) is related to monthly rhythms, and **seasonal affective disorder** (SAD) is related to seasonal changes in yearly rhythms. SAD may affect as many as one-fourth of the population in northern latitudes but is rare near the equator. The pineal gland secretes melatonin at night and serotonin in daylight when signals elicited by light inactivate the enzyme that converts serotonin to melatonin. According to one theory, when carbohydrate is eaten, insulin secretion increases, causing most amino acids to move into cells. This leaves a high blood concentration of the amino acid tryptophan, which is not affected by insulin. Much of the tryptophan crosses the blood-brain barrier and is used to make the neurotransmitter serotonin. When serotonin is released at brain synapses it may induce both sleep and carbohydrate satiety. Some researchers report quite different responses to carbohydrate meals: Normal people get sleepy, but victims of CCO who are deficient in serotonin experience carbohydrate satiety and a sense of well being without becoming sleepy.

See Questions—Objectives 7 through 10

Getting Proteins from Plants

Protein is essential in the human diet, but it need not come from meat. What humans need is an adequate supply of essential amino acids and other amino acids to allow protein synthesis. This need can be met by plant proteins eaten in the right combinations at the same meal. Most plants contain **incomplete proteins;** that is, they have too little of some amino acids—lysine, isoleucine, tryptophan, and the sulfur amino acids, cysteine and methionine—to meet human needs. Various plants lack different amino acids and dairy products can easily supplement these deficiencies (Table 8.7). By learning which plant foods supply certain essential amino acids and where dairy products are needed, complementary combinations, such as beans and rice or cereal and milk, that supply all essential amino acids can be made.

Human tissues are similar to the tissues of other animals and need amino acids in similar proportions. Even so, our bodies can use only about 90 percent of the amino

Questions

Objective 7

(a) What are the main effects of malnutrition?

(b) How do kwashiorkor and marasmus differ?

(c) What are the main attributes of vitamin-responsive metabolic disorders?

Objective 8

(a) How might diet contribute to or prevent hypertension?

(b) How might diet contribute to or prevent atherosclerosis?

Objective 9

What nutritional problems are associated with
(a) xanthines?

(b) pesticides?

(c) food additives?

(d) food pollutants?

Objective 10

In what ways might diet relate to cancer and to behavior?

Table 8.7
Availability of Selected Essential Amino Acids in Plant Foods

Food	Tryptophan	Isoleucine	Lysine	Sulfur-Containing Amino Acids
Soybeans	+++	++	+++	+
Peas, beans, and other legumes	+	++	+++	+
Nuts and seeds	+++	+	+++	+
Grains and whole-grain foods	++	+	+	++
Corn	0	+	+	++
Green vegetables	++	+	++	+
Dairy products	++	+++	+++	++
Eggs	+++	+++	+++	+++
Seafoods	++	++	+++	++
Meat and poultry	++	++	+++	++

+++ = excellent source, ++ = good source, + = poor source, 0 = very poor source

Questions
Objective 11

(a) Which amino acids are most likely to be deficient in plant proteins?

(b) How can foods from plants be combined to provide proteins of high biological value?

(c) What are some advantages of limiting one's intake of meat?

(d) Why is a strictly vegetarian diet not recommended?

acids in an egg, 80 percent of those in milk, and 70 percent of those in meat, poultry, and fish. These percentages, called the **net protein utilization,** represent the proportion of amino acids that can be converted to human proteins. Net protein utilization is directly proportional to biological value: Higher biological value means more amino acids will be used in protein synthesis. Plant foods generally have lower net protein utilization—from 70 percent for rice to 40 percent for kidney beans. To meet protein needs, larger servings of plant than animal foods must be eaten. For example, 4 ounces of cooked hamburger provides 30 grams of protein, 67 percent or 20 grams of which is usable by the body. To obtain 20 grams of usable plant protein would require a combination of 8 ounces of rice and 6 ounces of soybeans with 32 grams of protein, 61 percent usable.

Eating less meat has nutritional, ecological, and economic advantages. Nutritionally, increased plant food intake greatly increases dietary vitamins, minerals, and fiber and minimizes the risk of biologically magnified pesticide residues in meats. Because plants lack cholesterol and contain mostly unsaturated fats, eating them lowers cholesterol and saturated fat consumption. Eating raw, unprocessed plant foods avoids some food additives and pollutants. Ecologically, using large grazing areas for growing grain allows the land to feed many more people. (Animals need about 20 pounds of plant proteins to make 1 pound of edible protein.) Grain used to feed cattle a few weeks before slaughter uses still more plant food and adds saturated fats to the meat. Economically, plant foods even in the quantities needed to meet protein needs, cost less than meats.

A strictly vegetarian diet is not recommended because plant foods are deficient in vitamin B_{12} and the amino acids already noted. Small amounts of dairy products supply vitamin B_{12} and increase the biological value of proteins. Meat is not essential to a nutritious human diet.

See Questions—Objective 11

Study and Review

Did you get the essentials?

The following summary contains the basic concepts from chapter 8 except for the key terms that have been omitted. Read the chapter. Then try to fill in the blanks from memory. If you cannot fill in all the blanks, review the chapter paying particular attention to boldface terms.

_____ is the providing of all nutrients needed for good health through food intake. Humans need six kinds of nutrients—_____ , _____ , _____ , _____ , _____ , and _____ . Carbohydrates provide _____ and _____ , but lipids provide only _____ and a single _____ _____ _____ linoleic acid. Proteins provide _____ _____ cells used to make their own proteins. The diet must contain adequate amounts of _____ _____ _____ because the body cannot make them. The _____ _____ of a protein is a measure of how nearly its proportions of amino acids match those of _____ _____ . _____ are needed because they form part of _____ or other essential molecules such as _____ and _____ pigments. _____ are needed because they serve as enzyme _____ , maintain _____ , and are incorporated into _____ . _____ maintains proper volumes of body _____ .

Excess heat the body releases as it digests food is called _____ _____ . It is highest for _____ .

Eating the daily recommended servings from the _____ _____ _____ groups insures good _____ for most people. _____ _____ _____ provide the best available guidelines for quantities of certain nutrients needed by most healthy people. Other dietary guidelines advise _____ .

_____ removes CO_2 from molecules, _____ forms part of FAD, _____

forms part of NAD, and _____ _____ forms part of coenzyme _____ .
_____ is needed for amino acid metabolism, _____ transfer single carbons, and
_____ activates _____ and is absorbed only when _____ factor is
available. _____ is required for fatty acid synthesis. An _____ _____
deficiency leads to defective _____ and poor wound healing. All the above vitamins are
_____ soluble.

Vitamin _____ maintains epithelial cells, vitamin D fosters absorption of _____ ,
vitamin _____ protects cell parts from _____ , and vitamin _____ is
needed for synthesis of _____ and other _____ _____ factors. These
vitamins are _____ soluble.

Among minerals, _____ and _____ are found in large quantities in bone.
_____ is required to activate many enzymes.

_____ (bad nutrition) can include overeating but especially includes _____ (protein
deficiency) and _____ (lack of all _____). The potbelly of children with protein
deficiencies is caused by _____ , or _____ accumulation in the _____ .
People with malnutrition often develop _____ , a craving for nonfood substances.

Disorders and diseases that may be caused by or worsened by what a person eats include _____
_____ _____ .

_____ foods have many calories and few nutrients. _____ _____ are
intentionally added. Additives _____ when they return a food to its original nutritional value and
_____ when they provide greater than normal _____ value. _____
_____ are unintentionally added to foods. They include pesticides, which undergo _____
_____ as they are passed up the food chain.

Excretion; Fluid, Electrolyte, and Acid-Base Balance

9

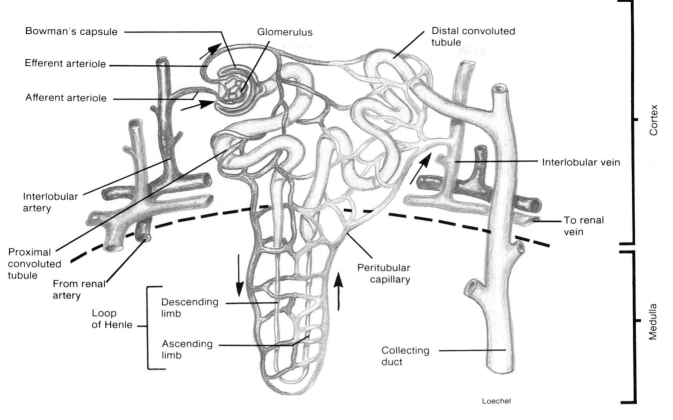

Bowman's capsule

Glomerulus

Distal convoluted tubule

Efferent arteriole

Afferent arteriole

Cortex

Interlobular vein

Interlobular artery

To renal vein

Proximal convoluted tubule

From renal artery

Peritubular capillary

Loop of Henle

Descending limb

Ascending limb

Medulla

Collecting duct

Loechel

Figure 9.1

The structure and location of cortical and juxtaglomerular nephrons.

Nephrons

Nephrons, the functional units of kidneys, perform two major functions—excretion of nitrogenous wastes and adjustment of body fluid volumes and solute concentrations. Each of the million nephrons in a human kidney consists of a tubule and its associated capillaries. **Cortical nephrons** are located solely in the cortex, whereas **juxtamedullary nephrons** extend into the medulla (Figure 9.1). All nephrons have a long, coiled, hollow tube attached at one end to a collecting duct with the other end modified into a cuplike **glomerular** (glo-mer′u-lar) **capsule** (Bowman's capsule). Inside the cup is a tuft of capillaries called a **glomerulus** (glom-er′u-lus). The capsule itself has an inner layer of epithelial cells called **podocytes** (pod′o-sītz) that are adjacent to the glomerulus and an outer layer of squamous epithelial cells. Substances filtering out of blood in the glomerulus pass through filtration slits in the podocytes and accumulate in the space between the two layers of the capsule.

Portions of the tubule between the glomerular capsule and a collecting duct are differentiated into the **proximal convoluted tubule,** the descending and ascending limbs of the **loop of Henle,** and the **distal convoluted tubule.** The loop of Henle is short in cortical nephrons and long in juxtaglomerular nephrons. These structures form a continuous tube through which kidney filtrate passes and is processed to become urine. Each collecting duct receives urine from several nephrons.

The capillaries surrounding the tubule also are part of the nephron, and in addition to the glomerulus, each nephron has **peritubular** (per-e-tu′bu-lar) **capillaries.** Many **afferent arterioles** branch from each interlobular artery and each goes to a glomerulus, where many substances are filtered out of the blood. From a glomerulus, blood enters an **efferent arteriole** and is carried to the peritubular capillaries. Peritubular capillaries surround the tubule and supply its parts in the following order: proximal convoluted tubule, distal convoluted tubule, and the loop of Henle. The capillaries surrounding the loop of Henle are called **vasa recta** (va′sah rek′tah). Many substances move between the fluid in the tubule and the blood in the peritubular capillaries. Blood leaving the peritubular capillaries enters venules that lead to the interlobular veins.

See Questions—Objective 1

Questions

Objective 1

(a) What are the major functions of nephrons?

(b) Describe the structure of a juxtamedullary nephron and explain how a cortical nephron differs from it.

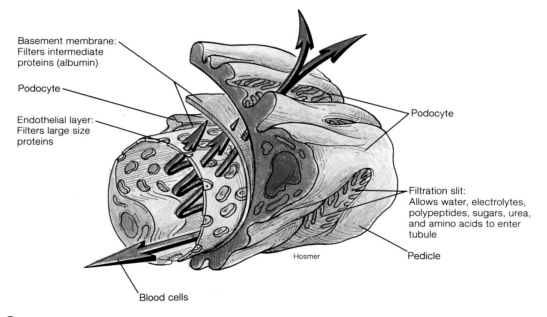

Basement membrane:
Filters intermediate
proteins (albumin)

Podocyte

Endothelial layer:
Filters large size
proteins

Podocyte

Filtration slit:
Allows water, electrolytes,
polypeptides, sugars, urea,
and amino acids to enter
tubule

Hosmer

Pedicle

Blood cells

Figure 9.2

Movement of substances and pressures in a glomerular capillary.

Glomerular Filtration

As blood flows through glomerular capillaries, plasma enters glomerular capsules by filtration (Figure 9.2). Water and low molecular weight substances pass through the endothelium, basement membrane, and filtration slits between podocytes, accumulating between the two layers of the capsule. The sievelike endothelial layer holds back blood cells, the basement membrane holds back large protein molecules, and filtration slits hold back smaller ones. The few small protein molecules that reach the filtrate are reabsorbed and digested by kidney tubule cells. Consequently, urine contains no protein unless glomeruli are damaged or tubules have impaired reabsorptive capacity.

Filtration depends, not only on the above anatomical properties, but also on relative pressures in glomeruli and capsules. Hydrostatic pressure in blood entering the glomeruli is about 55-torr—more than twice that in most capillaries. This pressure is derived partly from the heart's pumping and partly from efferent arterioles having smaller diameters than afferent arterioles. Blood oncotic pressure of about 30 torr and hydrostatic pressure of about 15-torr within the glomerular capsule together exert a 45-torr pressure against filtration. Thus, the **net filtration pressure** pushing substances out of a glomerulus is about 10 torr—about ten times that in most capillaries. Fluid leaving the glomeruli is **kidney filtrate,** sometimes called an **ultrafiltrate** because it is formed under pressure.

In normal human kidneys, plasma moving from glomeruli to capsular spaces forms glomerular filtrate at 115 ml/min to 125 ml/min—the **glomerular filtration rate** (GFR). GFR, which is proportional to body size and decreases with age, is determined mainly by net filtration pressure. A volume equal to the total adult plasma volume of about 3 liters enters the glomerular capsules once every 25 minutes. Over a 24-hour period, glomerular filtration produces about 180 liters of filtrate, which only produces about 1 liter of urine. The other 179 liters, including many dissolved substances, are returned to the blood by reabsorption in other parts of the kidney tubules. Except for proteins, dissolved substances have the same concentration in the filtrate as in the plasma.

See Questions—Objective 2

Tubular Reabsorption

Kidney filtrate flows from the glomerular capsule into the proximal convoluted tubule and on through the renal tubule to a collecting duct. During this passage, water and solutes are returned to the blood by tubular reabsorption. Water, waste urea, and ions such as Na^+, K^+, and Cl^- can be reabsorbed from all parts of kidney tubules and collecting ducts, but most reabsorption takes place in the proximal tubules. There metabolically active epithelial cells have a brush border composed of many tiny microvilli that greatly increase the absorptive surface area. Active transport of solutes out of the filtrate creates a small osmotic gradient and water leaves the filtrate by osmosis in the same proportion as the solutes. At the end of the proximal tubules, 60 to 70 percent of the filtrate's water and solutes have been reabsorbed.

Reabsorption of Na^+ involves both active transport and diffusion. To reach the blood plasma from the tubule lumen, Na^+ diffuses down electrical and concentration gradients across the brush-border membrane into the epithelial cells. Na^+ is then actively transported into **peritubular** (interstitial) **fluid** as K^+ is transported into the cells by a sodium-potassium pump. Carrier molecules, ATPase, and energy from ATP carry out this process,

Questions
Objective 2

(a) Where does filtration occur and what forces produce it?

(b) What determines which substances leave the blood during filtration and which are held back?

which accounts for more than 5 percent of basal metabolism. Once in peritubular fluid, Na^+ diffuses through peritubular capillaries into blood at a rate proportional to the concentration gradient.

Several other substances also are actively reabsorbed. Glucose is reabsorbed by **cotransport** like that in the digestive tract in which energy expended by the sodium-potassium pump indirectly transports glucose along with Na^+. Amino acids are transported carrier molecules. Each of four carriers transports a particular group of amino acids according to their shapes and charges. A portion of certain wastes such as uric acid and creatine (a product of muscle metabolism) also are actively reabsorbed.

The reabsorption rate of any actively transported solute is the product of the solute's plasma concentration and GFR. For a plasma concentration of 80 mg/dl of glucose and GFR of 125 ml/min, reabsorption is directly proportional to filtration and occurs at a rate of 100 mg/min. The theoretical maximum reabsorption rate, determined by the availability of ATP, ATPase, and carriers, can be much higher—up to 375 mg/min for glucose and sufficient to remove all but traces of glucose from the filtrate even when the plasma concentration exceeds 300 mg/dl. The plasma concentration at which no more of a solute can be returned to the blood and is retained in significant amounts in the urine is called the **renal threshold.** According to these data, the theoretical renal threshold for

Questions

Objective 3

(a) Where does reabsorption occur and what produces it?

(b) What determines which substances are returned to the blood during reabsorption?

(c) What kinds of defects impair reabsorption and how do they exert their effects?

glucose is about 300 mg/dl, but the actual threshold is closer to 200 mg/dl because not all kidney tubules are working to maximum capacity to bind and transport glucose.

See Questions—Objective 3

Tubular Secretion

Whereas large quantities of water and nearly all solutes are reabsorbed from the kidney filtrate, only small quantities of certain plasma solutes are secreted into it. Tubular secretion occurs mainly in the distal tubules, which are impermeable to water. Secretion is a fine-tuning mechanism—it makes final adjustments in the chemical composition of plasma beyond what is accomplished by glomerular filtration and tubular reabsorption.

Like other active transport processes, tubular secretion requires carriers, ATPase, and energy from ATP. H^+, often combined with ammonia, is secreted as a part of the acid-base regulating mechanism. K^+ is secreted and Na^+ is reabsorbed under the influence of aldosterone. This hormone stimulates synthesis of carriers and enzymes for active transport in the distal tubule epithelium. Na^+ reabsorption creates an osmotic gradient that conserves water. Penicillin, certain other drugs, and some toxic substances are removed from the blood by tubular secretion.

Filtration, reabsorption, and secretion of plasma constituents are summarized in Table 9.1 and Figure 9.3. In addition to the 60 to 70 percent of solutes and water reabsorbed in proximal tubules, another 20 percent is reabsorbed in the loops of Henle, and the remainder is reabsorbed in distal tubules and collecting ducts.

See Questions—Objective 4

The Countercurrent Mechanism

The **countercurrent mechanism** concentrates urine by increasing osmolarity in peritubular fluids around collecting ducts. It involves the loops of Henle as **countercurrent multipliers** and the vasa recta (capillaries around the loops of Henle) as **countercurrent exchangers.** In a **countercurrent system,** outflow runs parallel to and in the opposite direction of the inflow for a significant distance, as illustrated by straight and U-shaped tubes with water flowing though them (Figure 9.4). Applying heat to the middle of a straight tube warms the outflow but has little effect on the inflow. Applying heat to the bend in a U-shaped tube warms the outflow and heat from the outflow warms the inflow. Thus, outflow affects inflow in a countercurrent system, whether the system is a heated U-shaped tube or a system that moves Na^+, Cl^-, and water in a loop of Henle.

The loops of Henle of kidney tubules are **countercurrent multipliers.** The loop shape allows passive countercurrents to develop, and active transport of substances out of the system's outflow portion multiplies the

Table 9.1

Renal Handling of Various Plasma Constituents in a Normal Adult Human on an Average Diet

Substance	Per 24 Hours				Percentage Reabsorbed	Location
	Filtered	*Reabsorbed*	*Secreted*	*Excreted*		
Na⁺ (mEq)	26,000	25,850		150	99.4	P,L,D,C
K⁺ (mEq)	600	560*	50*	90	93.3	P,L,D,C
Cl⁻ (mEq)	18,000	17,850		150	99.2	P,L,D,C
HCO₃⁻ (mEq)	4,900	4,900		0	100	P,D
Urea (mmol)	870	460†		410	53	P,L,D,C
Creatinine (mmol)	12	1‡	1‡	12	. . .	. . .
Uric acid (mmol)	50	49	4	5	98	P
Glucose (mmol)	800	800		0	100	P
Total solute (mOsm)	54,000	53,400	100	700	87	P,L,D,C
Water (ml)	180,000	179,000		1,000	99.4	P,L,D,C

Reprinted with permission from Ganong, W. F.: *Review of Medical Physiology,* 14th Edition, 1989, by Appleton & Lange, Norwalk, CT and San Mateo, CA.

Note: P, proximal tubules; L, loops of Henle; D, distal tubules; C, collecting ducts.

*K⁺ is both reabsorbed and secreted.

†Urea diffuses into as well as out of some portions of the nephron.

‡Variable secretion and probable reabsorption of creatinine in humans.

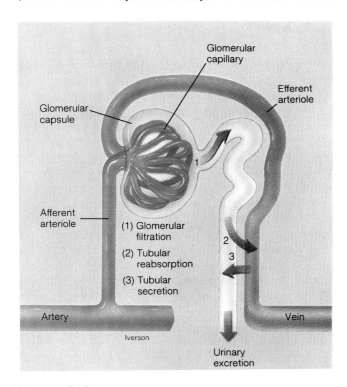

Figure 9.3

The three processes by which substances enter or leave the renal tubule are glomerular filtration, tubular reabsorption, and tubular secretion.

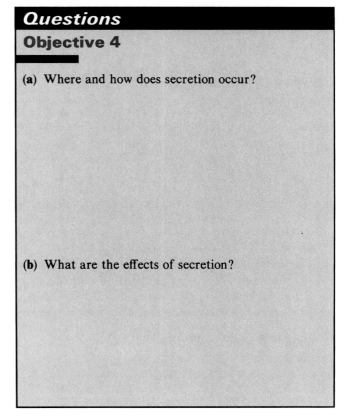

Questions

Objective 4

(a) Where and how does secretion occur?

(b) What are the effects of secretion?

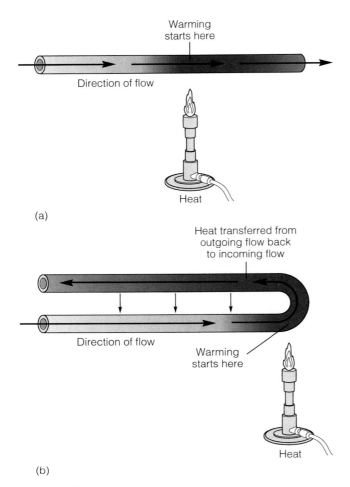

Figure 9.4

(*a*) Heat is transferred from the straight tube to the environment from the point of heat application in the direction of the outflow. (*b*) Heat is transferred from the bent tube to the environment as in (*a*) *and* from the outflow toward the inflow. The latter constitutes a countercurrent system.

countercurrent. Countercurrent multiplication greatly increases the filtrate concentrations of Na^+ and Cl^- at the tips of the loops. More significantly, it greatly increases the osmolarity of peritubular fluids, thereby drawing water from filtrate in collecting ducts and concentrating the urine. Countercurrent multipliers create an osmotic gradient between extracellular fluids and filtrate, and countercurrent exchangers prevent solute loss that could reduce the gradient as blood flows through the medulla. The significant effect of the countercurrent mechanism is to create an osmotic gradient that draws water out of the filtrate and concentrates the urine.

See Questions—Objective 5

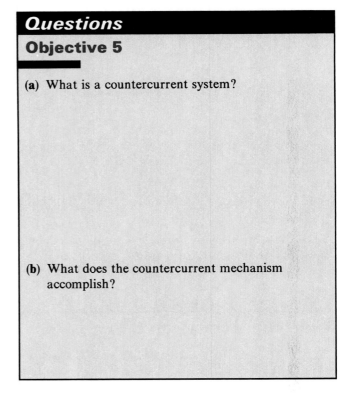

Questions
Objective 5

(a) What is a countercurrent system?

(b) What does the countercurrent mechanism accomplish?

Regulation of Urine Formation

Various regulatory mechanisms in urine formation respond quickly to changes in the composition and volume of body fluids and thereby maintain fluid homeostasis. The kidneys excrete nitrogenous wastes, mainly urea, regardless of urine volume, conserve water when fluid volume is low, and excrete excess water when fluid volume is high. They also regulate blood electrolyte concentrations and assist in maintaining acid-base balance.

Synthesis (chapter 4) and excretion of urea are the body's main means of disposing of amino groups from protein catabolism. On a moderate protein diet, the body produces 25 to 30 grams of urea per day—more on a high-protein diet and less on a low-protein diet. Urea is normally excreted at the same rate as it is synthesized. If excretion is impaired, as in some kidney disorders, toxic levels can accumulate in body fluids.

The rate of urea excretion is determined by the product of plasma urea concentration and GFR—the amount of urea entering the filtrate. For example, at a normal plasma urea concentration of 15 mg/dl to 35 mg/dl and GFR of 125 ml/min, urea enters the filtrate at 20 mg/min to 40 mg/min. Kidney tubules are somewhat permeable to urea and under normal physiological conditions about 40 percent of urea diffuses back into the

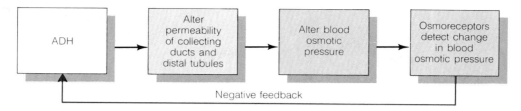

Figure 9.5

The role of ADH in regulating blood osmotic pressure and, to some extent, blood hydrostatic pressure.

blood and 60 percent remains in the filtrate and is excreted. If the urine volume is small and flow through tubules is slow, or the plasma urea level is high, more urea leaves the filtrate and only 10 to 20 percent is excreted. Conversely, if urine volume is large and flow through tubules is relatively rapid, or the plasma urea level is low, less urea leaves the filtrate and up to 70 percent is excreted. Some urea that leaves the filtrate diffuses back into the blood, but a portion remains in peritubular fluids where it contributes to osmolarity, as noted earlier. As a practical matter, high protein intake leads to concentrated urine because it increases peritubular osmolarity, whereas a high fluid intake leads to dilute urine because rapid flow allows less time for urea to contribute to osmolarity.

Among nitrogen-containing wastes in human urine, urea accounts for about 85 percent. Another 4.5 percent is creatinine, a product of muscle metabolism excreted at a constant rate. About 3 percent is ammonia and 2 percent is uric acid from the catabolism of nucleic acids. The remainder consists of very small quantities of amino acids and other nitrogenous substances.

If the entire glomerular filtrate were excreted as urine without concentration, body fluids would be depleted in less than an hour. To remove excess solutes while maintaining normal blood volume, the kidneys must concentrate the urine. We have already seen how the countercurrent mechanism and urea accumulation help to concentrate urine and have noted that aldosterone returns Na^+ to the blood with water following by osmosis. Antidiuretic hormone is another extremely important factor.

Interstitial fluid becomes hyperosmotic when blood is hyperosmotic because of too much solute or too little water. Such interstitial fluid causes hypothalamic osmoreceptors to respond by stimulating secretion of **antidiuretic hormone** (ADH). ADH activates an enzyme by the cAMP mechanism, and the enzyme phosphorylates a protein that makes the cell membrane much more permeable to water and a little more permeable to urea. ADH acts on luminal membranes of epithelial cells mainly in collecting ducts and to a lesser degree in distal convoluted tubules. Much water and a little urea move passively into the peritubular fluids. Most of the water then diffuses into the blood, lowering blood osmotic pressure, maintaining

blood volume, and preventing dehydration. Lowering blood osmotic pressure prevents further stimulation of osmoreceptors, thereby completing the loop in a negative feedback mechanism (Figure 9.5).

ADH indirectly removes urea from the filtrate. As water is reabsorbed from the collecting ducts, the urea concentration in the ducts increases. When urea becomes more concentrated in the filtrate than in the peritubular fluid, its net diffusion is toward the peritubular fluid where it increases osmolarity.

An excess of ADH temporarily increases reabsorption and decreases urine volume. However, sustained excessive ADH secretion increases blood volume and arterial pressure, and the higher pressure increases filtration, counteracting the ADH.

When the blood is hyperosmotic, such as after one drinks a large quantity of fluid, a less concentrated urine is excreted because osmoreceptors receive no stimulation and ADH is not secreted. In the absence of ADH, cells in distal tubules and collecting ducts remain relatively impermeable and water is excreted.

In ADH deficiency, or **diabetes insipidus** (not to be confused with diabetes mellitus), distal tubules and collecting ducts remain impermeable, and large volumes of fluid—up to 15 liters per day—are excreted. The patient is always thirsty but usually cannot drink enough fluid to compensate for the water loss. This condition is treated by administering ADH either by nasal spray (absorbed through membranes) or injection.

We have already seen that increased urine production, or **diuresis** (di-u-re′sis), can be caused by blood hyperosmolarity and by ADH deficiency. **Osmotic diuresis** is caused by filtrate hyperosmolarity, in which a high solute concentration (such as glucose in untreated diabetes) holds fluid in the filtrate. Osmotic diuresis can raise urine volume to 4 or 5 times normal, severely dehydrating tissues and disturbing electrolyte balance. **Diuretics** (di-u-ret′ikz) include alcohol, caffeine, and certain drugs, which can be classified by site and mechanism of action (Table 9.2). Diuretics are commonly used to prevent fluid accumulation in hypertension and conditions that decrease cardiac output. They can have undesirable side effects and should be used with caution.

Table 9.2
Properties of Diuretics

Class of Diuretic	Site and Mechanism of Action	Examples
Osmotic diuretics	Increase osmotic pressure in kidney filtrate and cause water to be excreted instead of reabsorbed.	Mannitol, sorbitol
Carbonic anhydrase inhibitors	Inhibit the enzyme carbonic anhydrase in tubule cells and reduce formation of carbonic acid. This prevents secretion of H^+ into filtrate and diffusion of HCO_3^- into the blood and allows increased excretion of Na^+, K^+, and HCO_3^-. Water follows these ions by osmosis. Excess HCO_3^- excretion can cause acidosis.	Acetazolamide (Diamox)
Thiazides	Inhibit reabsorption of Na^+ and Cl^- mainly in distal tubules and reduce osmotic pressure in interstitial fluids. This reduces the kidneys' ability to concentrate urine. It also increases excretion of various electrolytes and can lead to K^+ depletion.	Chlorothiazide (Diuril), hydrochlorothiazide
ATPase inhibitors	Inhibit ATPase that normally provides energy for Na-K pumps in the thick ascending loop of Henle (and proximal and distal tubules). This reduces Na^+ reabsorption and greatly increases Na^+ and water excretion. It can cause HCO_3^- retention and lead to contraction alkalosis—decreased volume and increased pH of body fluids.	Furosemide (Lasix), ethacrynic acid (Edecrin)
Potassium-sparing natriuretics	Inhibit Na^+-K^+ exchange in distal tubules and collecting ducts by inhibiting aldosterone (spironolactone) or by inhibiting Na^+ reabsorption (triamterene).	Spironolactone (Aldactone), triamterene (Dyrenium)
Vasopressin (ADH) inhibitors	Inhibit action of vasopressin (ADH) on collecting ducts.	Water, ethyl alcohol
Xanthines	Probably decrease tubular reabsorption of Na^+ and increase glomerular filtration rate.	Caffeine, theophylline

A specialized structure called the **juxtaglomerular apparatus** consists of juxtaglomerular cells and the macula densa. **Juxtaglomerular cells,** modified smooth muscle cells in afferent arteriole walls that touch macula densa cells, are filled with an inactive form of the enzyme **renin** (ren′in). The **macula densa** (mak′u-lah den′sah) consists of secretory cells in distal tubule walls. When macula densa cells detect a decrease in blood pressure or GFR, they probably secrete a messenger substance (as yet unidentified) onto the juxtaglomerular cells, causing them to activate and release renin. Decreases in Na^+, in distal tubules, and in sympathetic signals also can release renin.

Renin release activates the **renin-angiotensin** (an-je-o-ten′sin) **mechanism** (Figure 9.6), which regulates blood pressure and protects against sodium depletion. Activated renin acts on the plasma protein **angiotensinogen** (an-je-o-ten-sin′o-jen) and releases **angiotensin I.** Almost as rapidly as it is formed, angiotensin I is converted to **angiotensin II** by **plasma-converting enzyme** acting mainly in small blood vessels of the lungs. Angiotensin II remains active in the blood for only a few minutes until degraded by **angiotensinase** (an-je-o-ten′sin-ās).

Angiotensin II, a powerful vasoconstrictor, causes rapid constriction of arterioles in the kidneys and elsewhere in the body and mild constriction of venules and veins. Blood vessel constriction increases peripheral resis-tance and thereby elevates blood pressure. Because angiotensin II has a greater effect on efferent than on afferent arterioles, it increases glomerular pressure and thus GFR. Angiotensin II also decreases sodium and water excretion by stimulating aldosterone secretion.

Several similar and small **atrial peptides** are released from atrial muscle tissues when blood pressure increases and the heart's atria become distended. Atrial peptides are also called atrial natriuretic hormone (ANH) or **atrial natriuretic factor** (ANF) because in early experiments they were observed to cause **natriuresis,** or sodium excretion. Natriuresis is significant only under special conditions such as extremely high sodium loads, and the peptides have other actions such as decreasing blood pressure and cardiac output and possibly acting as brain neurotransmitters.

The mild natriuresis from atrial peptides is probably caused by a slight alteration in Na^+ transport in collecting ducts, but atrial peptides have other renal effects. They inhibit angiotensin II and reduce sodium reabsorption and may inhibit antidiuretic hormone and increase water excretion. They also relax renal blood vessels, increasing glomerular filtration, filtrate, and urine volumes. Finally, they increase sodium and potassium excretion.

See Questions—Objectives 6 and 7

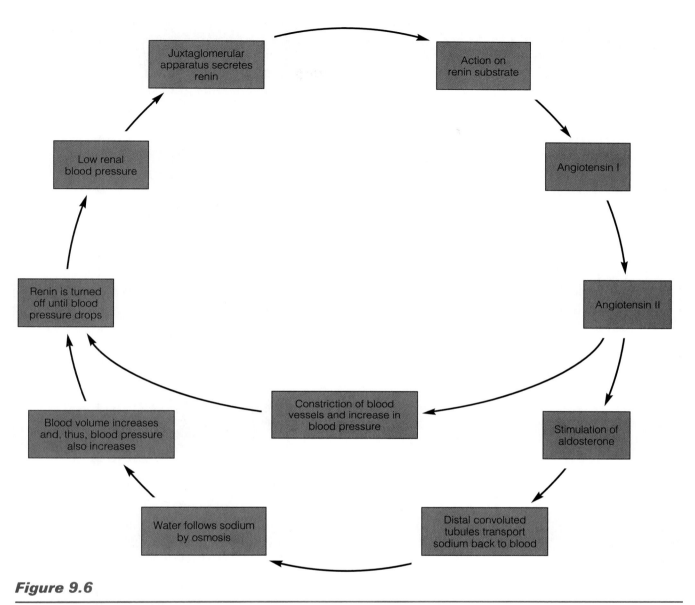

Figure 9.6

The effects of the renin-angiotensin system on blood pressure. A decrease in blood pressure turns this system on, and an increase in blood pressure turns it off.

Measurement of Kidney Function

An important measure of kidney function is **clearance,** the volume of blood that can be cleared, or freed, of a substance per minute. Clearance of a substance is proportional to the sum of the amounts filtered and secreted less the amount reabsorbed. As a substance is cleared from the blood, the amount excreted (E) in mg/min is the concentration in the urine (U) in mg/ml multiplied by the rate at which urine volume (V) forms in ml/min.

$$E = U \times V$$

The amount filtered (F) in mg/min is the concentration of the substance in plasma (P) in mg/ml multiplied by GFR in ml/min.

$$F = GFR \times P$$

If the substance is neither reabsorbed nor secreted, the amount filtered (F) equals the amount excreted (E), as shown in this equation, $GFR \times P = U \times V$. Rearranging this equation to solve for GFR is shown below.

$$GFR = \frac{U \times V}{P}$$

In this case GFR equals the **renal clearance rate,** or clearance in ml/min.

Questions

Objective 6

(a) How and to what extent is urea excreted?

(b) What other nitrogenous wastes are excreted by humans?

(c) What conditions can cause diuresis to occur in humans?

(d) What steps occur in the renin-angiotensin mechanism and what are its effects?

Objective 7

(a) How does antidiuretic hormone regulate urine volume?

(e) How do atrial peptides act on the kidneys?

(b) What is diabetes insipidus and how is it treated?

These formulas can be applied to the clearance of **inulin** (in'u-lin), a fructose polysaccharide made by onion and garlic plants. Suppose a person is given inulin intravenously until the plasma concentration (P) reaches 1 mg/ml. Later, the urine inulin concentration (U) is 125 mg/ml and urine volume (V) forms at a rate of 1 ml/min. Clearance (C) = 125 ml/min, as shown below.

$$C = \frac{125 \text{ mg/ml} \times 1 \text{ ml/min}}{1 \text{ mg/ml}} = 125 \text{ ml/min}$$

Clearance is equal to GFR because inulin is filtered but neither secreted nor reabsorbed. For the inulin concentration to reach 125 mg/ml, filtrate must form at the normal rate of 125 ml/min. In patients with glomerulonephritis, in which glomeruli are damaged or destroyed, clearance studies show significant decreases in GFR.

Clearance can be used to study other kidney functions. Whereas inulin clearance measures GFR, **para-aminohippuric** (par-ah-am"in-o-hip-ur'ik) **acid** (PAH) clearance estimates blood flow through the kidneys because PAH is filtered and secreted but not reabsorbed. In fact, 91 percent of plasma PAH is cleared in one pass through a kidney. Suppose we administer PAH to a person, obtaining plasma PAH (P) of 0.01 mg/ml. We later find that urine PAH (U) is 4.4 mg/ml and urine volume (V) forms at 1.3 ml/min.

$$C = \frac{4.4 \text{ mg/ml} \times 1.3 \text{ ml/min}}{0.01 \text{ mg/ml}} = 572 \text{ ml/min}$$

If this calculation of clearance accounts for 91 percent of plasma, the total plasma volume is 629 ml/min (572 ml/min divided by 0.91). As plasma comprises about 55 percent of the total blood volume, blood flows through the kidneys at 1,144 ml/min (629 ml/min divided by 0.55).

Clearance also can be used to determine what proportion of the nitrogenous waste urea is actually cleared from the blood. Recall that some urea is reabsorbed and some is trapped in peritubular fluids. Suppose we determine that a person has a plasma urea (P) of 0.2 mg/ml and a urine urea (U) of 12 mg/ml when urine volume is formed at 1.3 ml/min.

$$C = \frac{12 \text{ mg/ml} \times 1.3 \text{ ml/min}}{0.2 \text{ mg/ml}} = 78 \text{ ml/min}$$

We see that urea clearance occurs at a rate equal to removing all the urea from 78 ml of plasma per minute, and we have determined the GFR of 125 ml/min. Then we can reason that 78/125 or 62 percent of the urea is being removed from the blood.

The plasma concentration of **creatinine** (kre-at'in-in), a product of muscle metabolism, provides a quick estimate of kidney function. Plasma creatinine is reported with other blood constituents in routine blood analyses. Muscles release creatinine at a constant rate regardless of activity, and the plasma creatinine concentration is nearly constant throughout life. Creatinine clearance also remains constant if kidney function is normal. It is filtered and secreted but is not reabsorbed. Elevated plasma creatinine suggests that glomerular filtration has decreased and warns of possible kidney disease.

See Questions—Objective 8

Renal Failure and Dialysis

The inability of the kidneys to perform their functions, **renal failure,** can be acute or chronic. Tissue edema occurs because sodium and water excretion is impaired. Hypertension, waste retention, tissue ischemia, and acidosis soon develop. If renal shutdown is complete, death will occur in 1 to 2 weeks without treatment. With appropriate treatment and tissue repair, nearly normal function may be regained.

Questions
Objective 8

(a) What kind of information about kidney function is provided by clearance of inulin, PAH, or urea?

(b) What might be the significance of an increase in plasma creatinine?

Kidney functions can be performed by **dialysis** (di-al'is-is), the transfer of wastes to dialysis fluids, in people suffering from renal failure until or instead of a kidney transplant. **Hemodialysis** (he-mo-di-al'is-is) draws wastes from the blood into dialysis fluid as the blood passes through selectively permeable tubes outside the body. **Peritoneal dialysis** draws wastes into a sterile dialysis fluid located in the abdominal cavity. Dialysis fluid has the same concentrations of electrolytes and nutrients as normal plasma, but lacks waste products. Waste substances such as urea and ammonia freely diffuse into the dialysis fluid but only excesses of glucose, amino acids, and electrolytes do so. Nutrients in dialysis fluid diffuse into the blood.

See Questions—Objective 9

Regulation of Body Fluids

Body fluids are more than 90 percent water and contain a variety of solutes (dissolved substances). Such fluids are found in cells, between cells of most tissues, and in blood and lymph. With respect to water itself, the bodies of adult males of normal weight contain 55 to 60 percent water and those of females have relatively less water and more fat. Because fatty tissue contains almost no water, extremely obese bodies may be no more than 40 percent water and extremely lean ones as much as 75 percent

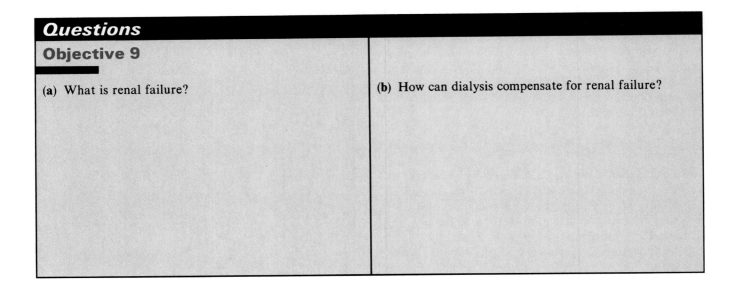

Questions

Objective 9

(a) What is renal failure?

(b) How can dialysis compensate for renal failure?

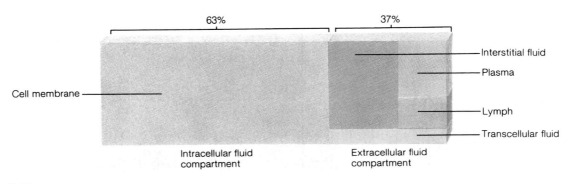

Figure 9.7

Volumes of fluid compartments.

water. Total body water declines with age from about 70 percent at birth to 60 percent at six months and 40 percent in old age.

About two-thirds of body fluid is **intracellular fluid** found within cells. The remaining third, **extracellular fluid,** consists mostly of **interstitial fluid** between cells and **plasma.** Small quantities of extracellular water also are found in lymph, cerebrospinal fluid, and fluid in joint cavities and inside the eyeballs. Typical volumes of the major fluid compartments—the intracellular compartment, the interstitial fluid compartment, and the plasma compartment—are shown in Figure 9.7.

All body fluids contain **electrolytes** (e-lek'tro-lītz), substances that ionize and can conduct electricity, but their concentrations differ by compartments (Figure 9.8). Electrolyte concentrations are small and are expressed in "milli" units, milliequivalents or milliosmols. For electrolytes with a 1^+ or 1^- charge, milliequivalents equal milliosmols. A milliequivalent (mEq) equals the number of

charges in a millimole (1/1,000 mole) of hydrogen ions. A milliosmol (mOs) is 1/1,000 mole of a substance multiplied by the number of particles it forms in solution.

Electrolyte concentrations in fluids of different compartments vary significantly (Table 9.3). Of these differences, high interstitial Na^+ concentrations and high intracellular K^+ concentrations are especially important in membrane excitability, particularly in nerve and muscle cells.

Within any one compartment the cations (positively charged ions) normally equal anions (negatively charged ions)—about 155 mEq/l of each in plasma and interstitial fluids and about 200 mEq/l of each in intracellular fluid (Figure 9.8). All fluid compartments have an osmolarity of about 282 mOsm/l—higher than electrolyte concentrations because uncharged (nonelectrolyte) substances also contribute to osmolarity.

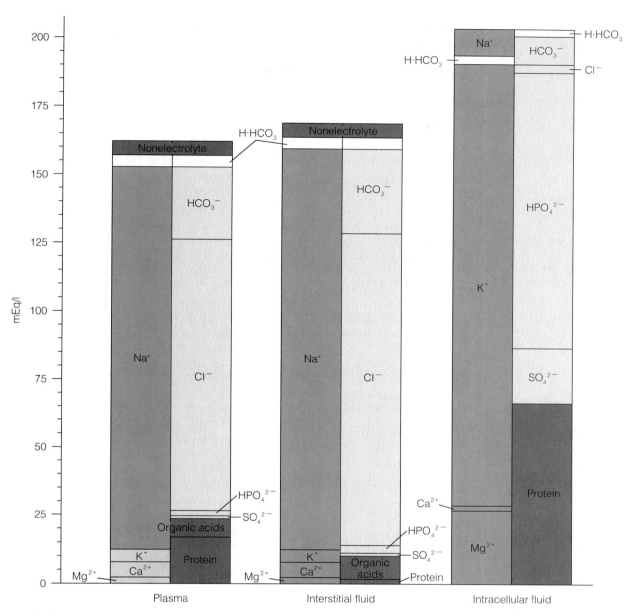

Figure 9.8

Composition of plasma, interstitial fluids, and intracellular fluids.

Water is essential for the operation of metabolic cycles. It is ingested in food and beverages, released in metabolic reactions (metabolic water), and lost through urine, feces, sweat, and as water vapor in exhaled air. Water in air and sweat during rest is referred to as **insensible water loss;** it cannot be measured directly, but is nonetheless important clinically. When water intake equals water output the body is in **water balance.**

Under normal physiological conditions osmotic gradients never persist across cell membranes. Whenever small, local differences arise between extracellular and intracellular osmotic pressures, water moves to equalize the pressures in less than a minute. When more extensive disturbances occur, more time is needed to reestablish equilibrium. For example, suppose that a person loses a large volume of water and electrolytes in sweat while working hard outdoors on a hot day. Even when the person drinks sufficient electrolyte replacement fluid, such as Gatorade, it takes 4 to 6 hours for the fluid to be absorbed and properly distributed throughout fluid compartments.

Table 9.3
Variation in Electrolyte Concentrations in Body Fluids

Differences between Extracellular and Intracellular Fluids

Extracellular Fluid	*Intracellular Fluid*
High in Na^+ and Cl^-	Low in Na^+ and Cl^-
Low in K^+ and HPO_4^{2-}	High in K^+ and HPO_4^{2-}
Relatively high in HCO_3^-	Relatively low in HCO_3^-
Relatively low in Mg^{2+}	Relatively high in Mg^{2+}
Relatively low in protein	High in protein

Differences between Plasma and Interstitial Fluids

Plasma	*Interstitial Fluid*
Relatively high in protein	Very low in protein
High in Cl^-	Extremely high in Cl^-

Though osmosis is directly responsible for maintaining equilibrium between fluid compartments, at least four mechanisms regulate the amounts of fluid entering or leaving the body.

1. Antidiuretic hormone conserves water by increasing the permeability of collecting ducts and distal tubules of the kidneys.
2. Atrial peptides increase water and sodium loss, probably by various effects on the kidneys.
3. Aldosterone stimulates sodium reabsorption causing water to follow by osmosis. It also stimulates potassium excretion as will be explained later in this chapter.
4. The thirst mechanism initiates drinking when the osmoreceptors in the hypothalamus detect increased plasma osmotic pressure. **Thirst** is a conscious desire for liquid that results in drinking. The brain's **thirst center** has **osmoreceptors** (oz″mo-re-cep′torz) that respond to increased osmolarity of tissue liquids, probably by detecting increases in sodium. When these liquids have high osmotic pressure (are too concentrated), osmoreceptors send signals to the cerebrum that cause a person to drink liquids. They also initiate release of antidiuretic hormone, which decreases urine volume and conserves liquid.

Fluid volume disturbances include **negative water balance** in which output exceeds intake and **positive water balance** in which intake exceeds output. Negative water balance is caused by decreased fluid intake or in-creased fluid loss, such as excretion of much dilute urine, burns that leak plasma, vomiting, diarrhea, excessive sweating, and hemorrhage. These conditions lead to **dehydration,** a lack of body water, and usually to losses of electrolytes. Dehydration first affects extracellular compartments. When relatively more water than solute is lost, the fluid becomes hyperosmotic. Intracellular fluid moves by osmosis to extracellular compartments. All compartments then have less than their normal volume but are only slightly hyperosmotic. Regardless of the cause, dehydration is made worse by **obligatory fluid loss** in urine. Though the kidneys can conserve fluid and excrete a concentrated urine, they cannot function without excreting some water along with waste solutes.

Severe dehydration—in untreated diabetes insipidus, burns over large body areas, and persistent vomiting and diarrhea—can reduce blood volume to a life-threatening degree. This decreases blood pressure, cardiac output, and tissue perfusion, and can damage the brain and other vital organs. Conditions that cause dehydration also can cause electrolyte and acid-base imbalances. For example, excessive sweating can deplete sodium and prolonged vomiting can deplete hydrochloric acid and raise body fluid pH.

Compared to adults, infants are especially prone to dehydration because their bodies have a higher surface-to-volume ratio and thus a greater evaporative surface, relatively more water than the bodies of adults, and immature kidneys with limited urine concentrating ability. Fluid regulating mechanisms must keep more water in tissues in spite of evaporative loss and immature kidney function. It is not surprising then, that an infant might take in 20 percent of its weight in fluid daily.

Positive water balance leads to **water intoxication,** or too much body water and a corresponding decrease in osmolarity of fluid compartments. Water intoxication can occur when people quench their thirst with pure water, tea, coffee, soft drinks, or other beverages hyposmotic to sweat. Failure to replace lost electrolytes leads to diarrhea and abdominal and leg cramps, which can occur many hours after the sweating. Though hypotonic beverages increase extracellular fluid volume, they make the fluid even more hyposmotic and some water moves by osmosis into cells, decreasing intracellular osmolarity and causing cellular swelling.

Edema, another water balance disorder, is an increase in extracellular fluid volume without a change in osmolarity. Characterized by a puffy face and swollen abdomen, ankles, and fingers, edema usually is caused by excess sodium. Excess sodium in a body fluid draws water into a compartment until osmotic balance is reached.

See Questions—Objectives 10 and 11

Questions

Objective 10

(a) What is the composition of fluid in each of the body's three main fluid compartments?

(b) Where are these compartments located?

(b) What are the causes and effects of dehydration?

(c) of water intoxication?

(d) of edema?

(e) How does the body compensate for these conditions?

Objective 11

(a) What mechanisms contribute to regulating fluid volume and how do they work?

Regulation of Specific Electrolytes

Having considered the volumes and compositions of fluids themselves, we can now focus on mechanisms that regulate the concentrations of their electrolytes—sodium, potassium, calcium, magnesium, phosphates, chloride, and bicarbonate.

Sodium is the most abundant electrolyte in extracellular fluids, and potassium is the most abundant in intracellular fluids. More than half of the body's sodium is deposited in bone and other connective tissues, whereas 95 percent of potassium is confined to intracellular fluids. The imbalance of sodium and potassium concentrations across membranes is maintained by sodium-potassium pumps in cell membranes. Sodium salts account for more than 90 percent of body fluid osmolarity, so sodium regulation largely determines body fluid volumes. Potassium regulation helps to maintain normal cell volumes. Both must be precisely regulated to maintain normal membrane excitability.

Sodium and potassium levels and fluid volumes are regulated mainly by **aldosterone,** which causes sodium reabsorption and potassium secretion in kidney tubules. Aldosterone acts by combining with a receptor protein in cells of the distal convoluted tubules and collecting ducts. The aldosterone-receptor complex activates DNA, which in turn initiates synthesis of proteins for the sodium-potassium transport process. Whether proteins are carriers, enzymes, or both is not entirely clear, but experiments on animals show that the sodium-potassium transport rate increases only after about a 45-minute lag.

Proteins synthesized from the above process act at the membrane surface of sensitive cells. Active transport of sodium out of cells creates a low intracellular sodium concentration and a negative potential (-70 mV). The low cellular sodium and the negative potential both draw sodium from the tubule into the cell. At the same time, aldosterone causes active transport of sodium to interstitial fluid from which it diffuses into plasma; it also causes transport of potassium from interstitial fluid into tubule cells.

Extracellular potassium initiates and maintains aldosterone secretion in proportion to the potassium concentration. Thus, both potassium and aldosterone are homeostatically controlled by negative feedback. When blood pressure or fluid volume drop, the renin-angiotensin system is activated and angiotensin stimulates aldosterone release. Sympathetic stimulation and ACTH release, which cause cortisol secretion, also can cause aldosterone secretion. Renal hypotension, shock, and heart failure elevate renin, angiotensin, and aldosterone.

Severely restricting dietary sodium intake increases aldosterone release, which in turn increases sodium reabsorption and partially defeats the effect of dietary restriction. Possible mechanisms for this effect are as follows:

1. Reducing sodium intake reduces extracellular volume, which in turn reduces cardiac output and renal blood flow. Reduced renal blood flow activates the renin-angiotensin system and stimulates aldosterone release.
2. Reducing sodium intake lowers potassium excretion and excess potassium increases aldosterone release.

Failure of regulatory mechanisms to maintain normal sodium or potassium levels can lead to the disorders listed in Table 9.4.

Though calcium and phosphate are found in body fluids as electrolytes, as much as 99 percent of the total body content of these minerals is deposited in bone as the crystalline salt **hydroxyapatite** (hi-drox″e-ap′ah-tīt). Here we will consider calcium and phosphate as electrolytes. Total plasma calcium is maintained between 8.5 and 10.5 mg/dl, with 40 percent bound to plasma proteins and not ionizable at normal blood pH, 50 percent as Ca^{2+}, and 10 percent in ionizable inorganic salts such as calcium citrate and calcium phosphate. Though protein-bound calcium remains in plasma, Ca^{2+} freely diffuses between plasma and interstitial fluid maintaining interstitial calcium at about 6 mg/dl—60 percent of that in the plasma.

In plasma, an equilibrium between protein-bound calcium and free calcium is influenced by the plasma pH:

$$Ca\text{-protein} + 2H^+ \rightleftarrows Ca^{2+} + 2H\text{-protein}$$

Decreasing plasma pH drives the reaction to the right as excess H^+ replaces Ca^{2+} on plasma proteins; increasing plasma pH drives the reaction to the left as H^+ is released from proteins and Ca^{2+} attaches to them. Plasma contains relatively more free Ca^{2+} when slightly acidic and relatively less Ca^{2+} when slightly alkaline.

Intracellular fluids normally contain low concentrations (10^{-7} M) of free Ca^{2+}. A variety of cellular processes are activated by the influx of Ca^{2+} into stimulated cells.

Phosphate concentration is relatively high (15 mEq/l) in intracellular fluids and much lower (2 mEq/l) in extracellular fluids. In cells, phosphates participate in phosphorylation of glucose and nucleotides such as ADP. In extracellular fluids, phosphates, which are about 80 percent dibasic phosphate (HPO_4^{2-}) and 20 percent monobasic phosphate ($H_2PO_4^-$), help to maintain a normal blood pH. As with calcium, the relative proportions of phosphates are affected by the pH of extracellular fluids:

$$HPO_4^{2-} + H^+ \rightleftarrows H_2PO_4^-$$

Decreasing plasma pH drives the reaction to the right as H^+ from plasma binds to HPO_4^{2-}, whereas increasing plasma pH drives the reaction to the left as H^+ enters

Table 9.4
Some Causes and Effects of Sodium and Potassium Imbalances

Imbalance	Causes	Effects
Hyponatremia (plasma Na+ less than 136 mEq/l)	Sodium loss Kidney disease, Addison's disease, vomiting, diarrhea, excessive sweating, use of diuretics Water excess Excessive water intake, excessive aldosterone release, oliguria from renal failure	Decreased osmolarity of extracellular fluids leads to excess excretion of fluid until normal osmolarity is restored; can lead to dehydration and shock. Decreased response of nerve and muscle cells to stimulation because of decreased sodium gradient. Water intoxication leads to dilution of intracellular fluids, edema, and decreased concentration of nearly all plasma constituents.
Hypernatremia (plasma Na+ greater than 150 mEq/l)	Sodium accumulation Intake of excessively salty fluids or foods Water deficit Deficient water intake, increased insensible water loss, excessive use of diuretics, osmotic diuresis because of high concentration of wastes in urine	Increased osmolarity of extracellular fluids and fluid retention; fever due to increased metabolic rate to supply ATP for Na+-K+ pump; hyperexcitability of nerve and muscle cells because depolarization is faster and requires only a weak stimulus. Dehydration, water retention to the extent possible, oliguria, shrinkage of cells.
Hypokalemia (plasma K+ less than 3.5 mEq/l)	Use of diuretics, diarrhea, vomiting, Cushing's syndrome, renal disease	Hyperpolarization of membranes leads to decreased excitability of nerve and muscle cells; this leads to lethargy, muscle weakness, and reduced gut motility; abnormalities in the heart conduction system.
Hyperkalemia (plasma K+ greater than 5.5 mEq/l)	Renal failure, hypertonic dehydration, massive tissue damage, Addison's disease	Partial depolarization of membranes leads to temporary increase in excitability of nerve and muscle cells; further reduction in polarization causes cells to lose ability to respond to stimuli; impairment of the heart conduction system and eventual cardiac arrest.

plasma. Unlike many changes in electrolyte concentrations, changes in extracellular phosphate concentrations—from one-fourth to four times the normal—cause no problems.

When Ca^{2+} increases, phosphate ions decrease, and vice versa, but the product of their concentrations remains constant (at about 35). Should the total exceed that which can dissolve in plasma, $CaHPO_4$ will precipitate and deplete body fluids of the ions. People who frequently drink carbonated beverages, which contain large amounts of phosphate, may fail to absorb Ca^{2+} because it binds to phosphate in the gut before being absorbed.

In contrast to phosphates, changes in extracellular calcium ion levels have profound effects on many physiological processes. Ca^{2+} regulates cellular processes, muscle function, neurotransmitter release, blood clotting, and cardiac function. Very small Ca^{2+} concentrations are effective, and excesses or deficiencies seriously interfere with regulation. The effects of excesses and deficiencies of calcium are shown in Table 9.5.

Body fluids contain numerous other electrolytes—magnesium and chloride (discussed here), sulfate, nitrate, lactate, urate, and trace minerals. The magnesium ion concentration is high in intracellular fluids, low

in extracellular fluid, and highly variable in urine depending on diet. Fruits, whole grains, and plant foods contain more magnesium than foods from animals. Both magnesium and Ca^{2+} are actively transported in opposite directions by the same membrane pump, which maintains a high intracellular magnesium concentration and a high extracellular calcium concentration. Magnesium has important functions in both intracellular and interstitial fluids. In cells, it is a cofactor for many enzymes, especially those that catalyze reactions involving ATP. In interstitial fluid, it blocks acetylcholine release at synapses and neuromuscular junctions, decreasing excitability of both nerve and muscle cells. This property is applied medically by administering magnesium sulfate to prevent seizures. In addition to active transport at cell membranes, magnesium is regulated by the kidneys, probably via aldosterone and negative feedback (similar to potassium regulation). Magnesium usually is well regulated. Imbalances are rare and their possible causes and effects are listed in Table 9.6.

The chloride ion concentration is low in intracellular fluids and high in extracellular fluids. It varies in urine according to the amount ingested but is usually

Table 9.5
Some Causes and Effects of Calcium Imbalances

Imbalance	Causes	Effects
Hypocalcemia (total plasma Ca^{2+} less than 8.5 mg/dl or ionic Ca^{2+} less than 4.0 mg/dl)	Excess of phosphates can precipitate calcium; lack of plasma proteins fails to carry adequate bound calcium, and plasma protein deficiencies result from liver or kidney disease or starvation; decreased parathormone secretion, vitamin D deficiency, hyperventilation by increasing plasma pH.	Partial depolarization of nerve and muscle cell membranes leads to hyperexcitability. Hyperexcitability is displayed by muscle cramps, convulsions, mental confusion, and tetany (continuous muscle contraction). Tetany prevents normal breathing movements and can cause severe dyspnea and death. Increased cardiac excitability causes a delay between depolarization and repolarization of the ventricles; it also reduces the strength of contractions.
Hypercalcemia (total plasma Ca^{2+} greater than 12 mg/dl or ionic Ca^{2+} greater than 5.6 mg/dl)	Increased parathormone secretion, malignant tumors (that are thought to secrete parathormone), excessive intake of vitamin D, hyperthyroidism, Addison's disease, certain bone diseases, and immobilization (that leads to excessive bone reabsorption).	Excessive active transport of Ca^{2+} out of cells, which interferes with entry of Na^+ when cells are stimulated and leads to hyperpolarization of nerve and muscle cell membranes. Decreased excitability leads to fatigue, listlessness, sleepiness and even coma, muscle weakness, reduced gut motility, sometimes abdominal pain and anorexia. Dehydration may also occur as Ca^{2+} interferes with the reabsorption of Na^+ and the concentration of urine. $CaHPO_4$ precipitates in tissues and especially in the collecting ducts of the kidneys and the walls of the arteries.

Table 9.6
Some Causes and Effects of Imbalances in Magnesium and Chloride

Imbalance	Causes	Effects
Hypomagnesemia (plasma Mg^{2+} less than 1.5 mEq/1)	Excessive loss from excess aldosterone, thiazide diuretics, or polyuria; inadequate intake from malnutrition or alcoholism; inadequate absorption from excess calcium that competes for transport, diarrhea, vomiting, hypoparathyroidism, and malabsorption disorders.	Muscle cramps, tetany, muscle weakness; tachycardia and heart arrhythmias; psychological changes, ataxia, and convulsions; hypertension.
Hypermagnesemia (plasma Mg^{2+} greater than 2.5 mEq/1)	Renal failure, hyperparathyroidism, inadvertent overdose of magnesium in therapy.	Bradycardia and hypotension as Mg^{2+} blocks sympathetic stimulation of heart and blood vessels; nausea and vomiting; lethargy, depressed respiration, and decreased reflex activity; death from respiratory or cardiac arrest.
Hypochloremia (plasma Cl^- less than 95 mEq/1)	Prolonged vomiting, and in infants prolonged diarrhea; excessive use of diuretics; metabolic alkalosis	No specific effects known. Hypochloremia is usually associated with alkalosis.
Hyperchloremia (plasma Cl^- greater than 115 mEq/1)	Depletion of bicarbonate in extracellular fluids as a result of metabolic acidosis.	No specific effects known. Hyperchloremia is usually associated with acidosis. Some recent evidence suggests that high chloride intake may contribute to hypertension.

higher than in plasma. Most cell membranes are relatively impermeable to chloride ions; the ions can be actively transported with sodium or can follow electrical gradients created by sodium ion transport. In chloride shift, negatively charged bicarbonate ions move across a membrane and chloride ions move in the opposite direction, equalizing the charge. Chloride concentrations are maintained by membrane phenomena (impermeability, active transport, and chloride shift) and by kidney excretion. When aldosterone causes sodium reabsorption, chloride ions follow along the electrical gradient and are indirectly regulated. Excess chloride ions are excreted. Like magnesium, chloride is well regulated. Imbalances are rare and their possible causes and effects are listed in Table 9.6.

See Questions—Objective 12

Questions

Objective 12

(a) What mechanisms are involved in regulating sodium and potassium concentrations in body fluids?

(b) What disorders result from improper regulation?

(c) What mechanisms are involved in regulating calcium and phosphate?

(d) What disorders result from improper regulation?

(e) What mechanisms are involved in regulating other electrolytes?

(f) What disorders result from improper regulation?

Acid-Base Regulation

The slightly acidic to neutral pH of intracellular fluids (6.7 to 7.0) is due to acids produced during cellular metabolism. In spite of cells releasing carbon dioxide and other wastes that move from interstitial fluids to plasma, the pH of such fluids stays within a narrow range of 7.35 to 7.45 because of quick-acting buffers and acid-base regulating mechanisms.

A **buffer** is a substance that resists pH change, usually by taking up or releasing H^+. Recall that pH is a measure of H^+ concentration. The human body has three important buffer systems—bicarbonate-carbonic acid, phosphates, and proteins. The **bicarbonate-carbonic acid buffer system** consists of a weak acid called carbonic acid

(H_2CO_3) and its salt called sodium bicarbonate ($NaHCO_3$). The salt of an acid is formed by replacing hydrogen with another cation, such as sodium. This buffer system acts mainly in plasma. The **phosphate buffer system** consists of two salts of phosphoric acid (NaH_2PO_4 and Na_2HPO_4). It acts mainly in intracellular fluids and in urine, not in plasma. The **protein buffer system** consists of carboxyl and amino groups of amino acids not involved in peptide bonds and therefore free to donate and accept H^+s, respectively. Hemoglobin and plasma proteins are important buffers that act in the blood.

Buffers can resist pH change even in the presence of a strong acid or strong base. For example, hydrochloric acid (HCl) is a strong acid. It completely ionizes into H^+ and Cl^-. Sodium hydroxide (NaOH) is a strong base. It completely ionizes into Na^+ and OH^-. The OH^- greatly increases the pH of a solution because it acts as a H^+ acceptor, that is, it removes H^+ from solution. Carbonic acid is a weak acid—only a small proportion of carbonic acid molecules ionize to form H^+ and HCO_3^-. Sodium bicarbonate acts as a weak base—it ionizes in solution to form Na^+ and HCO_3^-. The Na^+ does not affect pH, but bicarbonate ions are weak H^+ acceptors that remove some H^+ from solution.

In the bicarbonate-carbonic acid system (Figure 9.9), when a strong acid and sodium bicarbonate react, a new salt and a weak acid are formed:

$$HCl + NaHCO_3 \rightarrow NaCl + H_2CO_3$$

The salt has no effect on pH, and carbonic acid releases far fewer H^+ into the solution than did hydrochloric acid. The number of free H^+ is reduced and the solution's pH is unchanged or held in a narrow range. Similarly, when a strong base and a weak acid react, water and a weak base are formed:

$$NaOH + H_2CO_3 \rightarrow H_2O + NaHCO_3$$

The water has no effect on pH, and the sodium bicarbonate accepts far fewer H^+ than the strong base. Fewer H^+ are removed from the solution and the pH is held within a narrow range. For the pH to remain stable, the solution must contain enough of the buffer to react with all the strong acid or base in the solution.

In the phosphate buffer system, sodium dihydrogen phosphate (NaH_2PO_4) behaves as a weak acid and disodium hydrogen phosphate (Na_2HPO_4) behaves as a weak base. The effects of this system on strong acids and bases are shown in the reactions below.

$$HCl + Na_2HPO_4 \rightleftharpoons NaCl + NaH_2PO_4$$

(weak base)   (weak acid)

$$NaOH + NaH_2PO_4 \rightleftharpoons H_2O + Na_2HPO_4$$

(weak acid)   (weak base)

Proteins buffer by taking up or releasing H^+, thereby stabilizing the H^+ concentration in a solution. The carboxyl group of an amino acid can gain or lose H^+:

The amino group of an amino acid can gain or lose H^+ or it can gain or lose OH^-:

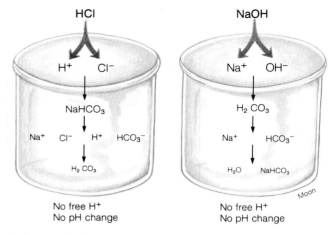

Figure 9.9

How the bicarbonate buffer system works.

All buffering reactions are easily reversible—they can go in either direction. Because the globin of hemoglobin and plasma proteins can hold or release either H^+ or OH^-, they act as a powerful buffer system in the blood.

Buffers are extremely important in regulating the pH of body fluids. They assure that excesses of acids or bases exert minimal effects on body fluids until they can be removed by the lungs or kidneys. The actions of buffers are summarized in Table 9.7.

Body fluids are in **acid-base balance** when their pH is in a normal range, and they contain neither an excess nor a deficiency of H^+. Acid-base balance is maintained by both respiratory and nonrespiratory mechanisms.

The main threat to acid-base balance is CO_2, which is continuously produced by cells. When it dissolves in plasma it produces carbonic acid as the following reaction goes to the right:

$$CO_2 + H_2O \rightleftharpoons H_2CO_3 \rightleftharpoons H^+ + HCO_3^-$$

To prevent the plasma from becoming acidic, CO_2 must be removed continuously, driving the reaction to the left. Under normal physiological conditions, CO_2 is removed from the plasma as blood passes through lung capillaries and diffuses into alveoli before being exhaled. This causes HCO_3^- and H^+ to form more CO_2 (and water), and more CO_2 diffuses out of the plasma. After CO_2 leaves the blood and H^+ leaves blood buffers, the blood can again circulate

Table 9.7
The Actions of Buffer Systems

System	Actions
Bicarbonate/carbonic acid	Bicarbonate reacts with a strong acid to form a salt and a weak acid: $HCl + NaHCO_3 \rightarrow NaCl + H_2CO_3$ Carbonic acid reacts with a strong base to form water and a weak base: $NaOH + H_2CO_3 \rightarrow H_2O + NaHCO_3$
Phosphate	Sodium monohydrogen phosphate reacts with a strong acid to form a salt and a weak acid: $HCl + Na_2HPO_4 \rightarrow NaCl + NaH_2PO_4$ Sodium dihydrogen phosphate reacts with a strong base to form water and a weak base: $NaOH + NaH_2PO_4 \rightarrow H_2O + Na_2HPO_4$
Proteins	Amino acids in the globin of hemoglobin or in plasma proteins can gain or lose either H^+ or OH^-. By doing so they can have significant effects on the blood pH.

The major problem of maintaining acid-base balance under normal physiological conditions is to prevent acidosis from CO_2 and H^+ accumulation. Both the lungs and kidneys can respond to either decreases or increases in the blood pH. When the blood pH begins to drop, the respiratory rate and the rate of H^+ excretion increase. Substances that acidify the blood are removed more rapidly. Conversely, when the blood pH begins to rise, the respiratory rate and the rate of H^+ excretion decrease. CO_2 and H^+ accumulate, counteracting the pH increase. By continuously adjusting the rate of removal of acidic substances, the body maintains the blood pH in the normal range.

Buffers and lung and kidney functions operate together to maintain acid-base balance but have different capacities and different limitations. Buffer systems react rapidly but can become overloaded by the accumulation of acid or base. The lungs can respond relatively rapidly to either an increase or decrease in plasma CO_2 and act to return it to normal. They, too, can be overloaded especially when respiratory disorders slow gas diffusion or interfere with regulation of the rate and depth of breathing. The kidneys provide the slowest but most complete regulation of acid-base balance. They respond slowly because they must remove H^+ by active transport, but they are thorough because they can adjust the blood concentrations of many plasma solutes.

The combined actions of buffers, lungs, and kidneys maintain the arterial blood pH between 7.35 and 7.45 and the HCO_3^- to CO_2 ratio near 20:1. Venous blood has a lower pH because it carries more CO_2 and H^+.

See Questions—Objectives 13 and 14

through tissues, pick up new CO_2, and carry it to the lungs with no significant change in the blood pH. This constitutes respiratory regulation of acid-base balance.

In nonrespiratory regulation of acid-base balance, kidney tubules secrete H^+. CO_2 passes from peritubular fluids into the tubule wall epithelium. Within the epithelial cells, carbonic anhydrase catalyzes the combination of CO_2 and water to form carbonic acid. Some of the acid ionizes into HCO_3^- and H^+. H^+ is actively transported from the epithelium into the tubule lumen. HCO_3^- moves in the opposite direction to interstitial fluid and some of it diffuses on to the blood.

Once in a kidney tubule lumen, H^+ can combine with ammonia from amino acid deamination or with phosphate ions. Kidney filtrate typically contains phosphates that form a buffer system in urine. When H^+ combines with any substance, it no longer contributes to acidity. Secretion of H^+ into the kidney filtrate removes acid from the blood. Binding of H^+ to ammonia or a phosphate buffer prevents urine acidity. Only on a very acidic diet does enough H^+ enter the filtrate to overwhelm buffers and allow the urine pH to drop as low as 5.

Acid-Base Imbalances: Causes, Consequences, and Compensation

Disturbances in acid-base balance can cause **acidosis** (as-id-o'sis), an excess of H^+, or **alkalosis** (al-kah-lo'sis), a deficiency of H^+. Either can be respiratory or metabolic (nonrespiratory), depending on its cause. Disorders of the respiratory system, including control centers, can cause respiratory acidosis or alkalosis. Metabolic disorders such as accumulation of ketone bodies in untreated diabetes or ammonia in hyperammonemia can cause metabolic acidosis or alkalosis, respectively. Because such disorders can be caused by vomiting, diarrhea, kidney disorders, and other problems, the term nonrespiratory is more descriptive than metabolic.

Whenever an acid-base imbalance develops, the body's regulatory mechanisms operate to stabilize or *compensate* for the imbalance. **Complete compensation** occurs

Questions

Objective 13

(a) What is a buffer?

(b) What kinds of buffers are found in the body?

(c) What are the specific effects of each of the body's three main buffer systems?

Objective 14

What mechanisms are involved in maintaining acid-base balance?

when the body completely counteracts the imbalance: pH, HCO_3^- and CO_2 concentrations, and the ratios among them are restored to normal. In most acid-base imbalances, the condition that caused the imbalance impairs normal regulation and **incomplete compensation** occurs: the HCO_3^- to CO_2 ratio and pH are restored to normal at the expense of maintaining both HCO_3^- and CO_2 concentrations above or below normal. The mechanism and the effect of incomplete compensation differ depending on the nature of the imbalance. Acidosis, alkalosis, and compensation are summarized in Table 9.8.

Impaired ability of the lungs to remove CO_2 allows it to accumulate in the blood in **respiratory acidosis.** Part of the CO_2 combines with water to form H_2CO_3, some of which ionizes, releasing H^+ and lowering the blood pH. The acid-base imbalance in respiratory acidosis is excess CO_2 that leads to excess H^+.

Any factor that interferes with gas exchange in the lungs can cause respiratory acidosis, including fluid accumulation in the alveoli in pneumonia, impaired exhalation in emphysema, obstruction of respiratory passageways in asthma and bronchitis, and reduction in respiratory surface area in any disorder that destroys alveoli. Damage or narcotic suppression of respiratory centers can reduce the respiratory rate and lead to acidosis.

When lung disease causes respiratory acidosis, the kidneys create what amounts to compensating metabolic alkalosis. They excrete H^+ and return enough HCO_3^- to the blood to balance the excess CO_2 that remains. The 20:1 HCO_3^- to CO_2 ratio is restored and the blood pH returns to 7.4 (according to the Henderson-Hasselbalch equation), but *both HCO_3^- and CO_2 are above normal.*

If the lungs remove CO_2 from the blood too rapidly, **respiratory alkalosis** occurs as more carbonic acid

Table 9.8
Acidosis, Alkalosis, and Compensation

	Acidosis	Alkalosis
General	Assume pH = 7.1 Given pK = 6.1, then $\log \dfrac{[HCO_3^-]}{[CO_2]} = 1.0$ antilog of 1.0 = 10 $\dfrac{[HCO_3^-]}{[CO_2]} = \dfrac{10}{1}$ (half the normal ratio)	Assume pH = 7.7 Given pK = 6.1, then $\log \dfrac{[HCO_3^-]}{[CO_2]} = 1.6$ antilog of 1.6 = 40 $\dfrac{[HCO_3^-]}{[CO_2]} = \dfrac{40}{1}$ (twice the normal ratio)
Respiratory	Caused by impaired ability of lungs to rid the body of CO_2. *Problem: excess CO_2 and, thus, excess H^+.*	Caused by excessive removal of CO_2 as in hyperventilation. *Problem: deficiency of CO_2 and, thus, deficient H^+.*
Compensation	Accomplished by kidney because respiratory system malfunction caused the condition and is incapable of correcting it.	
	Kidney excretes more H^+ than normal and returns more HCO_3^- to blood until a blood pH of 7.4 is restored. The HCO_3^- is retained in blood until its concentration is 20 times the CO_2 concentration. Both HCO_3^- and CO_2 are present in excess in incomplete compensation.	Kidney excretes less H^+ than normal and returns less HCO_3^- to blood until a blood pH of 7.4 is restored. Only enough HCO_3^- is present in the blood to restore the ratio of 20:1 with CO_2. Because the CO_2 concentration is below normal so is the concentration of HCO_3^- below normal in incomplete compensation.
Metabolic	Caused by any of several metabolic disorders such as excesses of metabolic acids, but results in a decrease in bicarbonate. *Problem: deficiency of HCO_3^- and, thus, excess H^+.*	Caused by any of several metabolic disorders such as loss of acids but results in an increase in bicarbonate. *Problem: excess of HCO_3^- and, thus, deficient H^+.*
Compensation	Accomplished by the respiratory system because it is not impaired in metabolic disorders and can compensate the condition more rapidly than the kidney.	
	Lungs release more CO_2 into exhaled air until the blood pH is returned to 7.4 and the ratio of HCO_3^- to CO_2 is restored to 20:1. Because of the original deficiency of HCO_3^-, the concentrations of both HCO_3^- and CO_2 are below normal in incomplete compensation.	Lungs release smaller amounts of CO_2 into exhaled air until the blood pH is returned to 7.4 and the ratio of HCO_3^- to CO_2 is restored to 20:1. Because of the excess of HCO_3^-, the concentrations of both HCO_3^- and CO_2 are above normal in incomplete compensation.

forms CO_2 and water and more H^+ and HCO_3^- are used to make carbonic acid. Blood H^+ decreases and blood pH increases. The imbalance in respiratory alkalosis is deficient CO_2 that leads to deficient H^+.

Respiratory alkalosis, less common than acidosis, can be caused by anxiety, aspirin overdose, or a change from low to high altitude. Anxiety and aspirin act directly on respiratory centers, whereas altitude lowers blood oxygen. All of these conditions produce hyperventilation and the removal of greater than normal amounts of carbon dioxide.

Because the lungs cannot compensate, the kidneys compensate by secreting less H^+ and by reabsorbing less HCO_3^-. The 20:1 HCO_3^- to CO_2 ratio is restored, so the blood pH returns to normal, but both HCO_3^- and CO_2 are below normal.

Accumulation of acids and ketone bodies often causes **metabolic acidosis.** Plasma HCO_3^- is depleted by neutralizing H^+ and this taxes the blood's buffering capacity, allowing H^+ to accumulate and lowers the blood

pH. The imbalance in metabolic acidosis is deficient HCO_3^- that leads to excess H^+.

Metabolic acidosis includes all types of acidosis *not* caused by excess CO_2, but excessive fat metabolism in untreated diabetes mellitus is a common cause. Low carbohydrate diets, starvation, renal tubular defects that impair H^+ secretion and HCO_3^- reabsorption, and loss of HCO_3^- in severe diarrhea can cause acidosis. The combination of dehydration and acidosis from severe diarrhea causes many childhood deaths, especially where emergency medical care is not available.

The lungs usually act quickly in metabolic acidosis to compensate by removing CO_2 from the blood. The 20:1 HCO_3^- to CO_2 ratio is restored, the blood pH returns to normal, but both HCO_3^- and CO_2 are below normal.

Depletion of acids can lead to **metabolic alkalosis** as HCO_3^- binds most free H^+ and the blood pH increases. The imbalance in metabolic alkalosis is excess HCO_3^- that leads to deficient H^+.

Metabolic alkalosis can arise from taking too much antacid, loss of Cl⁻, and excess aldosterone secretion. Antacids used to treat ulcers or gastritis neutralize acids and increase blood HCO_3^-. Vomiting causes loss of stomach acid, which in turn depletes H^+ and Cl^- in extracellular fluids. H^+ remains depleted and Cl^- is replaced by HCO_3^-. Excess aldosterone causes Na^+ reabsorption and H^+ secretion. Metabolic alkalosis and magnesium deficiency may combine to cause the **delirium tremens** (de-lir′e-um tre′menz) of alcohol withdrawal. Diuretic use and abuse can lead to **contraction alkalosis,** in which much water is lost and H^+ secretion is excessive. This leaves an excess of HCO_3^- dissolved in a smaller water volume.

The lungs compensate for excess HCO_3^- by removing CO_2 from the blood more slowly. The 20:1 ratio of HCO_3^- to CO_2 is restored, the blood pH returns to normal, but both HCO_3^- and CO_2 are above normal.

All acid-base imbalances involve an excess or deficiency of H^+, but the causes of such imbalances differ. In respiratory acidosis, H^+ excess is caused by an excess of CO_2. In respiratory alkalosis, H^+ deficiency is caused by a deficiency of CO_2. In metabolic acidosis, H^+ excess is due to a deficiency of HCO_3^-. In metabolic alkalosis, H^+ deficiency is due to an excess of HCO_3^-. The HCO_3^- to CO_2 ratio is high in acidosis and low in alkalosis.

Incomplete compensation fairly quickly restores the 20:1 ratio and returns the pH to 7.4. Complete compensation, which restores normal HCO_3^- and CO_2 concentrations, occurs slowly and may be impossible if lung or kidney disease severely impairs compensating mechanisms. Though healthy lungs act quickly to compensate for acid-base imbalances of metabolic origin, healthy kidneys assist at a slower pace by appropriately adjusting the rates of H^+ secretion and HCO_3^- reabsorption.

Acidosis or alkalosis can be life-threatening. When the blood pH drops below 7.0, depressed central nervous system function leads to coma. If not treated the patient will remain comatose and die from respiratory failure. When the blood pH rises above 7.8, the whole nervous system becomes hyperactive. Peripheral nerve hyperactivity spreads from forearm muscles, to facial muscles, and then to all muscles. Central nervous system hyperactivity can cause seizures and convulsions. Death is a usual outcome from respiratory muscle tetany.

See Questions—Objective 15

Questions

Objective 15

(a) How do acidosis and alkalosis differ?

(b) How do respiratory and metabolic (nonrespiratory) causes of these disorders differ?

(c) What is the cause, nature, and likely means of compensation for respiratory acidosis?

(d) for respiratory alkalosis?

Questions Continued

Objective 15

(e) for metabolic acidosis?

(f) for metabolic alkalosis?

Study and Review

Did you get the essentials?

The following summary contains the basic concepts from chapter 9 except for the key terms that have been omitted. Read the chapter. Then try to fill in the blanks from memory. If you cannot fill in all the blanks, review the chapter paying particular attention to boldface terms.

The functional unit of a kidney is a _____ , which includes the _____ that contains _____ and _____ that contain blood. The cuplike structure at the open end of a nephron is the _____ _____ and it holds a network of capillaries called the

_____ .

From this structure, filtrate moves through the _____ _____ tubules, the _____ of _____ , the _____ _____ _____ and drains into a _____ duct. Blood from an _____ arteriole enters the _____ , then capillaries around the _____ and _____ _____ tubules and finally to the _____ _____ around the _____ of _____ , and finally to the _____ arteriole.

The three processes carried out in nephrons are _____ pushing substances out of the blood, _____ passively returning or actively transporting substances to the blood, and _____ actively transporting substances from blood to filtrate. Kidney filtrate is formed under _____ pressure. The _____ _____ rate is normally about 120 _____ per _____ , but it forms _____ liters per _____ . Of this volume, _____ liters are returned to the _____ and only about _____ _____ excreted. Concentrations

Continued on next page

Study and Review

of substances in the filtrate are _____ in blood. Substances reabsorbed in the _____ _____ tubule include _____ _____ . The upper limit of the concentration of a substance returned to the blood is the _____ _____ . Tubular _____

In a _____ system, outflow runs _____ and in the _____ direction of the inflow. In the kidneys, the _____ mechanism increases the concentration of _____ in _____ fluids. This draws _____ from the filtrate and _____ the urine. _____ synthesis and excretion is the body's main means of getting rid of _____ wastes. Other wastes include _____ .

An important factor in regulating fluid volume is _____ hormone, which increases permeability of _____ ducts and _____ tubules. A deficiency of this hormone leads to _____ _____ . _____ are substances that increase urine volume. The _____ apparatus of the kidneys regulates blood flow by releasing a signal from cells of the _____ _____ that causes _____ cells to secrete _____ when the blood pressure _____ . This activates _____ , which causes constriction of _____ and secretion of _____ (a hormone that causes _____ retention and water retention by _____). _____ peptide probably deals with excesses of _____ . _____ is a measure of what volume of blood can be freed of a substance in a given _____ .

The kidneys' inability to remove sufficient wastes from the blood is called _____ _____ and it can be _____ or _____ . This disorder can be treated with _____ (removal of wastes from blood outside the body) and _____ _____ (removal of wastes into fluid in the abdominal cavity).

Among the body's main fluid compartments, the largest contains _____ fluid, the next largest contains _____ fluid, and the smallest contains _____ . The latter two together comprise the _____ fluid. Potassium and phosphates are present in greatest concentrations in _____ fluid, and _____ and _____ are present in greatest concentrations in _____ fluid and _____ .

Water lost in exhaled air and sweat is called _____ _____ loss. When water intake equals water output the body is in _____ _____ . _____ , the conscious desire for liquid, is elicited when _____ in the _____ center of the _____ detect increased _____ of tissue fluids. The body is in negative water balance in _____ and in positive water balance in _____ _____ and _____ .

Among electrolytes, _____ and _____ are regulated mainly by aldosterone, and _____ is regulated by PTH and calcitonin. Membrane pumps exist for active transport of two pairs of ions, _____ - _____ and _____ - _____ .

The body's buffer systems include the _____ - _____ system, the _____ system, mainly in cells and urine, and _____ , especially hemoglobin. The function of a buffer is to _____ .

The greatest threat of acid-base imbalance is accumulation of _____ , and the best defense is healthy function of the _____ system. This system responds quickly to adjust blood pH but it can alter only the _____ concentration. The _____ system acts more slowly but adjusts the concentrations of _____ electrolytes, including _____ ions (which are responsible for acidic conditions). Respiratory acidosis is caused by too _____ _____ and can be triggered by _____ . Metabolic acidosis is caused by too _____ _____ and can be started by _____ . Respiratory alkalosis is due to too _____ _____ and can be caused by _____ . Metabolic alkalosis occurs because of too _____ _____ and can be caused by _____ . These conditions are said to be incompletely _____ , when blood _____ becomes normal and completely _____ , when the concentrations of _____ and _____ become _____ .

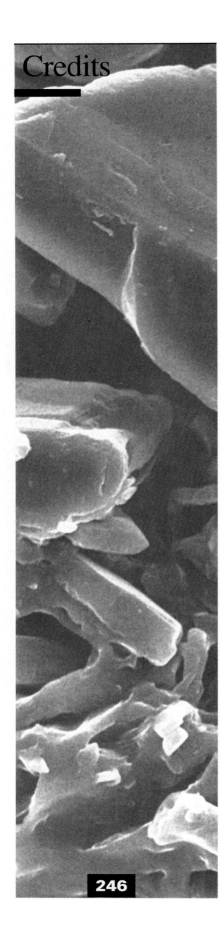

Credits

Illustrations

Unless otherwise indicated, all line art and tables are from Joan G. Creager, *Human Anatomy and Physiology,* 2d ed. Copyright © 1992 Joan G. Creager. All Rights Reserved.

Chapter 1

Figure 1.23 From Stuart Ira Fox, *Human Physiology,* 3d ed. Copyright © 1990 Wm. C. Brown Communications, Inc., Dubuque, Iowa. All Rights Reserved. Reprinted by permission.

Chapter 2

Figure 2.4 From Stuart Ira Fox, *Human Physiology,* 3d ed. Copyright © 1990 Wm. C. Brown Communications, Inc., Dubuque, Iowa. All Rights Reserved. Reprinted by permission. **Figure 2.12** From Kent M. Van De Graaff and Stuart Ira Fox, *Concepts of Human Anatomy and Physiology,* 2d ed. Copyright © 1989 Wm. C. Brown Communications, Inc., Dubuque, Iowa. All Rights Reserved. Reprinted by permission. **Figure 2.15** From Stuart Ira Fox, *Human Physiology,* 3d ed. Copyright © 1990 Wm. C. Brown Communications, Inc., Dubuque, Iowa. All Rights Reserved. Reprinted by permission.

Chapter 4

Figure 4.1*a* From John W. Hole, Jr., *Human Anatomy and Physiology,* 4th ed. Copyright © 1987 Wm. C. Brown Communications, Inc., Dubuque, Iowa. All Rights Reserved. Reprinted by permission. **Figure 4.1*b*** From John W. Hole, Jr., *Human Anatomy and Physiology,* 5th ed. Copyright © 1990 Wm. C. Brown Communications, Inc., Dubuque, Iowa. All Rights Reserved. Reprinted by

permission. **Figure 4.4** From John W. Hole, Jr., *Human Anatomy and Physiology,* 5th ed. Copyright © 1990 Wm. C. Brown Communications, Inc., Dubuque, Iowa. All Rights Reserved. Reprinted by permission. **Figure 4.12** From John W. Hole, Jr., *Human Anatomy and Physiology,* 5th ed. Copyright © 1990 Wm. C. Brown Communications, Inc., Dubuque, Iowa. All Rights Reserved. Reprinted by permission.

Chapter 5

Figure 5.5 From Kent M. Van De Graaff and Stuart Ira Fox, *Concepts of Human Anatomy and Physiology,* 3d ed. Copyright © 1992 Wm. C. Brown Communications, Inc., Dubuque, Iowa. All Rights Reserved. Reprinted by permission. **Figure 5.6** From John W. Hole, Jr., *Human Anatomy and Physiology,* 4th ed. Copyright © 1987 Wm. C. Brown Communications, Inc., Dubuque, Iowa. All Rights Reserved. Reprinted by permission. **Figure 5.17** From John W. Hole, Jr., *Human Anatomy and Physiology,* 5th ed. Copyright © 1990 Wm. C. Brown Communications, Inc., Dubuque, Iowa. All Rights Reserved. Reprinted by permission.

Chapter 6

Figure 6.6*a* and *b* From Stuart Ira Fox, *Human Physiology,* 3d ed. Copyright © 1990 Wm. C. Brown Communications, Inc., Dubuque, Iowa. All Rights Reserved. Reprinted by permission.

Chapter 9

Figure 9.1 From John W. Hole, Jr., *Human Anatomy and Physiology,* 5th ed. Copyright © 1990 Wm. C. Brown Communications, Inc., Dubuque, Iowa. All Rights Reserved. Reprinted by permission.

Many of the boldface terms in this glossary are followed by a phonetic spelling in parentheses. These pronunciation aids usually come from *Dorland's Illustrated Medical Dictionary*. The following rules are taken from this dictionary and will help in using its phonetic spelling system.*

1. An unmarked vowel ending a syllable (an open syllable) is long; thus *ma* represents the pronunciation of *may; ne*, that of *knee; ri*, of *wry; so*, of *sew; too*, of *two;* and *vu*, of *view*.
2. An unmarked vowel in a syllable ending with a consonant (a closed syllable) is short; thus *kat* represents *cat; bed*, *bed; hit, hit; not, knot; foot, foot;* and *kusp, cusp*.
3. A long vowel in a closed syllable is indicated by a macron; thus *māt* stands for *mate; sēd*, for *seed; bīl*, for *bile; mōl*, for *mole; fūm*, for *fume;* and *fool*, for *fool*.
4. A short vowel that ends or itself constitutes a syllable is indicated by a breve; thus *ĕ-fekt'* for *effect, ĭ-mun'* for *immune*, and *ŏ-klood'* for *occlude*.

Primary (') and secondary (") accents are shown in polysyllabic words. Unstressed syllables are followed by hyphens.

A

A cell Cell of the pancreas that secretes glucagon. 131
absorption (ab-sorp'shun) Taking in of substances. 167
absorptive (ab-sorp'tiv) metabolism Chemical reactions that take place as food is being absorbed. 180
acetyl-coenzyme A (as"et'il ko-en'zim a) A coenzyme needed to prepare molecules to enter certain metabolic pathways combined with a 2-carbon metabolite. 63
acetylcholine (as"e-til-ko'lēn) A neurotransmitter released from many axons, including those that control skeletal muscles. 101, 103

acid A substance that can donate hydrogen ions. 9
acid-base balance The process by which the pH of body fluids is maintained in a normal range. 238
acidosis (as"id-o'sis) The condition of having too low a blood pH. 238
actin (ak'tin) A contractile protein. 45, 87
action potential An altered electrical potential across the cell membrane of an excitable cell. 91
activation energy Energy required to start an enzyme-controlled reaction. 26
active site A region on an enzyme onto which the substrate binds. 27
active transport Transport of a substance against a gradient using an enzyme, carrier molecule, and cellular energy. 38, 40
acyl (as'il) A fatty acid. 72
adenosine (ad-en'o-sēn) triphosphate (ATP) The body's main energy storage molecule. 22
adenylate cyclase (ad-en'ilāt si'klās) A membrane-bound enzyme that converts ATP to cAMP. 117
adrenal (ad-re'nal) Above the kidney; a gland lying above the kidney. 128
adrenaline (ad-ren'ah-lin) A hormone secreted by the adrenal glands. 128
adrenergic (ad"ren-er'jik) Pertaining to an axon that releases norepinephrine (a neurotransmitter similar to adrenaline, or epinephrine). 103
adrenocorticotrophic (ad-re"no-kor"te-ko-trof'ik) hormone A hormone that causes the adrenal cortex to secrete hormones; corticotropin. 124
aerobic (a'ro-bik) With oxygen. 93
albumin (al-bu'min) A small protein found mainly in blood plasma. 145
alcoholism Disease associated with excessive intake of alcohol. 191
aldosterone (al-dos'ter-ōn) An adrenocortical hormone that increases sodium reabsorption. 129, 232
aliphatic Pertaining to organic molecules lacking ring structures. 11
alkalosis (al"kah-lo'sis) The condition of having too high a blood pH. 239

*From *Dorland's Medical Dictionary*, W. B. Saunders Company, Philadelphia, PA, 1988. Reprinted by permission.

D

deamination (de-am″in-a′shun) Removal of an amino group. 76

dehydration (de-hi-dra′shun) Removal of water. 8, 232

denaturation (de-nat-ur-a′shun) A change in the shape and properties of a protein molecule. 18

deoxyribonucleic (de-ok″se-ri′bo-nu-kla′ik) acid (DNA) A nucleic acid found in chromosomes that stores and transmits genetic information from one generation to the next. 22

diabetes insipidus (di-ah-be′tez in-sip′id-us) A disorder caused by a deficiency of antidiuretic hormone in which large quantities of dilute urine are produced. 225

diabetes mellitus (di-ah-be′tez mel-i′tus) A disorder caused by a deficiency or inactivity of insulin in which glucose accumulates in the blood and urine. 188

dialysis (di-al′is-is) The separation of small molecules from larger ones as smaller ones pass through a selectively permeable membrane. 229

diapedesis (di″ah-ped-e′sis) A process by which leukocytes squeeze between the cells of capillary walls. 159

diffuse endocrine system (DES) Different kinds of hormone-secreting cells in various locations throughout the body. 115

digestion The breakdown of large molecules into smaller ones. 167

dipeptidase (di-pep′tid-ās) An enzyme that breaks dipeptides into amino acids. 170

dipeptide (di-pep′tid) A molecule consisting of two amino acids held together by a peptide bond. 17

disaccharide (di-sak′ar-īd) A molecule consisting of two sugar (saccharide) units held together by a glycosidic bond. 14

diuresis (di″u-re′sis) Increased urine production. 225

diuretic (di″u-ret′ik) An agent that causes diuresis. 225

DNA polymerase (pol-im′er-ās) An enzyme involved in the synthesis of DNA. 47

dominant In genetics, a characteristic that appears in the phenotype whenever the allele for it is present in the genotype. 53

dopamine (do′pah-mēn) A neurotransmitter and precursor of norepinephrine. 101

dynorphin (din-or′fin) A polypeptide similar to an enkephalin. 109

E

edema (ĕ-de′mah) Accumulation of excess fluid in the tissues. 232

electrolyte (e-lek′tro-līt) Any substance that ionizes and conducts electricity. 230

electron A negatively charged particle that moves around the nucleus of an atom. 2

electron transport system A set of enzymes and coenzymes in the cristae of mitochondria that move electrons from substrates to oxygen. 66

element A fundamental unit of matter. 2

emulsification (e-mul″sĕ-fi-ka′shun) Process by which fat droplets in foods are broken into smaller particles by the action of bile salts. 170

endergonic (end″er-gon′ik) Using energy. 25

endocrine (en′do-krin) Pertaining to a ductless gland. 115

endoplasmic reticulum (en″do-plas′mik rĕ-tik′u-lum) A network of membranous vesicles within a cell. 44

endorphin (en-dor′fin) A brain peptide that binds to opiate receptors. 109

enteric (en-ter′ik) Pertaining to the intestine. 140

enterohepatic (en″ter-o-hep-at′ik) circulation Movement by which bile salts are returned to the liver and resecreted in bile. 175

enterokinase (en″ter-o-ki′nās) A proteolytic enzyme from the intestinal mucosa. 170

enzyme (en′zīm) A protein that catalyzes a chemical reaction. 21

eosinophil (e″o-sin′o-fil) A granular leukocyte capable of being stained with the dye eosin. 148

epinephrine (ep-e-nef′rin) Main hormone secreted by the adrenal medulla. 128

erythrocyte (er-ith′ro-sīt) Red blood cell. 144

erythrolabe (er-ith′ro-lāb) A cone pigment sensitive to red light. 109

erythropoiesis (er-ith″ro-poi-e′sis) Production of red blood cells. 140

erythropoietin (er-ith″ro-poi′et-in) A substance secreted by the kidney that stimulates production of red blood cells. 140, 147

essential amino acid An amino acid that must be present in the diet because the body cannot make it. 17, 74, 198

essential fatty acid A fatty acid that must be present in the diet because the body cannot make it. 72, 199

estrogen (es′tro-jen) A kind of female hormone that stimulates development of sex organs and secondary sexual characteristics. 135

excitability Ability to respond to a stimulus. 96

excitation-contraction coupling The connection between neural signals and muscle contraction. 91

exergonic (ek″ser-gon′ik) Giving off energy. 25

extracellular (ex″tra-sel′u-lar) Outside a cell. 230

F

facilitated diffusion Diffusion down a gradient aided by a carrier molecule but not requiring cellular energy. 39

fatty acid A long hydrocarbon chain with an organic acid group at one end. 16

ferritin (fer′it-in) A molecule consisting of the protein apoferritin and iron. 146

fibrin (fi′brin) A protein that forms a network in a blood clot. 151

fibrinogen (fi-brin′o-jen) Inactive fibrin. 151

fluid-mosaic model A description of the molecular structure of a cell membrane. 35

folacin (fol′ah-sin) A vitamin that helps transfer single carbon groups. 207

follicle-stimulating hormone (FSH) A hormone that stimulates maturation of ova and sperm. 123

functional group The part of a molecule that participates in a chemical reaction. 11

G

gamete (gam′ēt) An ovum or sperm. 132

gamma-aminobutyric (gam′ah ah-me′no-bu-tir′ik) acid (GABA) An inhibitory neurotransmitter of the central nervous system. 102

interstitial (in-ter-stish'al) Pertaining to spaces between cells. 230

ion A charged atom or group of atoms. 3

ionic bond A chemical bond through which atoms are held together by the attraction of unlike charges. 6

isomer (i'so-mer) A molecule having the same number and variety of atoms as another molecule, but arranged differently. 12

isosmotic (i''sos-mot'ik) Having the same osmotic pressure. 40

isotonic (i''so-ton'ik) Having the same tonicity. 40

isotope (i'so-tōp) An atom of an element having a different number of neutrons than certain other atoms of the same element. 4

J

juxtaglomerular (juks-tah-glo-mer'u-lar) Pertaining to or near a glomerulus. 226

juxtamedullary (juks''tah-med'ul-ar-e) Pertaining to or near the medulla, usually referring to the kidney. 219

K

ketone (ke'tōn) body Acidic molecule produced during excessive metabolism of fat. 71

kilocalorie (kcal) Amount of heat required to raise the temperature of 1 kilogram of water 1° C. 188

kinetic Energy of motion. 25

kinin (ki'nin) A substance that stimulates events in the inflammatory process. 158

Krebs cycle A sequence of reactions that oxidize acetyl-CoA; citric acid cycle; tricarboxylic acid cycle. 65

kwashiorkor (kwash-e-or'kor) Protein deficiency, usually in young children. 208

L

lactase (lak'tās) An enzyme that digests lactose (milk sugar). 170

lacteal (lak'te-al) A lymph vessel within a villus of the small intestine. 169

leukocyte (loo'ko-sīt) A white blood cell. 144

leukocytosis (loo-ko-si-to'sis) promoting (LP) factor A substance that increases the attraction of leukocytes to a site of injury. 159

ligand (li'gand) A substance that binds to a receptor. 21

lipase (li'pās) An enzyme that digests lipids. 167

lipid Fats and fatlike substances. 14

lipoprotein A molecule containing both lipid and protein. 73

loop of Henle A U-shaped segment of a nephron that helps to concentrate sodium chloride in peritubular fluid. 219

lutein (loo'te-in) A yellow pigment. 135

luteinizing (loo'te-in-īz''ing) hormone (LH) A hormone that stimulates ovulation and other reproductive processes. 123

lymphocyte (lim'fo-sīt) A leukocyte capable of participating in an immune response. 145, 148

lysosome (li'so-sōm) Membrane-bound organelle that releases digestive enzymes. 44

M

macula densa (mak'u-lah den'sah) Modified cells of the kidney's distal convoluted tubules associated with the juxtaglomerular apparatus. 226

major histocompatibility (his''to-kompat''ib-il'it-e) complex (MHC) proteins Proteins that account for antigenic individuality of a person's cells. 161

malnutrition Ill health due to an inadequate diet. 208

maltase (mawl'tās) Enzyme that digests maltose, a disaccharide derived from starch. 170

marasmus (mah-raz'mus) Near-starvation malnutrition. 208

matrix Inner cavity of mitochondrion. 44

matter That which has substance. 2

melanocyte (mel'ah-no-sīt) A cell that synthesizes melanin. 124

memory The process of storing and recalling previous experiences. 162

menarche (men-ar'ke) The onset of menstruation. 136

menopause (men'o-pawz) The cessation of menstruation. 136

menstrual (men'stroo-al) cycle A repetitive sequence of events in women of reproductive age, involving ovulation and preparation of the uterus for implantation. 136

menstruation (men-stroo-a'shun) The periodic discharge of blood, tissue debris, and fluid from the uterus. 136

messenger RNA (mRNA) A nucleic acid that carries information in the form of codons for the synthesis of a protein. 49

metabolic acidosis A lowering of blood pH because of a metabolic disorder. 241

metabolic alkalosis A raising of blood pH because of a metabolic disorder. 241

metabolic rate The rate at which nutrients are oxidized. 188

metabolic water Water derived from the oxidation of foodstuffs. 199

metabolism (met-ab'o-lizm) The sum total of all chemical reactions in the body. 2

micelle (mi-sel') A small fat droplet found in chyme. 171

microfilament A small, hollow protein fiber found within the cytoplasm of a cell. 45

microtubule A cylindrical organelle that contributes to a cell's mitotic spindle. 45

microvillus (mi''kro-vil'us) A tiny cytoplasmic projection from the surface of intestinal epithelial cells. 169

mineral An inorganic substance. 199

mitochondrion (mi''to-kon'dre-on) An organelle in which oxidative and energy-capturing processes occur. 44

mixture A combination of two or more substances in any proportions in which the substances do not lose their properties. 7

molecule The smallest quantity of a substance that retains its chemical properties. 2

monocyte (mon'o-sīt) A large, phagocytic, agranular leukocyte. 148

monosaccharide (mon-o-sak'ar-īd) A simple sugar. 12

motility Ability to move. 167

mutagen (mu'ta-jen) An agent that can cause changes in DNA. 55

mutation (mu-ta'shun) A change in the genetic information (DNA) in a cell. 54

myofibril (mi''o-fi'bril) A contractile fiber in a muscle cell. 86

myofilament (mi''o-fil'a-ment) A protein molecule that makes up part of a myofibril. 86

myoglobin (mi''o-glo'bin) A pigmented protein in muscle tissue that binds oxygen. 95

myokinase (mi″o-kin′ās) An enzyme that makes ATP and AMP from two molecules of ADP. 95

myoneural (mi″o-nu′ral) junction The joining of nerve and muscle tissue. 93

myosin (mi′o-sin) A protein that makes up the thick filaments of a myofibril. 87

N

naloxone (nal-oks′ōn) A drug that counteracts opium overdoses. 109

natriuresis (nat-re-u-re′sis) Stimulation of sodium excretion. 226

negative feedback A mechanism by which the output of a system suppresses or inhibits activity of the system. 115

nephron (nef′ron) The functional unit of a kidney. 219

net filtration pressure Pressure pushing substances out of a blood vessel. 221

net protein utilization The proportion of protein ingested that can actually be used by cells. 215

neuromuscular (nu″ro-mus′ku-lar) Pertaining to the association between the nervous and muscular systems. 93

neuroactive peptide (nu″ro-ak′tiv pep′tīd) A chain of amino acids having some influence on neural function. 102

neurotransmitter (nu″ro-trans-mit′er) A chemical substance released from one neuron that transmits a signal to another neuron at a synapse. 101

neutron (nu′tron) An uncharged particle in the nucleus of an atom. 2

neutrophil (nu′tro-fil) A granular leukocyte that fails to stain with either acidic or basic stains. 148

niacin (ni′as-in) A B vitamin used to make the coenzyme NAD. 204

nitrogen balance A condition in which nitrogen entering the body equals nitrogen leaving it. 74

norepinephrine (nor″ep-ē-nef′rin) A neurotransmitter of the sympathetic division of the autonomic nervous system and of some brain neurons. 101, 103, 128

nucleic (nu-kle′ik) acid A polymer of nitrogenous bases, 5-carbon sugars, and phosphates; DNA or RNA. 22

nucleolus (nu-kle′o-lus) An RNA-containing body within a nucleus. 42

nucleoplasm (nu′kle-o-plazm″) The substance of a nucleus. 42

nucleotide (nu′kle-o-tīd) A molecular unit containing a nitrogenous base, a 5-carbon sugar, and one or more phosphates. 22

nucleus (nu′kle-us) (1) A cell's control center; (2) the central part of an atom. 41, 42

nutrition The act of providing all substances needed for good health through food ingestion. 198

O

obesity An excessive amount of fat. 189

oogenesis (o-o-jen′is-is) The process by which an ovum is produced. 134

oogonia (o-o-go′ne-ah) Female germ cells. 134

opsin (op′sin) A protein associated with the response to light. 109

organelle (or″gah-nel′) A tiny functional unit within a cell. 35, 41

organic Pertaining to carbon-containing substances. 11

osmolarity The osmotic concentration of a solution determined by the number of osmotically active particles it contains. 40

osmosis Movement of water across a membrane from a region of lower to higher solvent concentration; diffusion of water from its own higher to lower concentration. 40

osmotic diuresis Increased urine volume because of the large number of osmotically active particles in the kidney filtrate. 225

osmotic pressure The pressure produced by osmosis. 40

ovulation (ov″u-la′shun) Release of an ovum from a follicle. 135

ovum (o′vum) A female gamete. 134

oxidation Addition of oxygen or loss of electrons or hydrogen. 12

oxidative phosphorylation Capture of energy in ATP as molecules are oxidized. 66

oxygen debt The amount of oxygen required to oxidize metabolites produced anaerobically during strenuous activity. 95

oxyhemoglobin (ok″se-he″mo-glo′bin) Hemoglobin-carrying oxygen. 154

oxytocin (ok″se-to′sin) A hormone that stimulates uterine contraction. 124

P

pancreas (pan′kre-as) A digestive gland that produces enzymes and hormones. 131

pantothenic (pan″to-then′ik) acid A B vitamin used to make coenzyme A. 207

parasympatholytic (par″ah-sim″pah-tho-lit′ik) Pertaining to substances that block or counteract sympathetic signals. 106

parasympathomimetic (par″ah-sim″pah-tho-mi-met′ik) Pertaining to substances that mimic the action of the parasympathetic nervous system. 106

parathormone (par″ah-thor′mōn) (PTH) A hormone from the parathyroid gland that decreases blood calcium. 128

parathyroid (par″a-thi′roid) glands Glands imbedded in the thyroid gland. 128

partial pressure The pressure exerted by one gas in a mixture of gases. 154

passive transport Any process that causes movement of substances without the expenditure of energy by the organism. 38

pentose phosphate pathway A metabolic pathway that metabolizes glucose, producing 5-carbon sugars and reduced NADP. 67

pepsin (pep′sin) An enzyme that digests protein in the stomach. 167

peptide (pep′tīd) bond A chemical bond that holds amino acids together. 17

peritubular (per″e-tu′bu-lar) Around a tubule, as in the kidney. 219

peroxisome (pĕ-roks′i-sōm) An organelle containing oxidative enzymes. 44

pH The negative logarithm of the hydrogen ion concentration. 9

phenotype (fen′o-tīp) The appearance of an individual with respect to one or all inherited characteristics. 53

phosphocreatine (fos″fo-kre′at-in) An energy storage molecule found in muscle. 94

phospholipid (fos″fo-lip′id) A lipid containing glycerol, fatty acids, and phosphoric acid. 16, 35

phosphorylation (fos″for-il-a′shun) Addition of a phosphate group to a molecule. 61

photon The smallest unit of light energy. 109

secretion (1) A cell product; the process of releasing a substance from a cell. (2) The active transport of substances from the blood to the kidney filtrate. 167, 178

selectively permeable A property of membranes allowing some substances to pass through while preventing the passage of others. 39

selenosis (se″len-o′sis) A disorder caused by too much selenium in the body. 208

serotonin (ser″o-to′nin) A substance secreted as a neurotransmitter in the brain and as a hormone in the gut. 102, 140

serum (se′rum) The fluid part of blood after formed elements and clotting factors have been removed. 144

sliding filament theory An explanation of how myofilaments move during muscle contraction. 87

sodium-potassium pump A mechanism that actively transports sodium ions out of cells and potassium ions back into them against natural gradients that cause the ions to move in the opposite direction. 41

sol The liquid state of a colloidal dispersion. 8

solute A dissolved substance. 7

solution A liquid containing dissolved substances. 7

solvent A substance in which other substances dissolve. 7

somatostatin (so″mat-o-stat′in) Growth-hormone-inhibiting hormone. 124

somatotropin (so″mat-o-tro′pin) Growth hormone. 123

specific heat The amount of heat needed to raise the temperature of a specific volume of substance 1° C. 8

specificity The quality of being specific. 28, 162

spectrin (spek′trin) A protein that contributes to the flexibility of erythrocyte membranes. 145

spermatogenesis (sper″mah-to-jen′es-is) The process by which sperm are formed. 134

stereoisomer (sta-re-o-is′om-er) A compound containing the same number and variety of atoms as another compound, but in a different spatial arrangement. 12

steroid (ste′roid) A lipid having a complex 4-ring structure. 16

streptokinase (strep″to-ki′nās) A substance used to digest blood clots in coronary arteries. 152

sucrase (soo′krās) An enzyme that digests sucrose. 170

surface-to-volume ratio The surface area of a structure divided by its volume. 39

sympatholytic (sim″pah-tho-lit′ik) Blocking or counteracting sympathetic signals. 106

sympathomimetic (sim″pah″tho-mi-met′ik) Mimicking the action of the sympathetic division, or a drug that does so. 106

T

T lymphocyte A thymus-processed lymphocyte that differentiates into one of several kinds of T cells. 160

target cell A cell that responds to a particular hormone. 115

taurine (taw′rēn) An amino acid derivative that conjugates with bile acids. 169

template Pattern. 47

testosterone (tes-tos′ter-ōn) A male hormone. 133

thiamine (thi′am-in) A water-soluble B vitamin. 204

thirst The desire for water or other fluid. 232

thirst center A nucleus in the hypothalamus that responds to osmotic changes in the blood, causing drinking behavior when the osmotic pressure increases. 232

thrombin An enzyme that converts fibrinogen to fibrin in the blood-clotting mechanism. 150

thrombocyte (throm′bo-sīt) A platelet. 148

thymosin (thi′mo-sin) A hormone secreted by the thymus gland. 139, 160

thymus (thi′mus) **gland** A gland that processes T lymphocytes before it regresses during puberty. 139, 160

thyroid gland A gland in the throat that produces hormones that regulate metabolism. 126

thyroid-stimulating hormone (TSH) A hormone that causes the thyroid gland to secrete hormones. 124

tissue plasminogen activator (tPA) A substance secreted by many tissues that converts plasminogen to plasmin. 152

tissue thromboplastin (throm″bo-plas′tin) A substance released by injured tissue that initiates the extrinsic blood-clotting mechanism. 150

tocopherol (to-kof′er-ol) A substance with vitamin E activity. 208

tonicity (to-nis′i-te) The degree to which conditions cause fluids to move into or out of cells. 40

trace element A chemical element normally found in the body in very small amounts. 208

transamination (trans-am-in-a′shun) Transfer of an amino group from one molecule to another. 76

transcription The process of transferring coded genetic information from DNA to mRNA. 47

transducin (trans-du′sin) Enzyme involved in visual process. 109

transduction The conversion of a signal from one type to another, as in converting a sensory signal from a receptor to an electrical signal in a neuron. 100

transfer RNA (tRNA) RNA that carries amino acids and places them in specific sites in a growing peptide chain. 50

transferrin (trans-fer′rin) A plasma protein that transports iron. 145, 146

translation The process by which information in mRNA codons is used to determine the sequence of amino acids in a protein. 47

transverse (T) tubule A tubule running crosswise of myofibrils in skeletal muscle that conveys a signal from the sarcolemma to the myofibrils. 87

triacylglycerol (tri-as″il-gli′ser-ol) A lipid consisting of glycerol and three fatty acids. 16

triad Any group of three. 87

triglyceride A triacylglycerol. 16

tropomyosin (tro″po-mi′o-sin) A muscle protein involved in regulating the contraction process. 87

troponin (tro-po′nin) A muscle protein involved in regulating the contraction process. 87

trypsin (trip′sin) A proteolytic enzyme released from the pancreas. 169

trypsinogen (trip-sin′o-jen) An inactive form of trypsin. 169

tubulin (tu′bu-lin) A protein found in intracellular microtubules. 85